S0-BXN-112

PHYSICAL METHODS IN MACROMOLECULAR CHEMISTRY

VOLUME 2

PHYSICAL METHODS IN MACROMOLECULAR CHEMISTRY

Edited by Benjamin Carroll

DEPARTMENT OF CHEMISTRY
RUTGERS—THE STATE UNIVERSITY
NEWARK, NEW JERSEY

VOLUME 2

1972

MARCEL DEKKER, INC. New York

MARCEL DEKKER, INC.

95 Madison Avenue, New York, New York 10016

LIBRARY OF CONGRESS CATALOG CARD NUMBER 69-12679

ISBN 0-8247-1085-1

PRINTED IN THE UNITED STATES OF AMERICA

PREFACE

The continuing rapid expansion of research in the macromolecular sciences has brought with it the inevitable "information explosion" particular to the field. This expansion of the literature compounds the difficulties of scientists keeping up with new developments in macromolecular chemistry.

This is the first series to bring together, in one place, information on methods in macromolecular chemistry that was previously scattered throughout the literature. It is the aim of the series to give the research worker a convenient and up-to date key to physical and physiochemical methods of particular significance. Physical Methods in Macromolecular Chemistry will provide the scientist with the understanding needed to select the most useful method for his research problem. By focusing on the interpretation of experimental data, the chapters point out important features and principal limitations of the methods treated. Critical assessments of the literature add further to the value of the discussion.

Synthetic as well as naturally occurring macromolecules are within the province of this series. The two groups of polymers may be considered separately or together, as best serves to illustrate the application of the methods.

Prepared by workers actively engaged in research, the chapters suggest new directions in the use of the techniques and also give hints of new horizons in the development of the methods themselves. Both new and routine applications of the various methods are covered. Instrumentation is emphasized for novel methods and those insufficiently covered in the literature, although lengthy discussion of standard hardware is avoided since this material is well covered in, for example, manufacturers' publications.

Among the new areas covered in this second volume of the series are gel permeation chromatography; molecular property evaluations in polymer chemistry; the interactions of polymers with small ions and molecules in solutions; dielectric properties of biopolymers; and thermal analysis methods used for solid state polymers.

The advance of a science is due in larger part to innovations in experimental techniques, as well as to the enhancement of the theoretical significance of the resulting data. It is hoped that Physical Methods in Macromolecular Chemistry will stimulate and encourage growth in both areas.

The editor is grateful to the authors for their efforts and to the publishers and their staff for their advice and gracious cooperation.

B. C.

Newark, New Jersey
1972

CONTRIBUTORS TO VOLUME 2

Donald D. Bly, E. I. du Pont de Nemours and Company, Inc., Wilmington, Delaware

Benjamin Carroll, Department of Chemistry, Rutgers University, Newark, New Jersey

E. O. Forster, Corporate Research Laboratories, Esso Research and Engineering Company, Linden, New Jersey

D. J. R. Laurence, Chester Beatty Research Institute, Fulham Road, London, England

Emmanuel P. Manche, Division of Natural Sciences and Mathematics, York College of the City University of New York, Jamaica, New York

A. P. Minton*, Polymer Department, Weizman Institute of Science, Rehovoth, Israel

*Present address: National Institute of Arthritis and Metabolic Diseases, National Institutes of Health, Bethesda, Maryland

CONTENTS OF VOLUME 1

*Present address: College of Pharmaceutical Sciences, Columbia University, New York, New York

CONTENTS OF VOLUME 2

PHYSICAL METHODS IN MACROMOLECULAR CHEMISTRY

VOLUME 2

Chapter 1

GEL PERMEATION CHROMATOGRAPHY IN POLYMER CHEMISTRY

Donald D. Bly

E. I. du Pont de Nemours and Company
Wilmington, Delaware

I. INTRODUCTION

A. The Influence of Gel Permeation Chromatography in Synthetic Polymer Chemistry

At the present time gel permeation chromatography is growing by leaps and bounds as did gas chromatography in the early 1950's. Over 100 publications appeared each year during the period 1967-1970 (1). Indeed, the literature explosion since 1964 has been so great that reviews and summaries (2-5) are already of much value.

Why the literature explosion? What happened in 1964 to capture the fancy of so many scientists and cause the big boom in gel permeation chromatography? The answer can be found most simply in the economist's terms of need, supply, and demand. Gel permeation chromatography was needed in polymer analysis. In recent years many new types of polymers have become commercially available with a tremendous variety of useful properties (6). Many of these properties depend on the molecular size distribution of the polymer or, for example, the block or other compositional distributions in copolymers.

The aim of polymer chemists has been to develop methods that will permit rapid analysis of the size distribution and, if possible, of the composition distribution. Examples of past methods include the turbidimetric titration and the Baker-Williams (7) modification of column fractionation. These and other rapid methods for estimating the molecular weight distribution (MWD) of polymers were reviewed in 1965 before the advent of gel permeation chromatography (8). Polymer chemists generally found in 1965 that very time-consuming bulk fractionation techniques were often required for fully determining the molecular size distribution of a desired polymer. Now, however, at its full potential gel permeation chromatography can reduce several weeks' work to a couple or hours of perhaps even a few minutes.

John C. Moore saw both the need for rapid fractionation processes in polymer analysis and the potential for satisfying this need by the technique known as "gel filtration chromatography." Gel filtration had been developed by biochemicst for separation of large, water-soluble molecules such as proteins.[1]

Moore was able to see that the gel filtration technique would be useful for synthetic polymer molecular size distribution analysis if a gel could be found that had the necessary properties. These properties included the right porosity range for fractionation, sufficient inertness to organic solvents, and structural rigidity so the gel would not collapse under moderate pressure. John Moore set out to prepare such a gel. He had some a priori chance of success in that local experience was available through colleagues, especially Wheaton and Bauman who had done extensive work on ion exchange resins and in ion exclusion from such resins (19-20). Moore's work was successful, and he reported the synthesis and use of certain crosslinked polystyrenes for molecular size separation at the Dallas, Southwest Regional Meeting of the American Chemical Society in 1962. In a feature article in Chemistry and Engineering News later in the year (21) details of the gel synthesis and the ability of the gel to fractionate by size were made more universally known. Moore put this new knowledge to work, and by 1964 he had prepared a substantial number of useful (various porosity) polystyrene gels and had used them to separate a number of organic monomers and polymers. Moore's classic publication in the Journal of Polymer Science (22) in 1964 was entitled "Gel Permeation Chromatography I. A New Method for Molecular Weight Distribution of High Polymers."

Now the problem of rapid polymer fractionation had a solution: gel permeation chromatography. And GPC, as gel permeation chromatography has come to be known, is a good solution. It is rapid, reproducible, and analytically descriptive of molecular size distributions. But these ingredients alone do not account for the explosive growth of GPC since Moore's paper in 1964.

Waters Associates, Inc., Framingham, Massachusetts, obtained a license from John Moore and the Dow Chemical Company to manufacture and use certain of the crosslinked polystyrene gels. Waters then began to manufacture gel permeation chromatographs which utilized these gels. The instrument, its use, and again the great potential of GPC for polymer analysis were described by L. E. Maley early in 1965 (23). Thus another ingredient was added to the fire: a comercially available instrument of demonstrated utility that made gel permeation chromatography a practical technique and easy to do. An aggressive technical (seminar) and marketing program by Waters fanned the flames and the explosion occurred.

B. Purpose and Limitations of this Chapter

The purpose of this chapter is to provide the reader with a knowledge of the present state of the art of gel permeation chromatography so that he can most productively apply it to his needs. It is intended primarily for the practicing synthetic polymer or macromolecular chemist, although GPC is certainly useful for small molecules too (24-25). The author has attempted to minimize analytical detail available elsewhere except in data handling which is often confusing. Separation theory, resolution, fractionation, efficiency, data interpretation and calculations, and equipment are discussed.

C. Description of the Technique and Terms

Gel permeation chromatography is a solid/liquid column chromatographic technique which sorts molecules according to their size in solution. The sample solution is percolated by solvent through the column which is packed with a rigid-structure, porous "gel." The size sorting takes place in the gel in the column. The very largest molecules exit first followed by smaller and smaller sized molecules. Because of this process, GPC can measure the molecular size distribution of polymers.

The molecules exiting from the column are detected and a plot of detector response versus volume is made. This process is much like that in gas chromatography. Detection may be continuous, using, for example, a differential refractometer, or discontinuous, viz., fraction collection followed by sample evaporation and residue weight analysis. Detailed description is found in Section IV. The volume/response plot itself is a measure of the size distribution.

Calibration can yield accurate, quantitative expressions of the size distribution from which information about the molecular weight distribution (MWD) or composition distribution can be obtained. Normal calibration is made with standards by which log molecular weight (log MW) or log hydrodynamic volume ($\log[\eta]M$) is plotted versus the retention volume (V_R) which is the peak position of the curve in volume units.

Many of the terms used in GPC and in this chapter have been listed in Appendix I. The symbols, units, and definitions have been taken in part from a list compiled by ASTM section D-20.70.04 recently published (25a). Frequently used terms and/or others defined in cited work are redefined in the text.

II. MOLECULAR PROPERTY EVALUATIONS

A. Qualitative Information

1. Value of Pictures

Because GPC sorts molecules according to size (see Section III-A), it inherently provides a picture of the size distribution of the polymer molecules. Much significant information can be obtained from this picture without further calculation, both with regard to the polymer under study and the GPC operating conditions per se. Several gel permeation chromatograms (pictures) are shown in Figs. 1 through 6 with interpretations to illustrate the utility of this qualitative approach. These figures represent actual chromatograms observed by the author. In the sketches the vertical axis is detector response, in this case differential refractive index (ΔRI) while the horizontal axis is retention volume (V_R), increasing from right to left. The curve peak (retention volume) is inversely proportional to log M which increases from left to right.

Figure 1 shows homopolymer being fully fractionated in that all the polymer molecules lie between the void volume (see Appendix I) and water peak with room to spare. The molecules are separated by size, and this picture describes the polymer MWD. Such a curve is obtained with Gaussian and normal polymer distributions. Atmospheric water and air are invariably detected and appear negative because of their very low refractive indices whereas the polymer refractive index is greater than that of the solvent in this case.

Figure 2 shows homopolymer at incomplete fractionation. In the figure, curve A blends into the water peak, and the low MW side of the distribution is accordingly obscured. Curve B begins near the void volume of the columns, indicating that an undetermined amount of high MW material has not been fractionated so that the high MW end of the distribution cannot be seen. Curve A can be improved by changing to, or adding additional, low porosity columns to improve the separation

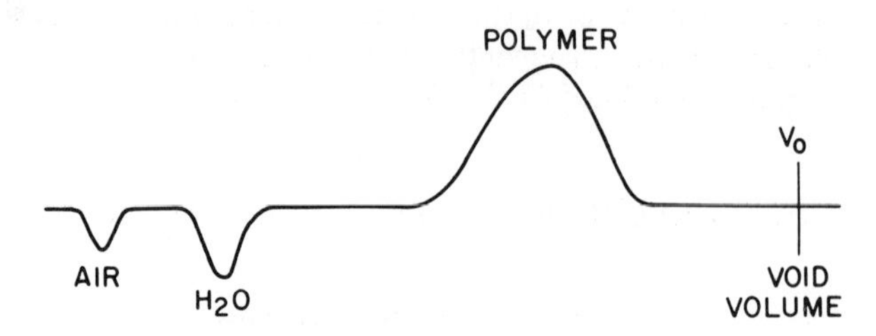

FIG. 1. Gel permeation chromatogram of fractionated homopolymer.

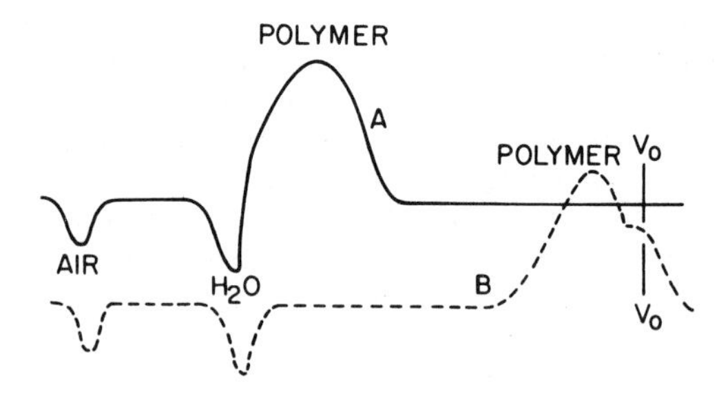

FIG. 2. Gel permeation chromatogram of homopolymer at incomplete fractionation.

of water and polymer, whereas in Curve B, additional high porosity is needed to separate the high MW material by size. Any gel in the polymer will invariably come out at the void volume, plug up the columns, or be filtered out in preanalysis preparation.

Figure 3 demonstrates an effect of catalysts on a polymer MW and MWD. The curves show vast differences in properties of polymers prepared under similar conditions but with different catalysts. Catalyst 1 gives a very broad distribution, from almost monomer to void volume, with a fairly high average molecular weight. Catalyst 2 gives a much narrower distribution but also a quite lower molecular weight. Catalyst 3 gives a polymer with fairly narrow molecular weight distribution and fairly high molecular weight.

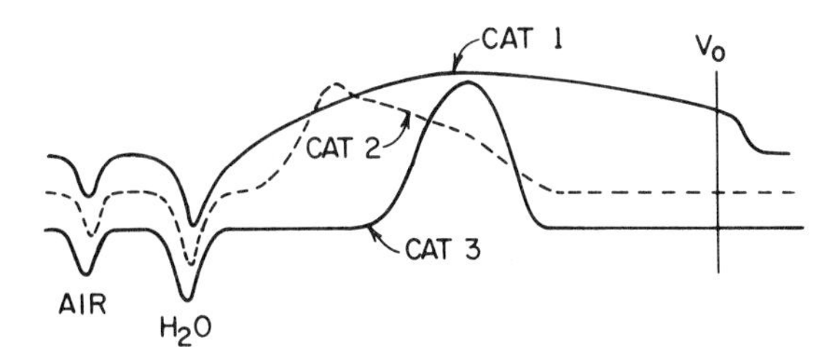

FIG. 3. Gel permeation chromatograms illustrating effect of a catalyst.

Figure 4 demonstrates how the fate of additives may be studied. In this case, it was desired to know whether an additive, dispersed in a polymer before extrusion, remained inert or chemically bonded to the polymer. Curve A represents the homopolymer with no additive, Curve B the test blend, and Curve C an unheated mixture of polymer and additive. It is immediately obvious that in sample B the components have either reacted or the additive has polymerized whereas in C no reaction has occurred.

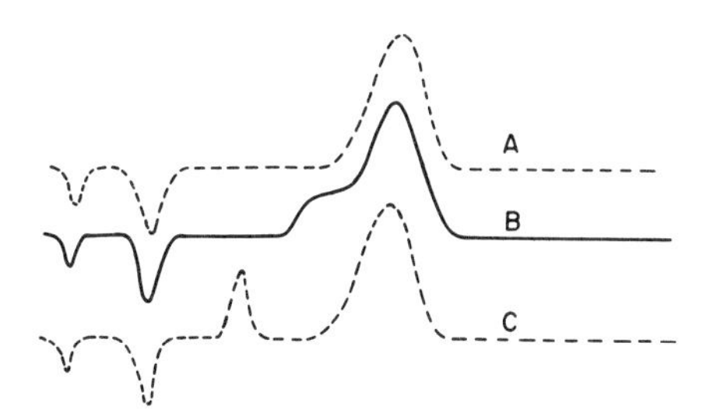

FIG. 4. Gel permeation chromatograms demonstrating fate of additives.

Figure 5 demonstrates a GPC study of the rate of a photolytic reaction. In the example shown, the ratio of the height of polymer to monomer in the curves (H/h) increased linearly with log T (photolysis time) over certain concentration ranges of each. Correlations

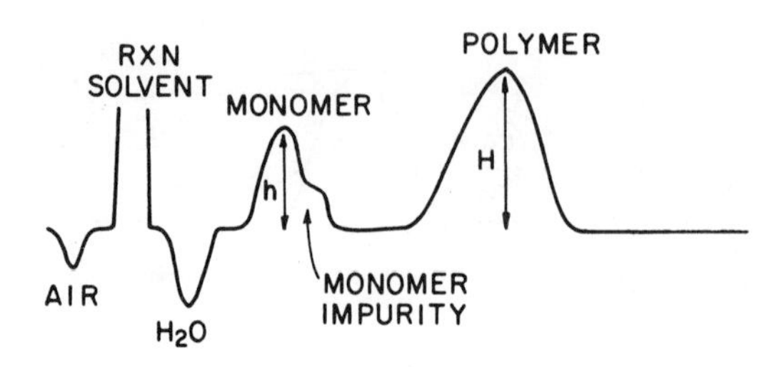

FIG. 5. Gel permeation chromatograms illustrating a study of reaction rate.

with $\sqrt{T}$ were also observed. Other correlations would be expected, for example, with I (photolytic intensity), catalyst, and solvent. This use of GPC has an additional advantage in that the molecular distribution of the polymer formed is also pictured. In this example a monomer impurity was also found that was heretofore not known to be present. Note the location of the water peak. Water is often anomalously located in a chromatogram. Since it is highly polar and hydrogen bonds readily, it may come off complexed with itself or with the chromatographic solvent and appear higher in MW than it is.

Figure 6 relates to the composition of co- and terpolymers. The chromatogram can be interpreted several ways, and which interpretation is correct cannot be known if the refractometer detector is employed. This chromatogram represents either a combination of at least two size distributions of molecules that overlap or a single size distribution of molecules whose composition varies with MW or a combination of these. It must be remembered that as composition changes, so does the refractive index that affects the height of a curve. Specific detectors are needed to sort the variables (see Section IV-B-2).

Besides the variable composition pitfall, the reader should also be cautioned to avoid long-term comparisons even with this qualitative approach. For example, if calibration procedures are not used, as in the above examples, a picture may not be useful for comparing to another obtained at some future date because the calibration of the GPC may have changed.

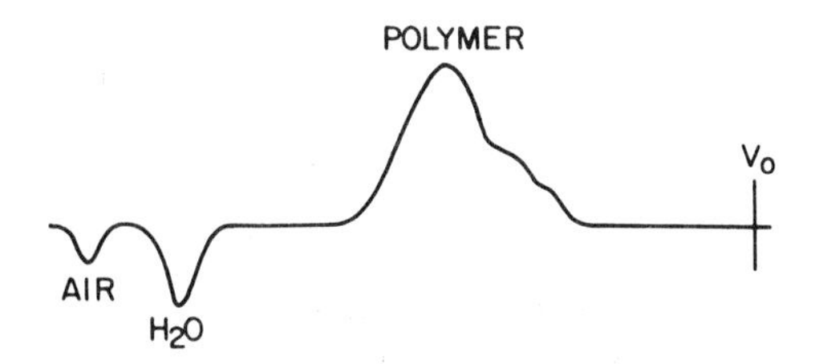

FIG. 6. Gel permeation chromatogram picturing possible effects of composition variation.

B. Semiquantitative Information

1. Comparative Techniques

Simple comparative techniques can often be used to obtain reliable estimates of molecular weight distribution (MWD) of many polymers. When the samples permit, the polydispersity, d, (M_w/M_n) can also be calculated.

In many research programs and scouting experiments, only the shape of the molecular weight distribution of a polymer is needed. It is sufficient to know if the MWD is bimodal, Gaussian, broad, narrow, long tailed, or weighted to the high or low MW side. To determine these features by GPC the problem narrows down to establishing that the chromatogram really represents the MWD of the sample. The author has studied this problem and has established that if a chromatogram falls in the linear region of the GPC calibration, the curve width and its features truly represent the sample MWD. This principle was demonstrated in the following way.

Using appropriate standards, it was shown that if a chromatogram falls in this linear region (log $\bar{M}_w$ versus V_R) and provided efficient columns are used, W/d is a constant (26-28); W is the GPC curve width and d is dispersity, that is, weight-average molecular weight, $\bar{M}_w$, divided by number-average molecular weight $\bar{M}_n$.[2] Since W/d is a constant, d can be calculated by comparing one curve width "a" for polymer "a" to another "b" for polymer "b" via Eq. (1).

$$W_a/d_a = W_b/d_b \tag{1}$$

Further, since d is but one measure of the MWD, Eq. (1) shows that the whole MWD of polymer "a" may be compared to "b" by comparing their gel permeation chromatograms. It should be noted that the comparison is not on a point for point MW for MW basis as in other calculation techniques, but rather via curve widths or features or both.

The advantages of the comparative technique are several and fulfill the needs stated above. The MWD features of a sample are observed simply by comparing the chromatogram obtained to that obtained for a standard. This is done visually, much like comparing two paintings of the same subject and with the confidence that differences are real. The principle applies even if d cannot be calculated because of the form of the MWD of the sample. However, in many cases d can be calculated rapidly if the above conditions are met. Bly illustrated the determination of d for PMMA (26) using Eq. (1) and a commercially available polystyrene as a standard.

$$\text{Sample 1, } d\ (\text{GPC}) = 2.3; \quad d\left(\frac{\text{light scattering}}{\text{osmometry}}\right) = 2.15 - 2.4$$

$$\text{Sample 2, } d\ (\text{GPC}) = 2.1; \quad d\left(\frac{\text{light scattering}}{\text{osmometry}}\right) = 1.8 - 2.0$$

To chromatograph and calculate the two samples and standard by GPC took about 5 hours. The light scattering and osmometry work together took about 3 1/2 days. It cannot be overemphasized that the GPC operating conditions must be identical for samples and standards and that the respective curves must fall in a linear log region.

2. Universal Applicability of the Comparative Technique

Perhaps it is intuitively obvious that valid comparisons can be made between various chromatograms that fall on the same linear calibration line. But what about polymers whose curves are on different linear calibration lines because of differences in the polymer types? Again Bly has shown that provided each GPC curve falls in the linear region of its respective linear calibration and regardless of the fact that one calibration line may differ considerably from another in location and slope, valid comparisons can be made between two chromatograms just as in the case above (27). In fact it is not even necessary to know the calibrations but only to know that the GPC curves would be in the linear region if the calibrations were known.

Under these conditions any standard can be compared to any unknown independent of polymer type via their chromatograms.

These conclusions were validated as follows. For any particular linear log calibration line, retention volume may be related to MW by

$$V_{R_2} = K - k \log \overline{M}_{w_2}$$
$$V_{R_1} = K - k \log \overline{M}_{w_1} \tag{2}$$

so that

$$k = \frac{V_{R_2} - V_{R_1}}{\log (M_{w_1}/M_{w_2})} \tag{3}$$

Furthermore, resolution R_s, normalized for sample polydispersity and MW, can be expressed for that line by Eq. (4) (see also Section III-B-2),

$$R_s = \frac{2(V_2 - V_1)}{(W_1/d_1) + (W_2/d_2) \cdot \log (M_{w_1}/M_{w_2})}$$
$$= \frac{2k}{(W_1/d_1) +} \qquad \frac{2k}{(W_2/d_2)} \tag{4}$$

Again, for that same particular linear log calibration $W_1/d_1 = W_2/d_2$, so

$$R_s = k/(W/d) \tag{5}$$

If we now compare two polymers ("a" and "b") whose curves lie on different linear log calibration lines, then

$$\frac{R_{s_a}}{R_{s_b}} = \frac{k_a/(W_a/d_a)}{k_b/(W_b/d_b)} = \frac{k_a \cdot W_b/d_b}{k_b \cdot W_a/d_a} \tag{6}$$

or

$$\frac{W_b/d_b}{W_a/d_a} = \frac{R_{s_a} \cdot k_b}{R_{s_b} \cdot k_a}$$

Since k is the reciprocal of the calibration line, that is

$$k = 1/\text{slope}$$

then

$$\frac{W_b/d_b}{W_a/d_a} = \frac{R_{s_a} \cdot S_a}{R_{s_b} \cdot S_b} \tag{7}$$

In words Eq. (7) states that the relationship between W/d's and consequently molecular weight distributions for two polymers that lie on different linear log calibrations, as for example a sample and a standard, depends inversely on the ratio of the products of their line slopes and resolutions, respectively.

A unique relationship was found between R_s and slop which provided the basis for a universal calibration in GPC because W/d was then uniquely defined by the slope. A variety of conditions, standard polymers, and solvents were studied with Styragel columns,[3] and the relationship was found to be independent of polymer type but dependent on each set of columns and operating conditions. Thus a family of curves was obtained (Fig. 7).

Each point in Fig. 7 represents the slope of a calibration line and a corresponding normalized resolution R_s for that line. So that the variety of polymers and conditions incorporated in Fig. 7 can be appreciated, details are provided in Table 1. The data show that the most dominant variable is the number of columns in a series. For any given number of columns, a change in operating conditions may cause some small change in the slope versus R_s relationship. It can easily be seen in Fig. 7 that the product $R_s \cdot$ slope is a constant for each set of columns and operating conditions independent of polymer type. This means that

$$R_a \cdot \text{slope}_a = R_b \cdot \text{slope}_b \tag{8}$$

and

$$W_a/d_a = W_b/d_b \tag{9}$$

even though

$$\text{slope}_a \neq \text{slope}_b \tag{10}$$

In words, Eq. (8-10) state that for the same columns and operating conditions, resolution varies inversely with line slope and that this variation is reflected almost entirely in the retention volumes rather than in curve widths. Note that Eq. (9) is the same as Eq. (1).

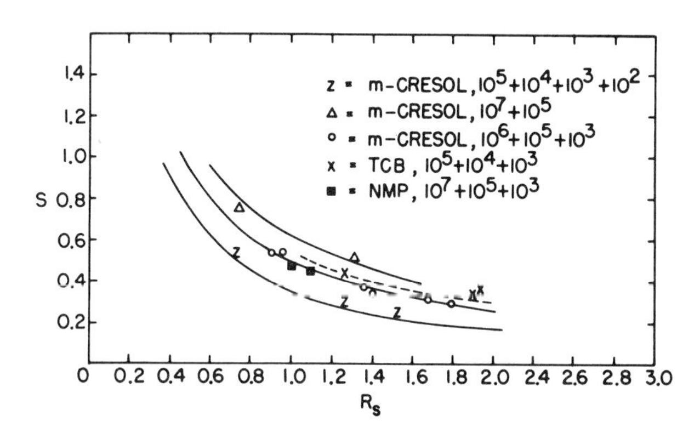

FIG. 7. Relationship between R_s and slope for linear calibrations in gel permeation chromatography as function of operating conditions. Reprinted from Ref. (27), p. 477, courtesy of American Chemical Society.

Thus two molecular weight distributions may be compared directly via their GPC widths even though their respective calibrations differ in slope, provided the above criteria are met. This author has observed that for samples which permit, d is directly calculable from Eq. (1) with an accuracy of about ±10% for $1.3 \leq d \leq 4.0$. For $d > 4$, accuracy has not been checked. For $d < 1.3$, results are quite sensitive to efficiency although the technique is applicable, especially if narrow d standards are run.

TABLE 1

Unique Relationship between R_S and Slope in Linear Log Calibrations in GPC as a Function of Operating Conditions

Conditions:	m-cresol, 100° $10^6 + 10^5 + 10^3$ [a]		Conditions:	TCB, 135°-150° $10^5 + 10^4 + 10^3$		Conditions:	m-cresol, 100° $10^7 + 10^5$	
Sample	Slope	R_s [b]	Sample	Slope	R_s	Sample	Slope	R_s
PS (Column Set 1)	0.36	1.40	PE	0.36	1.90	PS	0.52	1.30
N	0.55	0.95	PS	0.32	1.79	N	0.77	0.74
PS (Column Set 2)	0.32	1.67	BD	0.46	1.26			
N	0.55	0.90	HBD	0.35	1.93			
PMMA	0.40	1.36						
Conditions:	m-cresol, 90° $10^5 + 10^4 + 10^3 + 60$		Conditions:	m-cresol, 150° 10^4 only; 2.5 mg load		Conditions:	NMP, 40° $10^7 + 10^5 + 10^3$	
Sample	Slope	R_s	Sample	Slope	R_s	Sample	Slope	R_s
PS	0.25	1.50	PS	0.58	1.76	PS	0.48	1.00
N	0.56	0.71				PMMA	0.478	1.08

[a] All samples 5 to 10 mg loadings run at 1 ml/min on 3/8 in. x 4 ft. columns, one porosity/column as listed unless otherwise specified. PS is polystyrene, N is nylon, BD is poly(butadiene), HBD is hydrogenated poly(butadiene), and PMMA is poly(methyl methacrylate); PE is polyethylene.

[b] Calculated according to Eq. (4).

3. Coupled Columns

To obtain long linear-log regions for the comparisons described above, it is generally necessary to use coupled columns. The length and slope of the linear log region depend on many factors: solvent, polymer type, gel type, porosities, number of columns, etc. For Styragel, good combinations are three columns with $10^7 + 10^5 + 10^3$ or $10^6 + 10^5 + 10^4$ porosities for high molecular weights, viz., for polystyrene in the range 5×10^3 to 1×10^6. Only efficient columns should be used. The number designations are those used by Waters Associates, Inc. (1964-1969) for describing the gel pore sizes in polystyrene Å units consistent with the Q-factor concept (see Section II-C-b). A committee, ASTM D-20.07.04, now recommends designation of pore size by excluded hydrodynamic volume, $V_{h,\ max}$, expressed as $[\eta]$ M. It is anticipated that the gel manufacturers will follow this recommendation (see Appendix I).

The author has discussed the determination of column efficiency by means of plate count equations (28). J. Kowk et al. (29) have warned of some of the causes of decreased efficiency in using coupled columns. They suggest that it is generally unwise to couple columns which differ markedly in individual efficiency, column diameter, or composition.

Practical applications of chromatography generally require various compromises in operational procedure. Ultimate, or maximum attainable, resolution is usually sacrificed for increased speed of analysis by increasing solvent flow rate or reducing number of columns. In GPC highest resolution is usually obtained from single pore sizes or from a pore size distribution that matches the size distribution of the sample. The range of molecular sizes separable by a single pore size is narrow, and for polymers this range usually does not encompass all the molecular sizes in the sample. Therefore columns or varying individual pore sizes must be coupled or else consecutive runs made through individual columns of successively varying porosity followed by some sort of data compilation and summation using a complicated set of calibrations. Provided "sufficient resolution" is maintained, column coupling is easier to handle. The quantity "sufficient resolution" must be judged by the chemist. In our laboratory, for routine work we aim for long linear calibration regions in order to exploit the simplified data reduction and evaluation permitted by them and we judge the resolution as sufficient if it permits evaluation of the macroscopic details of MWD, $\overline{M}_w$, and $\overline{M}_n$. We normally do not exceed three columns in a series. If a problem warrants evaluation of microscopic properties, viz., a close look at a given molecular size region, maximum resolution over a more narrow range is sought.

C. Quantitative Data

1. Calibration for Molecular Weights

a. Real. Specific molecular weights in GPC can be determined only from a calibration curve. This is true whether simple or sophisticated calculations are used. In contrast to the polydispersity equation above, the calibration curve is very dependent on polymer type and on the molecular weight average used for the calibration, $\overline{M}_n$, $\overline{M}_w$, $\overline{M}_z$, $\overline{M}_v$, etc. (Certain functions such as $M \cdot [\eta]$ may not be so dependent; these are covered below.)

Calibration for a single molecular weight average such as $\overline{M}_w$ in GPC may be simple but for complex materials it is not, and obtaining proper standards is not. Often the required standards do not exist. Calibration requires chromatographing several samples of the specific polymer type in question (30) which have narrow molecular weight distributions and known, but significantly different, molecular weights covering the entire range of interest. The peak retention volumes are then plotted graphically against the known molecular weight average. For very narrow fractions, $\overline{M}_w = \overline{M}_v = \overline{M}_n$. Ideally part of the plot will be linear in log M versus V_R or its functionality easily defined (see Fig. 8). The molecular weight average of the unknown is determined from the calibration plot and the peak retention volume and is in the units of the calibration curve, $\overline{M}_w$, $\overline{M}_n$, or $\overline{M}_v$. Probably the variable used most often is $\overline{M}_w$ (22, 26).

In evaluating a GPC curve the M's obtained as above are quite accurate in the vicinity of the peak but become increasingly inaccurate with distance from the peak because of dispersion (31). For this reason other molecular weight averages calculated by curve integration and summation equations become more and more dependent on the number and quality of the standards used, the molecular weight average chosen ($\overline{M}_w$, $\overline{M}_n$, $\overline{M}_v$) for the calibration, and the correction techniques employed. Corrections are discussed below. For broad molecular weight distributions, dispersion affects $\overline{M}_w$, $\overline{M}_v$, and $\overline{M}_n$ to different extents. Errors in the calculation of molecular weight

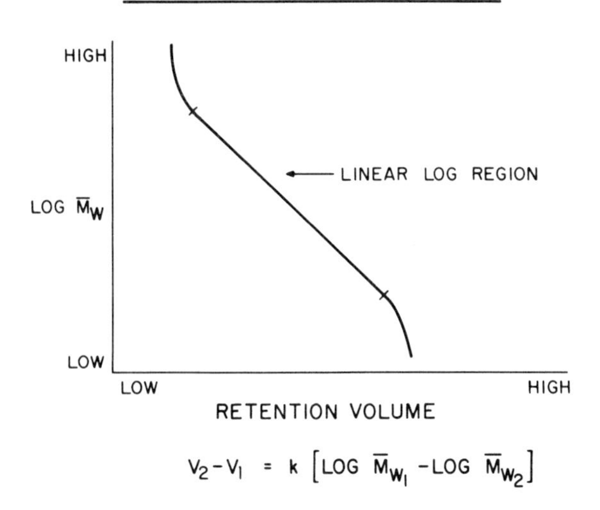

FIG. 8. Form of the general calibration in gel permeation chromatography.

averages of polymers from fractionation data have been analyzed by M. Bohdanecky et al. (32).

Well-characterized standards for a few polymer types have recently become available commercially (see folders from the National Bureau of Standards, the Pressure Chemical Company, and Phillips Petroleum Company). The ASTM (D-20.70.05) is also working on the problem, and at least two laboratories in the country provide service in both osmometry and light scattering (33, 34) for laboratories that do not have all the needed equipment. Nevertheless the shortage of proper standards may cause difficulty in obtaining quantitative data by GPC for some time.

b. The Q-Factor. The author would be remiss if he did not mention the "Q-factor" at this point. The concept of the Q-factor was introduced for polymers by L. Maley (23), D. J. Harmon (35), and L. W. Gamble (36) in 1965. It is also discussed by Waters Associates, Inc., in their GPC instruction manual and is employed by several service laboratories.

The Q-factor concept was employed by early workers in an attempt to find a universal calibration for GPC. The Q-factor is defined as "the molecular weight per angstrom length." The "angstrom length" is assumed to represent the size of the molecule in a fully extended configuration and is calculated from appropriate valence angles and bond lengths. Based on this assumption a calibration curve is prepared by chromatographing a series of well-characterized, narrow polydispersity standards. The logarithm of angstrom size (extended chain length) of the standard plotted against the chromatographic elution volume constitutes the calibration curve. It is assumed that a given elution volume corresponds to a given angstrom size and that if the molecular weight per angstrom length is known, the molecular weight for any polymer is easily determined.

However, the Q-factor can lead to very misleading MW values and its use is not recommended. Harmon immediately recognized discrepancies in his data when using the Q-factor and shortly recommended that calibration curves use only the same chemical species as those being studied and that the Q-factor not be employed (30). The problem with the Q-factor is not in the concept but rather in the model chosen, that is, the use of the fully extended length. Polymers vary in size in different solvents. This is perhaps most simply illustrated by the fact that dilute solution viscosity is dependent on polymer type and solvent. What is needed for a universal calibration is a model that will accurately relate molecular sizes in a given solvent independent of polymer type.

c. Universal Calibration. A number of publications have appeared relating V_h, hydrodynamic volume to retention volume, V_R, in GPC (37-41). These involve determining hydrodynamic volume in various ways and include studies with various polymer types, solvents, and temperatures. Z. Grubisic, P. Rempp, and H. Benoit have reported what is probably the most universal, descriptive, and easily used calibration, namely, a plot of $\log[\eta] \cdot M$ versus retention volume (41). They tell why $[\eta] \cdot M$ is a measure of the hydrodynamic volume and show a rather remarkable, nearly linear calibration line relating $\log[\eta] \cdot M$ to retention volume for a variety of polymers and molecular weights. The list includes linear polystyrene, poly(methyl methacrylate), polybutadiene, poly(vinyl chloride), block and graft copolymers of styrene and methyl methacrylate, star-shaped polystyrenes and poly(methyl methacrylates), and copolymers composed of a "three block sequence PMMA-PS-PMMA with PS grafts on the PMMA blocks."

There are several features associated with the use of the log $[\eta]\cdot M$ calibration. Under a fixed set of operating conditions (solvent, temperature, flow rate, etc.), one polymer may be used as a standard for another. Commercially available standards can be used to calibrate most systems. However, this approach has not been immediately accepted. For one reason the chromatogram is expressed in the $[\eta]\cdot M$ unit. People are not used to thinking in terms of this unit, and few if any correlations with physical properties have been established. To get molecular weights or molecular weight distributions expressed in molecular weight terms, intrinsic viscosities of the samples must also be measured. This, of course, is rather time consuming. Also, it is well known that the molecular expansion in solution changes at one or more molecular weights for many polymers; that is, "a" in the Staudinger equation, Eq. (11), is not a true constant,

$$[\eta] = KM^a \tag{11}$$

Therefore applicability to all polymers is not uniform. Recently Dawkins showed (42) that for the data of Meyerhoff (39, 78) use of unperturbed dimensions is superior to the plot of $[\eta]\cdot M$. More testing of both these approaches is needed (see Addendum).

2. Computations

a. The Pickett Program. To quantitatively reduce to common terms the wealth of information available in a gel permeation chromatogram a rather detailed and lengthy set of calculations must be made. Raw data are obtained as an analog tracing of retention volume, V_R, versus detector response, assumed to be proportional to concentration, X. Retention volume, V_R, is related to log M (or better hydrodynamic size), and the rate of change of retention volume with log M is $dV_R/d \log M$. It is desirable to normalize the raw data by converting operational variables such as sample loading, refractive increment, and instrument attenuation to unit sensitivity and then to make corrections for a sloping base line. For analyzing the molecular weight distribution, the quantity $d(X)/d(\log M)$ is needed (43-45). Therefore it is necessary to convert the normalized raw data dX/dV_R to $dX/d \log M$ by plotting dX/dV_R versus $dV_R/d(\log M)$. The resulting curve can be digitized and integrated to arrive at the classical sums describing $\overline{M}_n$, $\overline{M}_v$, $\overline{M}_w$, $\overline{M}_z$, and $\overline{M}_{z+1}$. A computer program for performing all these operations has been described by H. E. Pickett et al. (45-46). The program provides for printout of the MWD profile in differential and integral form and for a differential MW histogram. Some new interpretations on relative and absolute measures of heterogeneity of polymer sizes have recently been discussed by Carmichael (47) and might be useful additions to the computational programs above.

W. W. Yau et al. (44) have stressed that the only valid, completely unvarying basis for comparison (calculation) of MWD data is literally the true MWD curves themselves and that these can be obtained from GPC only through full use of the calibration curves. The use of differential calibration curves (dX/dV_R versus $dV_R/d(\log M)$) is also essential for nonlinear calibrations, and corrections for zone broadening (axial disperson) must be made when the magnitudes of such corrections are significant. A rigorous graphical procedure for obtaining and interpreting GPC data along these lines has been presented (44).

b. The Manual Method. Cazes (2) and Waters Associates, Inc. (48) have described the analysis of a gel permeation chromatogram on the basis of the principles discussed above. A small desk calculator is sufficient for the calculations and a computer is not necessary although these calculations can consume a great deal of time if done manually.

The method follows in outline form except that the complicating role of the Q-factor has been deleted (refer to Table 2).

(1) Draw a line across the base of the GPC curve, from the beginning of the curve until its end.

(2) Digitize this curve by drawing vertical lines from the base line to a point on the curve. Use 20 to 70 lines or digital points depending on the width of the curve and the accuracy required.

(3) Measure the height H of each vertical line or the height of the digital point from the base line to the point on the curve. This measurement must be quite accurate in inches or centimeters.

(4) Make a table listing the values of these heights versus the retention volume, V_i, Table 2, columns 1 and 2. The retention volume also must be measured very accurately.

(5) Add the heights and make a table of cumulative heights, ΣH_i, so that a second column is formed relating cumulative heights to elution volume. It is assumed that the cumulative height is retention volume. It is assumed that the cumulative height is related to the cumulative weight of polymer present at that point

just as any individual height measured is directly related to the amount of polymer present at that retention volume.

(6) Normalize the cumulative heights on a scale of 0 to 100 by dividing the individual cumulative height by the total cumulative height and multiply by 100; this is not shown in Table 2.

(7) Form the distribution curve by plotting the normalized cumulative height, from 0 to 100%, versus retention volume; from the calibration curve, retention volume can be related directly to molecular weight.

(8) The number-average and weight-average molecular weights can then be calculated readily using the other columns in Table 2 and the mathematical definitions for $\overline{M}_n$ and $\overline{M}_w$. These are the classical definitions where

$$\overline{M}_n = \frac{\Sigma N_i \text{ X } M_i}{N_i} \quad \text{and} \quad \overline{M}_w = \frac{\Sigma N_i \text{ X } M_i^2}{N_i \text{ X } M_i}$$

The symbols in Table 2 correspond to those in the summations.

(9) The reader must be cautioned that the calculated $\overline{M}_n$ and $\overline{M}_w$ values will not be accurate unless the chromatograms were corrected using the corrections discussed below, nor will the values be correct unless an absolute calibration curve is available.

TABLE 2

Arrangement for Calculation of MW from GPC

1 Retention volume	2 Digital height	3 Cumulative heights	4 MW of standard	5 Column 2 / Column 4	6 Column 2 X Column 4
V_i	$H_i(M_iN_i)$	H_i	M_i	N_i	$M_i^2N_i$

3. Chromatogram Correction Techniques

Strictly speaking, quantitative data may be obtained from GPC only if the calibration is fully and accurately known and if various biases on the chromatographic procedure are accurately determined and compensated. Bias may occur because of longitudinal diffusion and channeling in the column, mixing in the detector cell, column end fittings and connector tubing, viscosity drag, concentration overloading, and improper data interpretation. Complete knowledge of the magnitude of these variables is difficult to obtain because of variations between instruments, solvents, samples, operator technique, and so on. Nevertheless, good estimates of general utility have been made.

The H. E. Pickett et al. program (45) has done much toward providing a good scheme for GPC data reduction. It is incomplete, however, in not providing the corrections needed because of the curve-broadening operatives discussed above. Some corrections can be eliminated experimentally by careful planning, including the use of efficient columns, small-volume detector cells, small-volume end fittings, and dilute solutions. Various contributions to peak spreading have been studied by Hendrickson (49) who found that peak spreading conformed to the equation,

$$\overline{YV}^2 = \overline{YM}^2 + \overline{YA}^2 + \overline{YI}^2 + \overline{YD}^2 + \overline{YS}^2 \tag{12}$$

where $\overline{YV}$ is the peak width of a normal chromatogram, $\overline{YM}$ is the contribution due to the true molecular weight spread of the sample, $\overline{YA}$ is due to peak spreading in the apparatus, $\overline{YI}$ is spreading in the interstitial volume, $\overline{YD}$ is diffusional spreading due to time spent in the gel, and $\overline{YS}$ is due to sorption. Corrections for longitudinal broadening have been discussed by M. Hess and R. F. Kratz (31), who proposed a method for correcting for axial dispersion in GPC using experimentally available information from the GPC process itself and other analytical techniques. A computer program for the Hess and Kratz correction has also been made available for use in conjunction with the Pickett et al. program (50). The second Pickett program (46) is also a computational program but incorporates a reshaping method of correcting for bias and imperfect resolutions. The method looks promising, especially for broad MWD polymers.

Another approach to calculating molecular weight distributions from gel permeation chromatograms has been presented by L. Tung. In his first paper (51) Tung describes an integral equation to relate the chromatogram and the true MWD function and develops three solutions to this integral equation. The first is for the special case of the log normal MWD whereas the other two are numerical solutions for general distribution functions. The calculations were tested by Tung and others (52). A high-density poly(ethylene) sample was fractionated on a GPC unit. The cuts were rechromatographed and the chromatograms obtained were compared to those computed from the original curve. Agreement was very good. In further application of the method (53) Tung showed that correction for imperfect resolution, that is, curve broadening due to other than useful fractionation operatives, is important when the distribution is narrow or complex but minor for single-moded broad distributions. Actually, for the broad sample MWD is not much affected by correction for imperfect resolution, but the absolute values of $\overline{M}_w$ and $\overline{M}_n$ are. A small change in MWD integrated over a large retention volume range gives an appreciable change to the molecular weight averages. Although Tung has written two computer programs for the necessary computations (53) and the computations appear quite useful, they are somewhat formidable if one is mathematically unsophisticated (see Addendum).

Other more simplified techniques have been presented. Rodriguez, for example, has described techniques for comparing data via a set of distribution models (43), and Berger and Shultz (54) have investigated characteristics of gel permeation chromatograms for various molecular distribution functions.

It is necessary to be very careful in using computer programs to calculate and correct GPC data; the sophisticated technique easily gives the impression that all is well inside a rotten world. The data are certainly no better and sometimes worse than the calibration and corrections per se. The warning is given and the reader must beware. Pickett et al. have pointed out that the results obtained in their program (45) depend strongly on the number of digital points taken or their location on the curve, or both. In establishing a similar program at Du Pont, Potter concluded that it is a serious error to take data points from the raw GPC curve at equal units of retention volume (55-56). Rather, the points should be taken on a logarithmic basis because of the logarithmic form of the calibration. Points taken at equal retention volume intervals overemphasize the low molecular weight region and underemphasize the high molecular weights. Thus accurate and precise calculation and correction of the GPC data should not be overly simplified.

4. Method of Hamielec et al.

A. E. Hamielec and some of his students have recently published a new approach to evaluating gel permeation chromatograms (57-58).

The technique assumes a linear calibration curve of the form

$$V_R = C_1 - C_2 \log \overline{M}_w \tag{13}$$

Standards with known M values ($\overline{M}_n$, $\overline{M}_w$, or $\overline{M}_v$) are chromatographed under the conditions to be used for the unknowns, and the values for C_1 and C_2 are then guessed. A Rosenbrock search routine is used along with a computer. The estimated values of C_1 and C_2 define an estimated calibration curve, used to convert the chromatogram to a differential molecular weight distribution assuming infinite resolution. The differential distribution is then used to calculate the needed M values, which are compared to the true values and the magnitude of the error is noted. The routine is iterated until the calculated and true values agree within prescribed limits. The true, experimental calibration curve constants C_1 and C_2 are then explicit. The search is restricted to the desired range of possible C_1 and C_2 values by a high-failure technique. Of course the calibration curve is no better than the standards that are used and should cover no larger range of values than spanned by the standards.

The technique works best with broad MWD standards of known $\overline{M}_n$ or $\overline{M}_w$. Using an extension of this searching technique, it is also possible to find a nonlinear calibration curve, that is, the search program can be modified to search for N constants given N bits of information.

This versatile technique permits the use of one broad MWD standard with known number and weight average molecular weights, or two broad standards with known number average molecular weights, weight average molecular weights, or even viscosity average molecular weights. Therefore if two or more broad standards are used whose $\overline{M}_n$ and $\overline{M}_w$ are rather far apart and if these values are known, a long range of molecular weights is indeed encompassed by the standards.

It might be added that this technique is actually a more quantitative utilization of the comparative technique discussed above. It was assumed in the comparative techniqe that the processes which affect the breadth of the standards similarly affect the breadth of the unknowns so that direct comparisons can be made. If standards are used to obtain the calibration curve in terms of real operating conditions, the unknown samples can be directly calculated through that calibration curve because they experience the same biases as the standards.

Since Hamielec's technique involves no extra mathematical operations in calculating the distribution, none of the step-size problems or artificial oscillations are introduced which trouble many other calculation schemes. The technique has now been tested in several laboratories including the authors and has been found very useful. The work of Hamielec et al. is also significant in that it teaches about GPC theory and operating variables. It shows the importance of imperfect resolution and emphasizes for the first time the fundamental importance of skewing in GPC.

III. THEORY OF THE SEPARATION

A. Mechanism of the Separation

1. Summary of Theories

Until now we have considered what gel permeation chromatography does rather than how the job is done. GPC is a form of liquid chromatography which sorts molecules according to their size in solution. Many have sought to determine how this sorting is done.

Pecsok and Saunders (59), and more recently Altgelt (5), have reviewed studies on the mechanism of the GPC process. Discussions have been divided into three categories: steric exclusion, restricted diffusion, and thermodynamic considerations. The reviews cover most of the work through early 1967. With regard to the three mechanistic theories, Pecsok and Saunders state that the experimental data cited for each provide satisfactory support over limited ranges of molecular weights provided the column is precalibrated with known compounds of a similar type. They state further that although each author is able to produce data to substantiate his equation or method of plotting, essentially no independent data are available for rigorous testing. Recently independent work by Yau, Malone, and Fleming (60) has shown that at different times and under different conditions one or more of the stated mechanisms are operating in the GPC separating process.

The mechanism observed in a GPC experiment can be likened to the results obtained by a kineticist who is studying a complex chemical reaction consisting of more than one slow step and perhaps several fast steps. The kineticist observes only the slow steps as a function of experimental conditions and must infer the mechanism of the fast steps. Likewise in GPC under one set of operating conditions one mechanism is apparent whereas in other studies another mechanism or combination is apparent.

According to Altgelt (5), Flodin was the first to advance a theory of steric exclusion. Flodin (13) assumed that the separation process involves diffusional equilibrium; that is, the time for a molecule to diffuse into and occupy a pore is smaller than the residence time of the solute zone around that pore. If this is the case, the process would be neither diffusion controlled nor sensitive to diffusion and the distribution coefficient could be described by Eq. (14).

$$K = \frac{V_i(\mathrm{acc})}{V_i} = \frac{V_R - V_o}{V_i} \tag{14}$$

That is, the fractionation is described by the volume accessible in the pores, $V_R - V_o$, divided by the total volume in the pores, V_i. It follows that the retention volume of a material can be described by

$$V_R = V_o + (K \times V_i) \tag{15}$$

where V_o is the interstitial volume of the column. Various workers have tried to solve theoretically for K using different models. Porath (61), for example, studied flexible chains penetrating conelike volumes. Squire (62) made the same study as Porath but included cylindrical openings and crevices and considered only spherical solutes. Laurent and Killander (63) considered the gel to be a long network of randomly distributed rigid rods. The spaces between the rods represented the pores of the gel. Their calculations for exclusion are fairly well substantiated by their experimental data.

Ackers (64) on the other hand proposed a restricted diffusion mechanism for GPC. He assumed that the separation process did not involve diffusion equilibrium; that is, the time for a molecule to diffuse in and out of the pore was assumed significant relative to the residence time of the pore zone. Nevertheless, Ackers' experimental work, wherein static and dynamic K values were compared, showed that only for the highly swollen Sephadex 200 did restricted diffusion play a part. Yau and Malone (65) have extended both the theoretical and experimental approach to the role of diffusion in separation. Their work is discussed in the next section. See also Addendum, A. Mechanism.

The thermodynamic theories (as described by Altgelt) appear to be extensions of the steric exclusion theories. The principle is a comparison of experimental gel permeation chromatograms to theoretically predicted curves. DeVries (66, 67) calculated the elution curve of a polymer sample in terms of size exclusion, molecular weight distribution of the sample, size distribution of the pores (porous glass), and packing efficiency. He also assumed the molecules to have a rigid sphere conformation. The theoretical curves qualitatively agreed with his experimental results. However, the experimental calibration curves were considerably longer and more linear than the predicted curves.

Casassa (68-69) has considered the separation that would be obtained on a uniform gel pore size but with varying macromolecular chain conformation or coil size in solution. He sought to test Moore and Arrington's proposal (69a) that the entire range of probable domain dimensions of flexible coil macromolecules is important in the mechanism of size sorting in GPC. Casassa assumed that solute molecules within cavities are in quilibrium with those outside and that both phases are so dilute as to avoid polymer-polymer interaction. The major assumption, however, is a commonly made one, that conformation of the molecules can be described by random flight statistics. This implies adherence to the differential equation,

$$\frac{Pn(r)}{dn} = \frac{b^2}{6} \nabla 2 \quad Pn(r) \qquad (16)$$

where Pn(r) is the probability density for finding the n th step of a random flight at a point located by a vector, r, drawn from the coordinate origin and b^2 is the mean square length of a step in the polymer chain. Equation (16) is fundamental in dissipative physical processes and so has been solved for a variety of boundary conditions. Casassa develops an analogy to heat conduction and borrows this format for solution of the equation. He computes concentration ratios (inside the gel particle versus outside) and plots these versus R/a, where R^2 is the mean square radius of the molecule and a is the radius of the hollow cavity.

The precalculated concentration distribution agreed quite well with the observed experimental distribution obtained by GPC on a "single" pore size glass. In neither case is the discrimination according to molecular weight sharp and the region lying between nearly complete permeation of the gel and complete impermeability embraces the range $0.2 < R/a < 0.7$ which can be as much as a 12-fold range of molecular weight. Using a stochastic model for a monodisperse sample, Carmichael (70) calculated that the end-to-end distance of the polymer molecule coil and the pore size of the gel should correlate with V_R, the retention volume. The data calculated by Carmichael agreed at least semiquantitatively with Moore and Arrington's proposal and further are consistent with the experimental data of Cantow, Johnson, and Porter (71-73), shown in Figs. 9 and 10. Cantow et al.

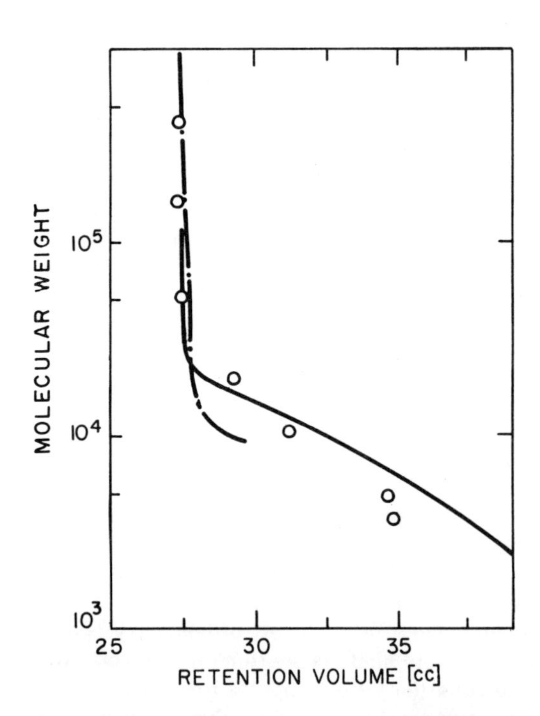

FIG. 9. Comparison of pore size distribution of Bio-Glas 200 to the fractionation of polystyrene. o, experimental points; ———, adsorptomat curve; —·—, mercury porosimeter curve. Reprinted from Ref. (71), p. 2836, courtesy of John Wiley and Sons.

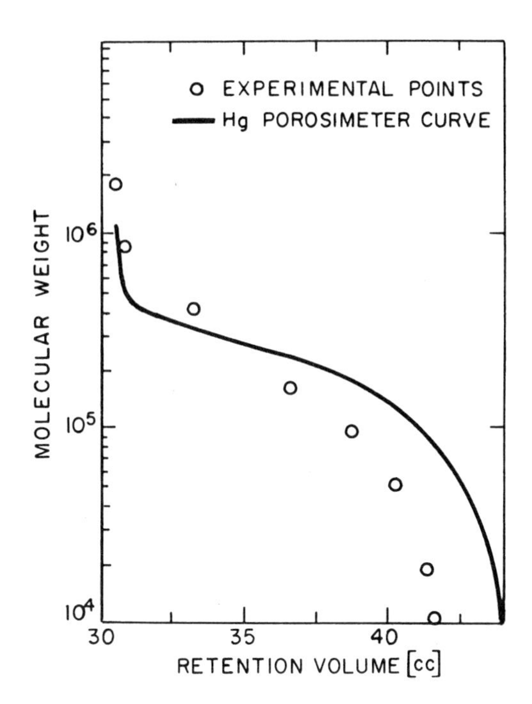

FIG. 10. Comparison of pore size distribution of Bio-Glas 1000 to the fractionation of polystyrene. o, experimental points; ———, mercury porosimeter curve. Reprinted from Ref. (71), p. 2838, courtesy of John Wiley and Sons.

( ) demonstrated that the sizes of the gel pores determined by mercury porosimetry correspond with calculated polymer coil sizes separated by the gel. The polymer size (in solution) was determined from viscosity measurements using appropriate theory and was then related to MW and V_R. The authors did not get a one-to-one correspondence of gel size to coil size as these figures alone might suggest, but the difference was a constant value. Figure 11 relates the mercury porosimetry data for a connected series of gel pore sizes to the separation of polymer coil sizes, showing a better match for the variable size gel. Figures 9 through 11 strongly support the contention that size exclusion is operative in the GPC separation process. However, in view of DeVries and Casassa's data it is likely that the

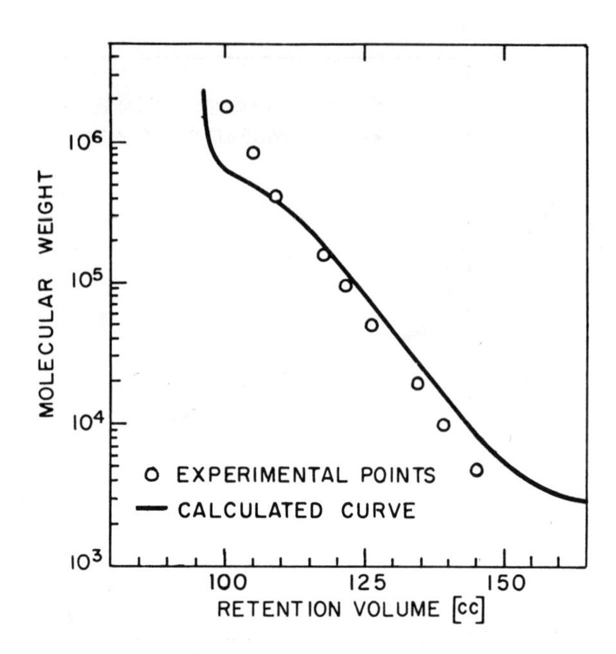

FIG. 11. Comparison of pore size distribution of multiple pore size Bio-Glas to the fractionation of polystyrene. o, experimental points; ———, calculated curve. Reprinted from Ref. (71), p. 2839, courtesy of John Wiley and Sons.

governing factor is not only the size of the gel pore but also the distribution of conformational sizes available to the polymer molecule in a given solvent.

2. Systematic Study by Yau

More currently Yau and co-workers have systematically investigated the diffusion and size exclusion theories in detail to determine the extent of their contributions to the gel permeation separation process. With Malone, Yau (65) tested the diffusion theory by assuming that the time a molecule takes to emerge from a GPC column is the sum of the time the molecule spends in the gel phase and in the flow stream. The authors assert that the fraction of the time that an average molecule spends in the flow stream is proportional to the

the probability of finding a molecule in the mobile phase. If there is no net flow through the gel phase, the elution time, t_e, can be written as

$$t_e = V_m/\bar{u} = V_m/u_m P \tag{17}$$

where $\bar{u}$ is the mean linear velocity of the molecules along the column, and V_m and u_m are the volume and the velocity of the fluid in the mobile phase, respectively. The probability, P, is approximated by a one-dimensional solution of Fick's diffusion equation,

$$C_g = \frac{2C_m}{(\pi)^{1/2}} \int_{x(4Dtd)^{-1/2}}^{\infty} \exp(-y^2)\, dy \tag{18}$$

where C_g is the concentration of solute molecules diffused inside the gel. Equation (18) can be written in terms of its error function complement,

$$\operatorname{erfc}[z] = \frac{2}{(\pi)^{1/2}} \int_z^{\infty} e^{-y^2}\, dy \tag{19}$$

or

$$C_g = C_m \operatorname{erfc}\left[\frac{x}{2(Dt_d)^{1/2}}\right] \tag{20}$$

where C_m is the solution concentration in the mobile phase; D is the diffusion coefficient; t_d is the diffusion time; and x is the distance from the gel surface. The number of molecules, N_g, diffused inside the gel phase can then be obtained by integration of Eq. (20) over the distance a of the gel phase and substituting N_m/V_m for C_m and N_g/V_g for the average C_g. After integration and substitution of identical terms, the authors arrive at their working equation,

$$V_R = V_o + V_g\left[\frac{k}{\left(\pi u_m M^b\right)^{1/2}} \; 1 - \exp(-u_m \cdot M^b/k^2)\right. \\ \left. + \operatorname{erfc}\left[\frac{U_m \cdot M^{b\ 1/2}}{k}\right]\right] \tag{21}$$

The authors note that in the intermediate molecular weight range Eq. (21) may be approximated by Moore's equation

$$\log M = A - B \cdot V_R \tag{22}$$

while at the larger molecular weights, Eq. (21) reduces to an equation equivalent to that of Meyerhoff,

$$V_R = KM^{-c} \tag{23}$$

where K and c are empirical constants.

Yau and Malone found that their calculated curve fit their experimental points remarkably well (Fig. 12). In the calculations V_g was

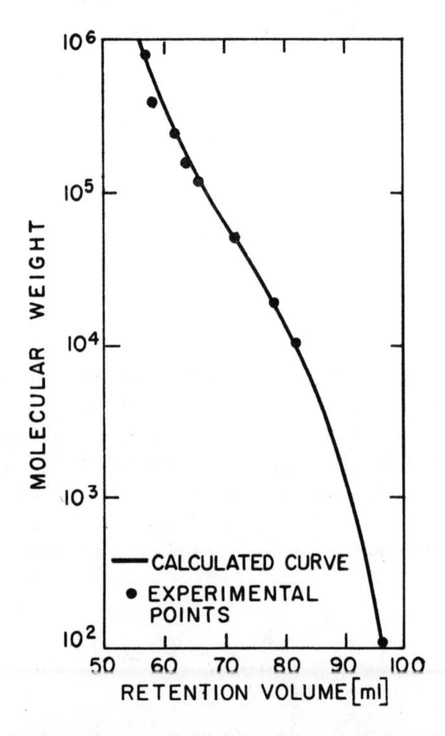

FIG. 12. Comparison of experimental points with calculated calibration curve (Eq. (21)). Reprinted from Ref. (65), p. 667, courtesy of John Wiley and Sons.

assumed to be the total volume of the gel phase by completely neglecting any exclusion effect, b was assumed to be 2/3, and the values of V_o and k were chosen to fit the experimental results. The retention volume approaches the void volume V_o for large molecular weight and approaches the total volume of the column ($V_o + V_g$) for small molecular weight. These limiting values of the retention volume were considered to be reasonable since the geometry of the column was expected to establish the limits for the retention volume of the molecule. The authors then compared the retention volumes at two different flow rates to the curves predicted by Eq. (21). Their results (Fig. 13)

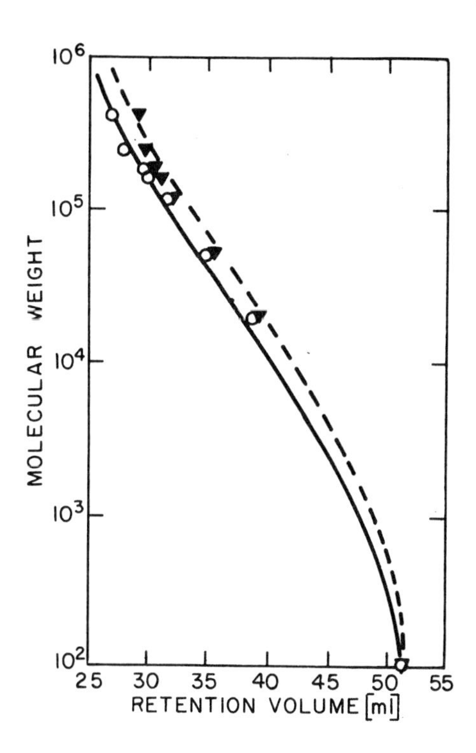

FIG. 13. Effect of flow rate on calibration curve. o, experimental points for 0.93 ml/min; ———, calculated for 0.93 ml/min; experimental for 0.67 ml/min.; ------, calculated for 0.67 ml/min. Reprinted from Ref. (65), p. 668, courtesy of John Wiley and Sons.

showed that although the flow rate data were in the predicted direction, quantitatively the prediction was rather poor. They concluded that the discrepancy was a consequence of neglecting the exclusion effect on the quantity V_g in Eq. (21). Nevertheless, the general agreement with the experimental results indicated that diffusion could play an important role in the GPC separation process.

Hermans (74) has since introduced a general theory for describing the diffusion of solute into the gel particles. Yau and Malone's approach described above is a special case of Hermans' generalized theory. The general theory solves the problem in principle but unfortunately leads to intractable equations. Hermans "solves" these equations for two limiting cases and uses them to consider some moments. The mathematics is rigorous and the reader is referred to the original paper for details. Hermans concludes by stating that caution should be used in the use of GPC data when diffusion into the gel particles plays an appreciable role. The diffusion results in relatively long chromatographic tails and these tails will tend to be longer the smaller the diffusion coefficient of the solute in the gel. Consequentely the high molecular weight species will show the most pronounced tails. These can overlap with the curves for the lower molecular weight species and reduce resolution. To improve resolution it may be necessary to use the smallest possible particle size.

With Malone and Fleming, Yau (60) next studied the role of exclusion in GPC. To eliminate the effect of diffusion it was necessary to extrapolate the gel permeation results to zero flow rate. This was done rather simply via a static mixing experiment analogous to that used by Ackers in his study of the biological system. A solution of known volume, V_i, and initial concentration, C_i, is mixed with a measured amount of dry, porous gel. The system is allowed enough time to reach the equilibrium distribution following mixing. The material balance of the solute molecules in this system was then expressed,

$$C_i \times V_i = C_o \times (V_i - V_g) + C_g \times V_g \tag{24}$$

where V_g and C_g represent the liquid volume and the equilibrium concentration inside the porous substrate, respectively, and C_o is the concentration outside the porous substitute. The following identities were then established:

$$(C_g/C_o) = K_x = (K_{GPC}/K_D) = K'_{GPC}/k \cdot K_D \tag{25}$$

By substitution and rearrangement, Eq. (24) becomes

$$1 - \frac{C_i}{C_o} = \frac{V_g}{V_i}(1 - K_x) = \frac{V_g}{V_i}\left(1 - \frac{K'_{GPC}}{kK_D}\right) \tag{26}$$

Thus if exclusion is important to the GPC separation, C_o should be greater than C_i and, in the case where K_D approaches unity and becomes independent of molecular weight, the plot of $(1 - C_i/C_o)$ versus K'_{GPC} should approximate a straight line.

The results of the experiments designed to test these predictions are shown in Fig. 14 (porous glass) and Fig. 15 (polystyrene gel). Although the solution concentration in the mixing experiment was somewhat higher than the actual concentration of the solute in the peak

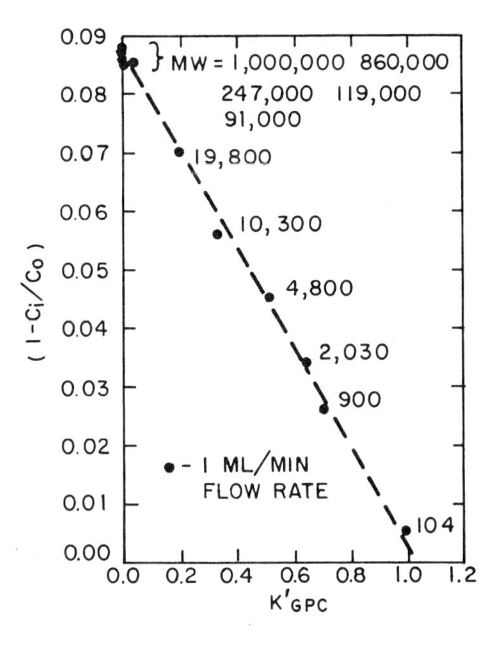

FIG. 14. Comparison of gel permeation chromatography [Eq. (28)] and static (mixing experiment) data for Bio-Glas 200. Reprinted from Ref. (60), p. 804, courtesy of John Wiley and Sons.

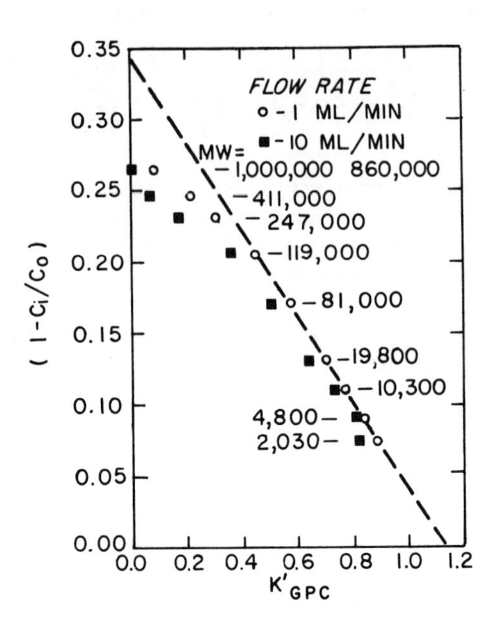

FIG. 15. Comparison of gel permeation chromatography as function of flow rate [Eq. (28)] and static data for polystyrene gel, 10^4 designation. Reprinted from Ref. (60), p. 805, courtesy of John Wiley and Sons.

eluting from the GPC column, this difference in concentration should not affect the data since $(1 - C_i/C_o)$ as determined from the mixing experiment is not a function of solute concentration. It is evident from the excellent agreement that exclusion does play a primary role in the gel permeation separation process. This, of course, agrees with other authors' views presented above. [Concurrent with Malone, Fleming, and Yau (60), Chang (75) carried out a series of similar experiments and came to essentially the same conclusion, that the solute molecule is excluded from part of the inner space of the gel particle which is entirely available to the solvent molecules and that the excluded volume increases with increase in the molecular size of the solute.]

Diffusion plays essentially no role in the GPC peak separation in the porous glass described above (60). The authors, however, point out that for larger pore size glass at a higher flow rate the data show small deviations from linearity at the high molecular weight end of the curve. The poor fit found with the polystyrene gel, particularly at the high flow rate and in the higher molecular weight region, shows that diffusion is also important in the separation mechanism with this gel.

Therefore Yau wished to examine further the diffusion contribution to the GPC separation observed in Fig. 13 in terms of additional theory. By linking several previous theories he formulated one (76) including both the steric exclusion and lateral diffusion effects in GPC. In laying the ground work he showed that the distribution coefficient, K_{GPC}, of Laurent and Killander (63), which was derived for rigid body solute molecules of definite dimensions, could also be applied to K_{GPC} for flexible polymer molecules; that is,

$$K_{GPC} = \frac{V_R - V_o}{V - V_o} = C_g/C_o \tag{27}$$

where C_g and C_o are the average solute concentrations per unit liquid volume inside and outside porous substrate, respectively. Equation (27) is a general representation of the relation between K_{GPC} and the GPC data irrespective of the conformation of the molecules in solution. Representation of GPC data in terms of K_{GPC} instead of V_R is generally preferred because data obtained in different laboratories or on different instruments can be compared. Unfortunately, values of V_R determined on different instruments will often vary.

In practice K_{GPC} is difficult to obtain because V is not readily available from GPC data. V should correspond to the peak retention volume of the solvent molecules at $K_{GPC} = 1$. However, experimental detection by GPC of the solvent retention peak where $K_{GPC} = 1$ is difficult if not impossible. Commonly V is replaced by the peak retention volume V' of some small monomer molecule. The distribution coefficient K'_{GPC} calculated in this way differs from K_{GPC} by a constant factor; that is,

$$K'_{GPC} = \frac{V_R - V_o}{V' - V_o} = k \times K_{GPC} \tag{28}$$

k is a constant independent of the molecular weight of the solute molecule.

As additional ground work for his combined theory, Yau defines K_{GPC} as the product of the contributions from the diffusion and exclusion effects as

$$K_{GPC} = K_D \times K_X \tag{29}$$

where K_D is the distribution coefficient that results from the non-equilibrium process of lateral diffusion between the mobile and stationary phase and K_X is the distribution coefficient that results from equilibrium exclusion of the solute molecules from the pores of the substrate. K_D should be flow rate dependent while K_X is not. At very slow flow rate K_{GPC} approximates K_X since K_D approaches unity.

Using these concepts Yau derived a mathematical theory (76) to include both restricted diffusion and steric exclusion. From the theory he predicts the effects of size exclusion and flow rate dependence and compares the predicted curves to experimental results. The equation used for the calculation is rather complex,

$$V_R = V_o + (V' - V_o) K_X' \left[(x/\pi)^{1/2} (1 - e^{-1/x}) + \operatorname{erfc} (1/x^{1/2}) \right] \tag{30}$$

Despite its complexity, this equation has been regorously derived in terms of the above variables. X is also rigorously defined by

$$X = \frac{g_1 (1 - g_2 M\beta)^2}{U_o M} \left[1 - 2.104\, g_2 M^{\beta} + 2.09\, g_2{}^3 M^{3\beta} - 0.95\, g_2{}^5 M^{5\beta} \right] \tag{31}$$

where U_o is the flow rate, M is the molecular weight of the polymer sample, $\beta = (a + 1)/3$, and a is the Mark-Houwink exponent. The constants g_1 and g_2 represent two groups of physical variables that are not functions of U_o and M. As shown in Fig. 16 the calibration lines calculated from this equation agree very well with the experimental values, both as a function of the flow rate and the molecular weight of polystyrene standards.

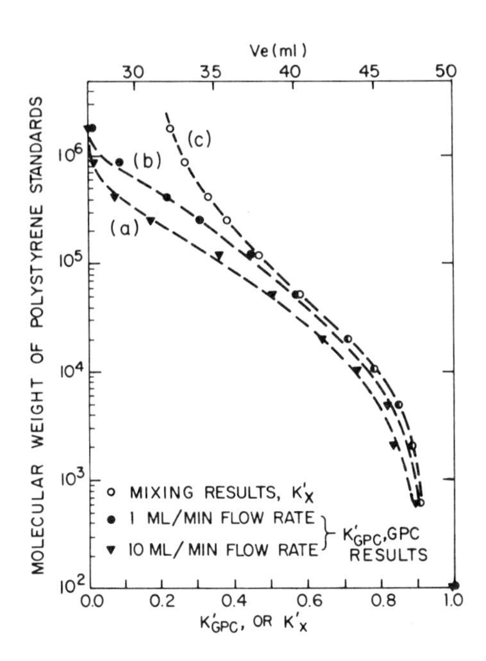

FIG. 16. Comparison of calculated calibration curves [Eq. (30)] to experimental curves for polystyrene gel. ▼ , experimental, (a) calculated for 10 ml/min flow rate; ●, experimental, (b) calculated for 1 ml/min flow rate; o, experimental for K^1x (c). Obtained from mixing experiment. Reprinted from Ref. (76), courtesy of John Wiley and Sons.

Yau concludes that in view of the complexities that remain to be clarified, the agreement of his experimental data with the theoretical predictions should be considered as additional evidence but not as proof for attributing the GPC separation mechanism solely to exclusion and diffusion effects. The complexities cited are adsorption of the polymer molecule on the surface of the substrate, possible deformation of the polymer molecule under the sheer strain existing in flow streams, deformation of gel substrate under the operating pressure, and perhaps variation of the degree of swelling of the gel substrate by the osmotic action between the eluting polymer solution and the solvent residing in the gel phase. In addition the effects of variations in temperature, concentration, and pore structure on GPC separation may need to be considered.

In summary, it is the opinion of the writer that in light of the experimental evidence the most expedient viewpoint is to consider size exclusion as the dominant separation mechanism in GPC. Size exclusion can be regarded as the sole mechanism operable in GPC only by defining any other effect, such as diffusion, to be a bias on this mechanism. This does not mean, however, that the size exclusion process is simple because organic gel pores can change size and shape and the polymer molecules can undergo size and conformational changes within the environment of the gel. Nevertheless, the dominant factor in the separation is based on certain size parameters. If the experimental conditions are such that equilibrium is not attained, that is, if diffusion, adsorption, or other mechanisms are involved, it can be considered that the basic mechanism for sorting in GPC has been to a certain extent perturbed. As was stated at the beginning of this section and as the data presented have shown, the mechanism observed in any permeation experiment of this type is a function of the limiting or slowest process. Under equilibrium conditions and without complexities, the size exclusion principle appears to be the basic mechanism operating in GPC.

B. Expressions of Performance

1. Separation Efficiency

In GPC as in all other areas of chromatography it is desirable to know the efficiency of the process. The manufacturer wants to know the quality of the columns he is selling. The GPC operator wants a simple way to test the separating power of the columns he buys and so on.

The problem of expressing efficiency has been faced in many different disciplines. Ultimately it is probably best expressed in terms of the number of theoretical equilibrations that a material undergoes in its transport through a medium. Thus the concept of theoretical plates or theoretical equilibrations has been introduced into distillation and into chromatographic theory. The theoretical plates can be expressed simply in terms of plates per unit length, by Eq. (32),

$$N = 16 \times (V_R/W)^2 \tag{32}$$

Still another way of expressing column efficiency is the height equivalent to a theoretical plate which is the length of the column divided by the total number of plates, that is, L/N. A detailed analysis of the theoretical plate concept is found in Purnell, Gas Chromatography, Wiley, New York, 1967.

Equation (32) shows that the efficiency depends on the location of the chromatogram in terms of V_R. In GPC it has been common practice to express N by Eq. (32). The technique has been to chromatograph a very low molecular weight monodisperse material, to measure V_R and W conventionally, and finally to calculate N. This technique provides the maximum plate count, N_{max}, available to the columns. However, N_{max} is not a realistic value for polymers (28), which exit from the columns at much lower V_R than the low molecular weight material. To calculate efficiency for a polymer, the chromatogram obtained for a polymer must be used in determining V_R. Since V_R is molecular weight dependent, N should also depend on the molecular weight.

There is also a problem in properly evaluating GPC curve width, W, which is a function of the molecular weight distribution of the polymer. Therefore the W obtained in order to use Eq. (32) is biased by the sample chosen. The author has studied the problem in detail (28) and has determined that the curve width for many polymers, especially standard materials, can be normalized for the molecular weight distribution by using the polydispersity of the sample. A standard polymer that has a chromatogram of regular gaussian shape is chosen. Normalization is obtained by dividing the curve width by the polydispersity of the standard sample. This technique is discussed in more detail in Section II-B where the quotient of W/d is shown to be a constant for a defined set of operating conditions. The restraints set by those conditions are not severe and the normalization is fairly broadly applicable. Since W/d is a constant, d/W is a constant, and Eq. (32), normalized for polydispersity, can be expressed as Eq. (33),

$$N = 16 \times [V_R/(W/d)]^2 \tag{33}$$

which rearranges to

$$N = 16 \times d^2 \times (V_R/W)^2 \tag{34}$$

Equation (34) shows that GPC efficiency can be expressed normally in terms of plate count N, but bias by the sample on the curve width has been removed and the efficiency is properly expressed as a function of retention volume. Equation (33) also says that for a given set of operating conditions $\sqrt{N}$ should equal V_R multiplied by a constant, that is,

$$N^{1/2} = C \times V_R \tag{35}$$

The constant C, of course, will vary with columns and operating conditions. Equation (35) was tested (28) and found to be valid by the data in Fig. 17. Note that the results are independent of polymer type. If $N^{1/2}$ is related to molecular weight, the plot is dependent on polymer type (Fig. 18).

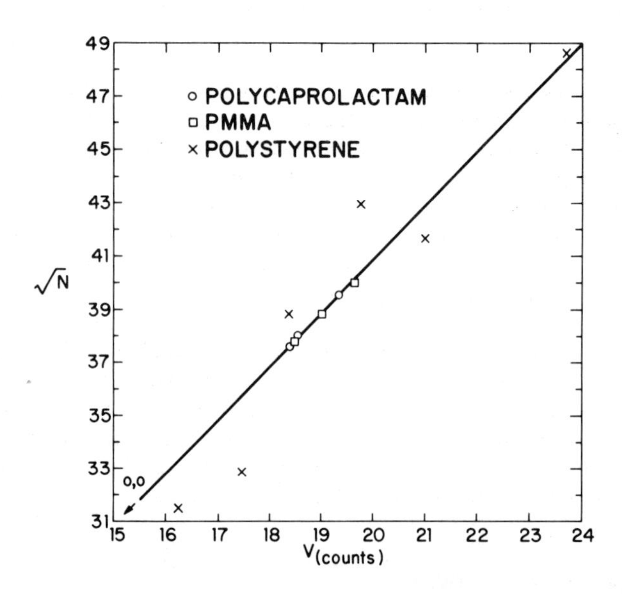

FIG. 17. Square root of theoretical plates (N) versus peak retention volume (V) as an independent function of polymer type in gel permeation chromatography. Reprinted from Ref. (28), p. 2088, courtesy of John Wiley and Sons.

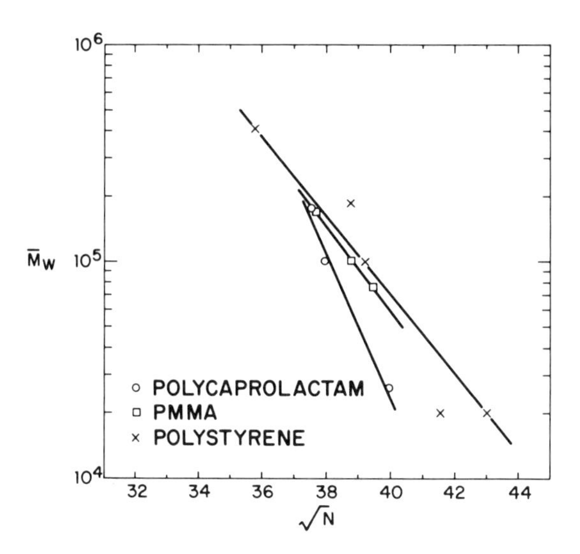

FIG. 18. Square root of theoretical plates (N) versus log $\overline{M}_W$ as function of polymer type in gel permeation chromatography. Reprinted from Ref. (28), p. 2088, courtesy of John Wiley and Sons.

The maximum number of components, n, resolvable by GPC has been discussed by J. C. Giddings (77), who has shown that Eq. (36) is a highly simplified expression for peak capacity of a gel filtration or gel permeation column.

$$n = 1 + 0.2 \times N^{1/2} \tag{36}$$

Comparing the actual capacity obtained for a given set of operating conditions to the predictions of Giddings' equation, using for N the data calculated by Eq. (34), further verifies that Eq. (34) is a valid way of calculating efficiency in GPC. From Table 3 it is assumed that polystyrene P-18M has come out at the void volume of the columns, 16.22 counts. The curve width divided by polydispersity is 2.05 counts and N is 993. Using this value for N, n is calculated to be 7. Thus the number 7 must be subtracted from all other calculations of n since the volume from zero to the void volume is not a useful separation. Using polystyrene 2A at 21.01 counts, N = 1733 and calculated n = 9.

TABLE 3

GPC Data for Standard Samples

Sample	$M_w \times 10^{-3}$ reported	d reported	V in counts	W in counts	W/D	N calc as $16d^2 (v/W)^2$
Polystyrenes[a]						
P-18M	1800	1.20	16.22	2.47	2.05	993.5
W-4190037	411	1.05	17.44	2.23	2.12	1079
N-705	185	1.10	18.35	2.08	1.89	1507
P-7a	51	1.06	19.73	1.93	1.82	1875
P-2a	19.8	1.06	21.01	2.14	2.02	1733
P-12a	2.03	1.10	23.63	2.14	1.95	2360
PMMA	M_w by L.S. $\times 10^{-3}$	from GPC				
1 (containing 4% ethyl acrylate)	169	2.00	18.39	3.90	1.95	1423
2 homopolymer	75	1.95	19.30	3.81	1.95	1561
3 (containing 4% ethyl acrylate)	100	1.86	18.95	3.63	1.95	1508
Nylons (anionically prepared polycaprolactam)						
1	175	3.51	18.38	6.86	1.95	1415
2	100	2.56	18.51	4.99	1.95	1443
3	26	2.5	19.66	4.91		1600
Acetone	-	-	25.77	1.22		7120

[a]P Stands for the Pressure Chemical Company, W for Waters Associates, Inc., N for National

The difference between 9 and 7 is the predicted number of peaks or curves resolvable in this region (between 16.22 and 21.01 counts); that is, it is predicted that two curves could be resolved in this distance and no more. Since the observed W/d values are about 2, it is verified that two curves are resolved in this region in practice under these operating conditions. Choosing as another example P-12a, where $V_R = 23.63$ counts and $N = 2360$, then $n = 11$ and by difference the number of resolvable peaks between the void volume and the location of sample P-12a is 4. Looking at the curve widths and summing indicate that indeed about four curves could be fully resolved in this region. If, however, the acetone value of 7120 is taken for N (again Table 3), then $n = 18$ and it is predicted that 11 curves could be resolved between the void volume and the acetone peak. In practice, however, using actual polymers as seen in Table 3, only five curves could be resolved in this region even when the polymers were low in dispersity and the narrowest W/d values were used. Thus n as calculated from Eqs. (34) and (36) agrees very favorably with Giddings' prediction, but n calculated from the maximum efficiency expected from the column using monodisperse low molecular weight materials, Eqs. (32) and (34), does not.

2. Resolution and Fractionation

Resolution in GPC has been expressed in a variety of ways. In his own work (26) the author has used common chromatographic expressions, normalized for samples that vary in molecular weight or polydispersity or both. Meyerhoff (78) and others have preferred to express resolution in terms of the width of the chromatogram by comparing the width obtained to the theoretical polydispersity, $\overline{M}_w/\overline{M}_n$. LePage, Beau, and DeVries (79) compared zone spreading as a function of column efficiency using the theoretical plate concept as a measure of column efficiency. They define resolution as the ratio of the difference between retention volumes of neighboring zones to the average base width of the zones. However, they do not normalize their equation for the molecular weight distribution of the sample used to calculate the resolution or for the difference in molecular weights between the samples. Therefore the resolution value obtained is not a function of the column alone but rather includes sample bias. Preferably, sample bias should be eliminated from calculations of resolution because the resolution values then obtained for any given column or sets of columns can be used to predict separation power for any set of samples.

Osterhout and Ray (80) and Feldman and Smith (81) discussed resolution in terms of changes in molecular weight per unit change in retention volume, d log M/dV_R (or the equivalent reciprocal term, change in retention volume per unit change in molecular weight). Feldman and Smith defined a Resolution Index as the ratio of molecular weights which would exhibit a displacement in their GPC peak positions equal to W, the curve width. The Resolution Index covers the range from zero, where there is no resolution, to unity, where there is a complete resolution of all species. This method is rather similar to that employed by Bly and is discussed below.

Purnell (82) in 1960 discussed the correlation of the separating power of gas chromatographic columns to efficiency where efficiency was calculated from the well-known theoretical plate equation (Eq. 32) discussed above. Many of Purnell's arguments are applicable to GPC. He says that apart from those interested in the theoretical aspects of chromatography, few workers use the theoretical plate number for determining the separation power of a column. Chemical analysts, for example, are more interested in the separation attainable with a given column and set of operating conditions than with apparent efficiency. According to Purnell, more satisfactory results have often been obtained for columns of lower "efficiency" than for others with higher plate numbers; GPC is analogous, for some workers have experienced equally good separations on both porous glass columns and organic gel columns even though the gel columns have much higher maximum plate efficiencies. After much deliberation in gas chromatography, an international body recommended Eq. (37) to define resolution of chromatographic columns (83). Equation (37) is also a good starting point for defining resolution in GPC,

$$R_{1,2} = (V_2 - V_1)/(W_1 + W_2) \tag{37}$$

There are actually three kinds of resolution to consider in GPC: theoretical resolution; limiting resolution; and practical resolution of chromatographic columns. Both theoretical resolution and limiting resolution in GPC are really dependent on the mechanism of the separation *per se* and discussion of these is included above. The reader is also referred to the work of Giddings and Mallik (84). The practical, or experimental, resolution is discussed further here.

Not much can be said about the resolution of a GPC column unless something is known about the samples that are injected into the column.

All the techniques described above require knowledge of the samples. Either their molecular weights or molecular weight distributions must be known. The technique of expressing the logarithm of the molecular weight versus retention volume amounts to defining the calibration curve. To use the Resolution Index d log M/dV_R to describe the resolution of a column, either the calibration curve must be known a priori and the chromatograms that are obtained properly corrected or else fractions must be collected and the molecular weight of those fractions determined. But again it must be emphasized that the true value of M for every V_R must be known. The value d log M/dV_R will not be a constant over the entire separation range because the calibration curve is not linear over the full range. Thus it is an incremental expression of resolution.

Bly (26) has preferred to work with resolution or the separating abilities of columns only in the linear region of calibration, log M versus V_R, because in this region many simplications are possible. Equation (37) can be normalized for the polydispersity of each sample by dividing the curve width by the polydispersity, whereas the retention volume V_R can be normalized by dividing by the molecular weight of the sample. See Sections II-B-2 and III-B. With proper normalization the resolution value, called the "specific resolution," R_s, is independent of the sample and is expressed,

$$R_s = \frac{2(V_2 - V_1)}{(W_1/d_1 + W_2/d_2) \cdot \log (M_{w_1}/M_{w_2})} = \frac{2k}{(W_1/d_1 + W_2/d_2)} \tag{38}$$

Equation (38) defines numerically the ability of a column and set of GPC operating conditions to separate two or more samples. In the linear log region of a calibration, R_s should be a constant because it has been corrected for the sample molecular weight and sample polydispersity and thus depends only on the columns and operating conditions. Figure 7 and R_s calculated from Table 3 provide support.

The reader must keep in mind that there are various causes for broadening of GPC curves. Polydispersity of the sample is a major cause, and longitudinal dispersion, viscosity drag, mixing in the end fittings, and so on, add additional breadth to the curve. Therefore this normalization technique will not be useful unless efficient columns are used and unless polydispersity is the dominant broadening function. This means that inefficient columns combined with narrow distribution standards will not provide good data for R_s. The normalization technique has been found generally applicable for polydispersities greater than 1.3 and less than 4.0 and for columns whose specific resolutions are greater than or equal to 1.0.

In practice it has been found that columns which give R_s values greater than or equal to 1.0 provide satisfactory GPC curves for obtaining macroscopic details of the molecular weight distribution of polymers. Columns which give R_s values less than 1.0 should not be used.

Fractionation. Once the specific resolution is known, and to obtain this only two samples of known molecular weights and molecular weight distributions are needed, it can be used to predict the fractionation of any two samples of any molecular weight distribution, but of the same polymer type, by Eq. (39).

$$\%F = R_s/1.5 \times \log (\overline{M}_{w_1}/\overline{M}_{w_2}) \times 100 \qquad (39)$$

The value 1.5 is used because two samples can be considered to be completely separated when the back and front of adjacent peaks reach the base line at $\pm 3\sigma$, respectively. This equation has been tested and verifying data have been published (26). Figures 19 and 20

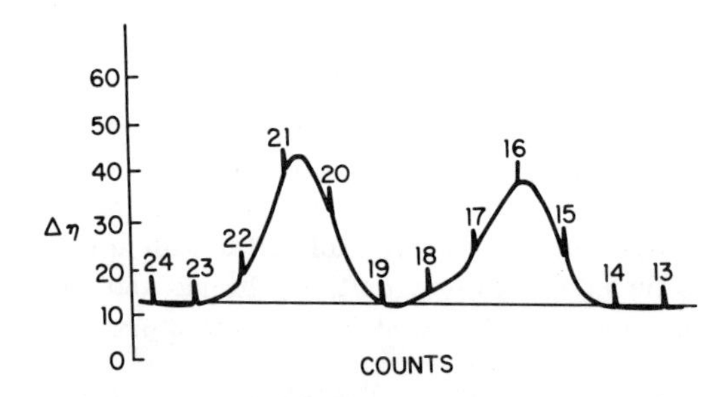

FIG. 19. Comparison of calculated fractionation of polystyrenes [Eq. (39)] to experimental values (1:1 mixture of 0.25% 7a + 4190038, run at 100°, 1 ml/min, 4 x m-cresol, $10^5 + 10^4 + 10^3$). Reprinted from Ref. (26), p. 18, courtesy of John Wiley and Sons.

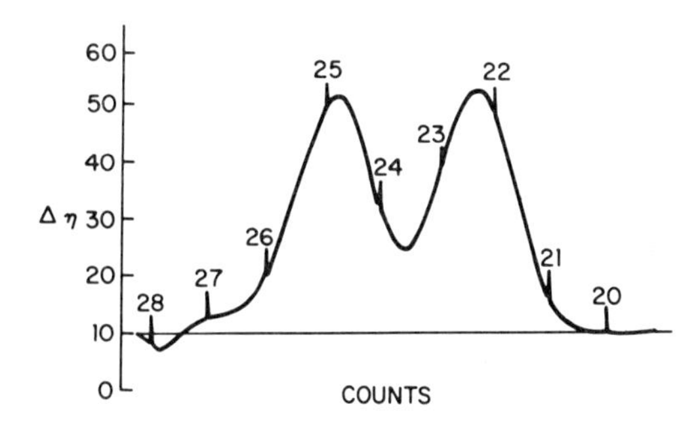

FIG. 20. Comparison of calculated fractionation of polystyrenes [Eq. (39)] to experimental values (1:1) mixture of 0.5% 11a + 2a, run at 100°, 1 ml/min, 4x, m-cresol, $10^5 + 10^4 + 10^3$). Reprinted from Ref. (26), p. 18, courtesy of John Wiley and Sons.

illustrate the fractionation of two different sets of polystyrenes. In Fig. 19 the specific resolution, R_s, is 1.42 and the predicted fractionation is 106%; that is, the equation predicts that the two curves will be slightly greater than 100% separated. Indeed it can be seen that this is the case. In Fig. 20 the same operating conditions and therefore the same R_s value of 1.42 were used, but the predicted fractionation is 71% whereas the measured value is 85%. The measured value was obtained simply as the percentage of the total curve area that does not overlap. The author has also used the specific resolution to indicate that valid comparisons may be made between molecular weight distributions of two samples whose chromatographic curves lie in the linear portions of their calibrations (see Section II-B).

3. Factors Affecting Separation or Resolution

There are many variables involved in a chromatographic experiment. Several of these may interlock in any test for the effect of one variable. For example, a study of the effect of flow rate must also consider the concentration of the sample used in making that study. The column is easily overloaded at the high MW end, a tendency that is accentuated as the flow rates increase. No fully systematic study of all the variables affecting separation has been made to date because of the many variables involved. Many independent studies have been made, but unfortunately they do not often link together the dependence of one variable on another. Variable sorting is often difficult.

Many of the variables that affect the GPC process are listed in Table 4. The table has been divided into two parts: static variables and dynamic variables. Static variables are those over which the experimenter has little or no control at the time he runs an experiment. In building a gel permeation chromatograph, the experimenter would have a choice of detector, pumping assembly, column design, and so on. However, once these are established, and most chromatographers today are using commercial, integrated equipment, few if any changes are made to conform to the needs of a single experiment. Thus these variables are static. On the other hand, the experimenter has immediate and direct control over most of the dynamic variables listed although he may often be faced with problems as to how these variables should be handled. For any given experiment he can choose and thus control the concentration, flow rate, inject time, temperature, solvent, solvent viscosity, and solution viscosity. Thus these are dynamic variables.

Table 4 also lists references to pertinent studies of the variables indicated.

Once the static variables have been optimized or are placed in acceptable working condition for an experiment, the most important and significant dynamic variable is probably concentration. Concentration should be kept to a minimum as much as is conveniently and detectably possible. The lower the concentration, the lower the chances of artificially biasing the chromatographic results. The higher the concentration, the greater the potential for trouble: viz., the higher the viscosity of the solution and the greater the pressure drop across a column, the less likely will be plug flow out of the sample injection loop and the more likely are wall effects, overloading of the column at high molecular weights, skewed chromatograms, and so on.

Unfortunately GPC today is limited in ability to detect minute quantities of material. Probably the most sensitive detector is the ultraviolet detector. However, it lacks universal applicability (see Section IV-B-2). The most used detector is the differential refractometer. Sometimes a lower concentration can be used by choosing a solvent whose refractive index is considerably different from that of the polymer. The reader should refer to Table 4 for additional references on other variables.

TABLE 4

Variables Affecting the Separation

Static Variables

1. Variable Name	Reference Number
Column design, packing configuration, size	22, 4, 87, 99, 101, 105, 84, 126, 125, 123, 127, 129
Instrument assembly	91, 87, 93, 99, 106, 107
Detectors, detection, cell design	106, 104, 103, 100, 109, 108, 107, 114, 113, 112, 111, 110, 115
Pumping assemblies, pumping design	116, 117, 118, 143, 110, 119
Pulsation damping	117, 118, 120

Dynamic Variables

2. Variable Name	Reference Number
Gel particle size	87, 99, 101, 105, 22
Gel porosity	87, 86, 71, 105, 124, 125, 126
Gel porosity distribution	71, 105
Concentration	95, 89, 92, 99, 97
Flow Rate	88, 89, 99
Inject Time	89, 102, 103
Temperature	72
Solvent, solvent viscosity	30, 96
Solution viscosity (skewness)	92
Dispersion	31, 101, 102, 103
Molecular conformation	94, 71, 30, 98
a. Solvent power	85, 30, 92, 96
b. Hydrogen bonding	85, 90
Adsorption	85

IV. INSTRUMENTATION AND TECHNIQUE

The efficiency of all practical chromatographic separations depends on operational and instrumental variables. In fact even the separation mechanism depends somewhat on these variables as discussed above. Therefore knowledge of the instrumental and operational scheme is required for a full understanding of the technique and its utility. Most of the studies reported in the literature have been performed on a Waters Associates Gel Permeation Chromatograph (48) so it is appropriate to discuss one in detail.

A. Waters Associates Gel Permeation Chromatograph

Figures 21-23 present the Waters' Model 200 instrument photographically and schematically. Operation, and indeed the experimental procedure, can easily be understood by following the solvent path through the instrument. Solvent from the 16-liter reservoir (Fig. 23) is directed through a heated, nitrogen-purged, degassing chamber from whence it passes through the pump, a ballast, and filters and is split into two streams. Flow is regulated by the flow control valves and pump stroke. A 250-lb pressure-relief safety valve is provided. The sample stream passes through the (closed) sample valve, working columns containing the porous gel, and one side of a differential refractometer. The four-way sample valve permits sample injection on opening by placing a tubular coil (2-ml capacity) previously filled with the sample solution into the solvent flow line. Plug flow is assumed. Even though this assumption is technically unrealistic, in practice serious errors are unlikely from this source if several columns are used in the sample line (103). Interruption of a light beam by solvent emptying from a 5-ml syphon allows a digital monitoring of the solvent flow rate. The interrupted signal activates a "pipper" on the recorder. The reference side of the solvent stream incorporates a dummy column to provide back pressure for better flow regulation after which the solvent passes through the other side of the differential refractometer cell. Some models incorporate a double bank of sample columns and a bank switching valve so that it is possible to switch from one column porosity series to another without a delay. Options available include an automatic injection system, a fraction collector, the R-4 Differential Refractometer, a pulseless pump, and a digital curve translator.

The Water's instrument has operated in the author's laboratory nearly continuously for 2 1/2 years using m-cresol at $> 100°$ for much of the time. It has proven to be rugged, reproducible, and dependable. The detector, the R-4 refractometer, is a proportional-type light meter. It is extremely sensitive, 10^{-7} R.I., and electronically stable. In chromatographic terms, detector sensitivity can be described as follows: A 50% scale deflection is obtained for nylon (R.I. = 1.54) in m-cresol (R.I. = 1.50 at 100°) using a 10-mg load (2 ml of 0.5% solution injected) at a 4X attenuator setting.

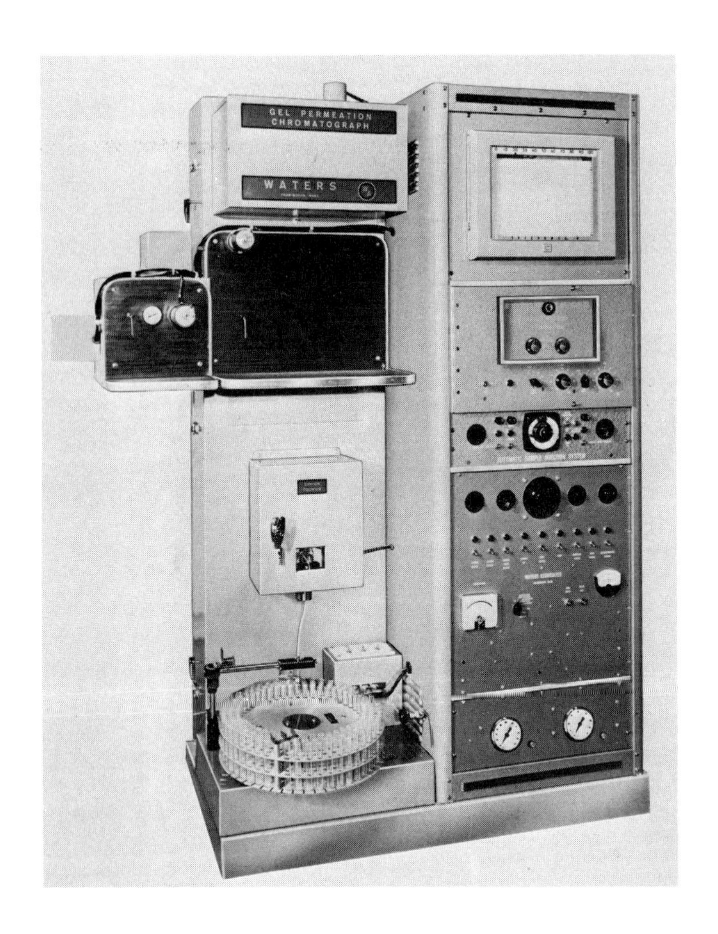

FIG. 21. Front view of Waters Associates, Inc., Model 200 Gel Permeation Chromatograph. Reproduced courtesy of Waters Associates, Inc.

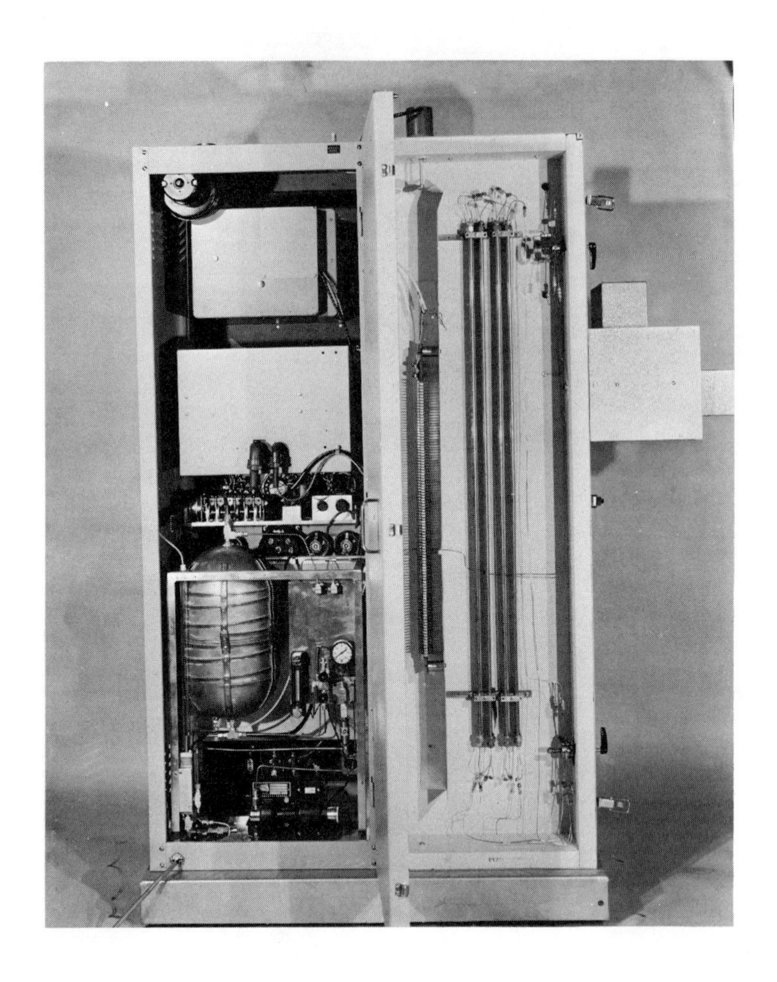

FIG. 22. Rear view of Waters Associates, Inc., Model 200 Gel Permeation Chromatograph. Reproduced courtesy of Waters Associates, Inc.

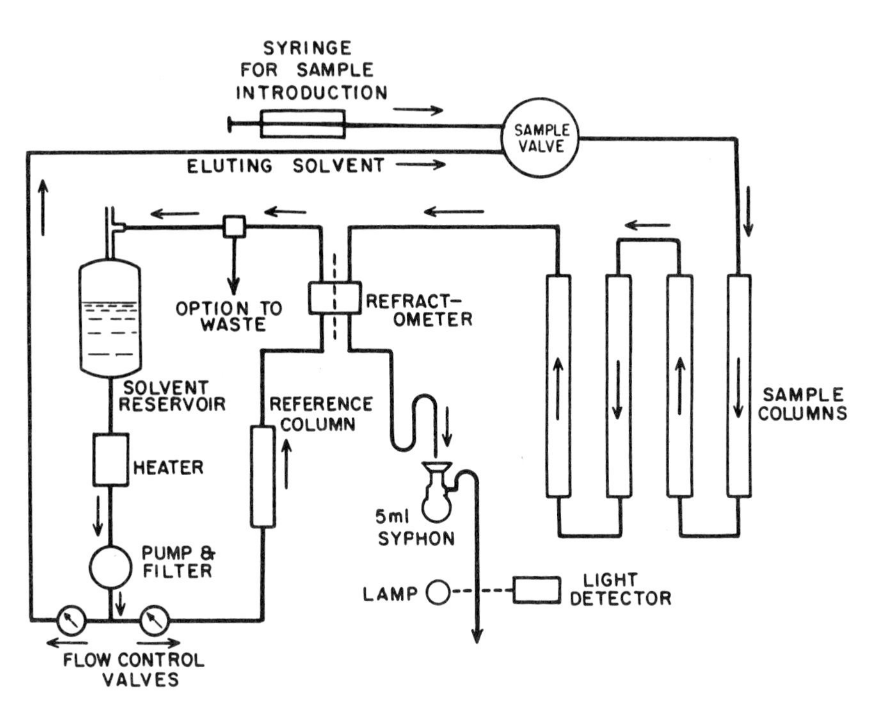

FIG. 23. Schematic drawing of solvent flow system in Waters Associates, Inc., Gel Permeation Chromatograph.

B. GPC Assemblies

The commercial Gel Permeation Chromatograph, Model 200, has two significant shortcomings: the initial investment cost is high and all parts are integrated into one unit. The latter lends itself to speed and convenience in repetitive type-analyses but concurrent utilization of different or multiple detectors and column rotation at high temperatures is difficult. This limitation can be overcome using compatible modular components in the major areas of the solvent-pumping assembly, column-oven area, the detector system and recorder, and/or the digital curve translator. In many cases quite simple modular equipment assemblies are sufficient for the experimenter's needs and the high investment is neither necessary nor justifiable. Also Waters

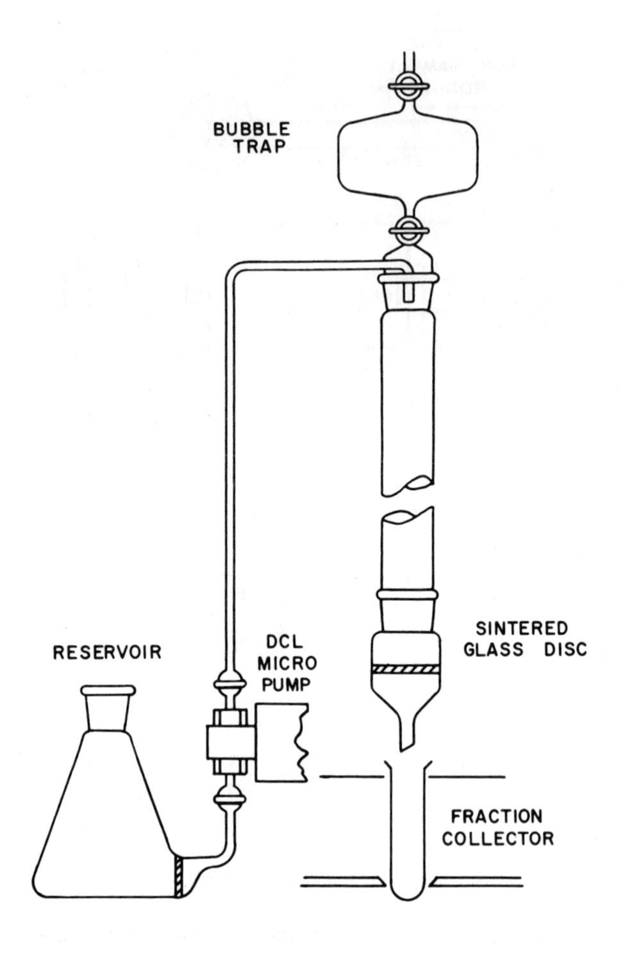

FIG. 24. Gel permeation chromatographic assembly of D. MacCallum. Reprinted from Ref. (106), p. 119, courtesy of Hüthig und Wepf Verlag.

Associates and others have recently introduced less expensive liquid chromatographs which can be used in GPC.

1. Simple Assemblies

Probably the least sophisticated GPC technique imaginable would be the gravity feed of solvent through a glass tube packed with "gel." The sample would be "poured in" on top and volume fractions of eluate taken, evaporated to dryness, and weighed. A somewhat more sophisticated approach, but nevertheless a simple and inexpensive one, has been discussed by MacCallum (106). His apparatus (Fig. 24) has all the elements for a successful experiment. With this apparatus MacCallum investigated the molecular weight distributions of polystyrene, poly(vinyl chloride), polychloroprene, and polyethylene. He used Santocel-A, a silica gel, as his fractionating medium and for detection measured the viscosity of the fractions collected. A comparison of this GPC method for polystyrene to other methods was shown by MacCallum (Fig. 25).

A more sophisticated "homemade" assembly (Fig. 26) has been described by Rodriguez et al. (107). This is a very versatile assembly and the reader can easily imagine it atop a laboratory bench. Parts, such as pumps, columns, and detectors, are easily interchanged. For our purpose it is worth while to quote Rodriguez' description of the assembly and its operation. The description covers the very essence of a chromatographic experiment.

"The apparatus consists of a solvent flow system, packed columns, a sample introduction valve, and a sample analyzer. The sample (0.2 ml of 2 to 5 wt% polymer in trichloroethylene) is introduced in about 20 sec by means of a 4-port, high-pressure valve. A reciprocating pump moves the solvent (trichloroethylene) from a reservoir through the sample valve and to the packed columns at rates up to 1.2 ml per minute. Pulsing is diminished by a bellows and spring arrangement. Four columns supplied by Waters Associates, Inc., Framingham, Massachusetts, are used. These are 0.300-in. i.d. stainless steel tubing, 4 ft long, capped with sintered-metal filters, and packed with a special copolymer of styrene and divinyl benzene. Two columns are '10^6 A' (large pore size) to assure resolution in the high molecular weight regions. A '10^5 A' column and a '10^4 A' column complete the unit. Several practical considerations limit the number of columns. A complete run requires an effluent volume of about 50 ml per column. A maximum rate for good resolution appears to be around 1 ml min; at this rate, the pressure drop per column is about 20 psi. With four columns, a run takes almost 4 h with a pressure drop

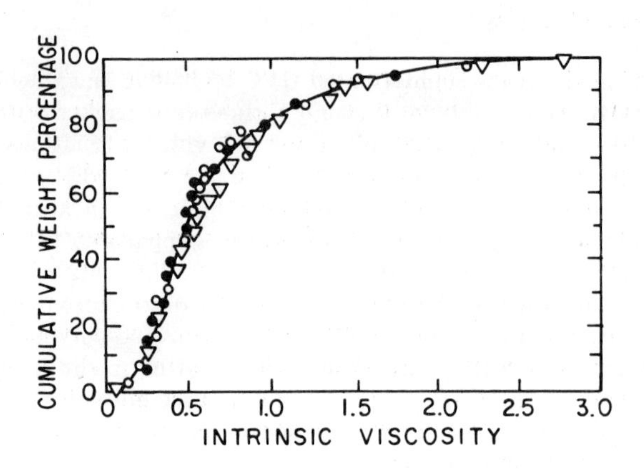

FIG. 25. Molecular weight distribution curves for a polystyrene sample obtained by ◁, volatilization, o, coacervation, •, gel permeation (by MacCallum). Reprinted from Ref. (106), p. 122, courtesy of Hüthig und Wepf Verlag.

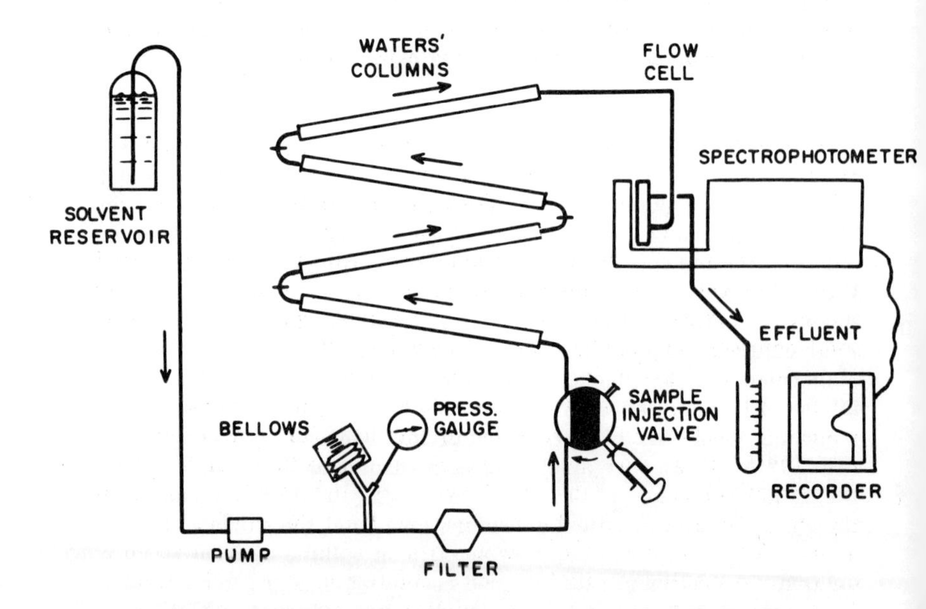

FIG. 26. Gel permeation chromatographic assembly of F. Rodriguez, R. A. Kulakowski, and O. K. Clark. Reprinted from Ref. (107), p. 122, courtesy of the American Chemical Society.

of 80 psi; the first 2 h merely flush the nonbead volume of the column (about 25 ml per column). Runs can be overlapped so that a complete analysis is obtained in each 2-h period.

"The concentration of polymer in the effluent can be determined on collected samples or continuously in a flow cell.

"For the analysis of proteins in water, flow cells using ultraviolet absorbance are available but for nonaqueous systems, differential refractive index has been used most often. In the present work a flow cell made from sodium chloride crystals inserted in an infrared (ir) spectrophotometer (Beckman IR-8, Beckman Instruments, Inc., Fullerton, California) has some advantages over differential refractive index. These are

"1. Limit of detectability for silicones is somewhat improved.

"2. Temperature has little effect on infrared absorbance.

"3. Solvent choice is not as critical when a cell containing the same solvent is used as a reference; almost any solvent which does not absorb greatly at the chosen wave length can be used.

"4. The measure of concentration can be very specific, that is, one group can be followed, such as Si-O-Si, -C-O-, etc. In fact, the distribution of monomers or other groups within a molecular weight distribution can be elucidated by successive runs at various wavelengths."[4]

Figure 27 illustrates calibration data for dimethylsilicones obtained by Rodriguez on this assembly

2. Components

a. Detectors. A GPC detector can be anything that senses or determines the presence of a given solute in a given solvent with required accuracy and precision. One can evaporate the solvent and weigh the residue or measure the viscosity of the solution as did MacCallum (106). The author and many others (4) have found the differential refractometer to be useful and practical but it has limitations. For example, Ross (108) cites its extreme sensitivity to

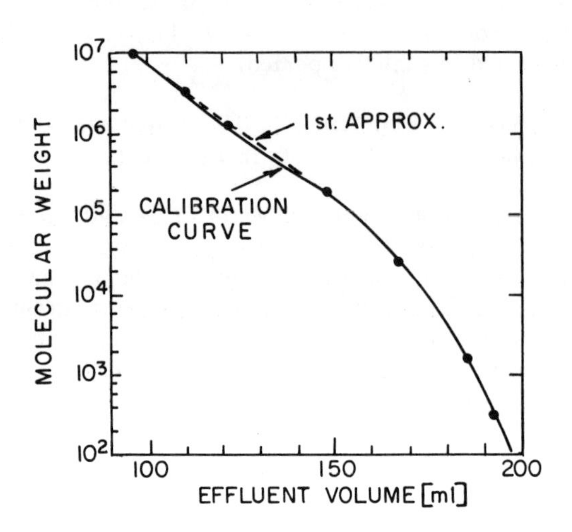

FIG. 27. Calibration curve for dimethyl silicones in gel permeation chromatography obtained by F. Rodriguez et al. Reprinted from Ref. (107), p. 124, courtesy of the American Chemical Society.

temperature and pressure change, variation of refractive index with molecular weight, and poor sensitivity with certain systems. Rodriguez (107) has used an infrared detector (Beckman IR-8) as have Ross and Casto (108) (Perkin-Elmer Model 12-C with a Carle Heated Microcell). Unfortunately the infrared detector suffers from poor sensitivity but its specificity offers certain advantages, especially in composition analysis. Terry and Rodriguez (109) have shown the combined utility of GPC and ir in interrelating functional groups, molecular weight, and composition. Others have used ultraviolet energy absorption with its high sensitivity (110-111), adsorption involving detection of the thermal energy of solute adsorption (112), and flame ionization (113-114). A variety of these detectors are available commercially. Recently a sophisticated "belt detector" operating on the flame ionization principle has been described by Coll et al. (115). The utility of a viscosity detector is discussed in the Addendum.

b. Pumps. In high-pressure liquid chromatography the working solvent is generally pumped through the columns and detector by means of a reciprocating piston pump. The Waters units employ controlled volume "Mini-Pumps" from the Milton Roy Company (116). These are excellent pumps for controlled low-volume flow and have a repeatable accuracy to ±0.3%. They are relatively trouble free but must be aligned very carefully or the sapphire plunger is likely to break.

Reciprocating actions will introduce pulsations into the liquid system. It has often been claimed that these pulses can cause noisy output from certain detectors and may also affect the resolving power of the columns. Much pulsation can be eliminated by employing spring-loaded bellows or surge tanks between the pump and columns, but even careful design does not remove all pulses from the flow pattern. For this reason high-pressure pulseless pumping assemblies have been developed (117-118). One design, offered by Chromatronix, Inc. (119), Models CMP-1 and CMP-2, utilizes two independent pump chambers. Two pistons are driven by two rotary cams 180° out of phase and the shape of the cams is such that one piston pumps at a constant rate while the other withdraws permitting solvent to fill the second chamber. Thus continuous, pulseless flow can be maintained. The pumps are useful to about 500 psi. However, little data published to date have shown significant improvement in resolution or fractionation in GPC because a pulseless pump was used in preference to a damped pulsing system. In fact Williams observed no changes in resolution when he compared a damped to nondamped system (120). Aris showed theoretically (121) that the effect of pulsation on peak dispersion in open tubes at laminar flow is insignificant unless the mean pressure gradient is smaller than the amplitude of the pressure gradient fluctuations. Horvath et al. (122) say that this theory is in good agreement with their findings which were that neither the pulsating nor the nonpulsating character of the flow had any demonstrable effect on the plate height or retention times in their liquid chromatography experiments.

c. Columns and Packings. It is well known that separating efficiency improves with an increase in column length (84), but longer columns supply excessive back pressure. Reducing column diameter leads to more rapid analysis but low volume columns are more easily overloaded. Ultimately the minimum detectable sample size and maximum usable pressure limit further changes in these variables. On the other hand columns used for preparatory scale fractionations generally employ high diameter-to-length ratios, 1/30 or less, in order to accommodate large sample dosages. It has also been shown that slower flow rates improve efficiency in terms of plates per foot (unit length) but reduce efficiency per unit time (plates per minute), so that a faster flow rate and longer column can do the same job more quickly (84). A multitude of such considerations and compromises must be applied to column construction. The experimenter must first establish a priority list for his needs (viz., sample detection limit, rapidity of analysis, efficiency of fractionation, and so on) and then tailor his column(s) accordingly (see Section II-B-3) and column coupling. Altgelt and Moore (4) have reviewed column designs and packing techniques specifically for gel filtration chromatography and for GPC. Their discussion is limited primarily to the handling of solvent-swollen gels (such as Styragel) in the solvent media. However, for dry "gel" simple technques can be employed and are often preferred (123). The reader is referred to chromatography textbooks and monographs for additional detailed information on column requirements and packing techniques.

Many different kinds of column packing materials have been employed for GPC but the only two that have found widespread acceptance to date are crosslinked polystyrene and porous glass. Polystyrenes were introduced by Moore whereas the impetus for developing porous glass was provided by Haller (124). Haller made his own porous glasses and using them demonstrated the size separation of a variety of viruses. A good characterization of the glass porosity is provided by Haller (124). Styragel, the crosslinked polystyrene described by Moore (22), and a porous glass under the name "Porasil" have for some time been available commercially from Waters Associates, Inc., Framingham, Mass. "Porasil" has been described by LePage et al. (125), and Bio-Glas, available from Bio-Rad Laboratories, Richmond, Calif., has been evaluated by Cantow and Johnson (126). A variety of other materials that may gain wide acceptance is now coming on the market. Pharmacia (127) now supplies Sephadex LH-20, prepared by alkylating Sephadex G-25, which is useful in polar organic solvents and which may be familiar to those involved in gel filtration. Corning (128) has a new porous glass, the GPC-10 series, and Du Pont a new controlled-surface porous glass (129). The Du Pont glass is used in conjunction with a new liquid chromatograph (see Ref. 111).

The polystyrene gel and "Porasil" are of good quality and are generally useful for polymer fractionation. There is little to choose between them, except that adsorption problems may require use of deactivated Porasil, now also available from Waters. The other glasses have not been extensively studied to date. However, glass is generally more stable and of fixed porosity whereas the pore size of of the swollen gel is solvent dependent. Likewise glass is more polar and thus more likely to perturb the size separation mechanism by adsorption, especially with more polar polymers. Published information and calibration curves (125-129) are useful for aiding the inexperienced chromatographer in selecting the right pore size(s) for his columns. Column parts and packed columns are available commercially from the vendors cited above.

C. Preparatory Scale Units

Preparatory scale gel permeation chromatographs are generally a little larger than their analytical counterparts. The columns are more massive, and provision must be made for introducing larger samples, handling much larger solvent volumes, and splitting the liquid stream prior to the detector. The basic components, or the job they do, remain unchanged. Simple assemblies for preparative scale work can be constructed just as the analytical chromatographs discussed above (Fig. 21 through 24 and 26) and they function in essentially the same way. The greatest problem in preparative work is polymer-fraction recovery from the large volume of solvent, followed by solvent recovery itself, or disposal. If a cheap, volatile solvent such as THF is used, this is no real problem, but, for example, if one is trying to isolate nylon from m-cresol, it is significant. The Waters Ana-prep. unit under normal continuous operation puts 20 gal a day through the columns. The reader can imagine the difficulty of working up 20 gal of m-cresol a day.

The operation and utility of the Ana-prep. have been described by Bombaugh et al. (130) and Maley et al. (131). Bombaugh (130) proved its utility for polystyrene by fractionating NBS 706 ($\overline{M}_w/\overline{M}_n$ = 2.1) into 12 fractions. The $\overline{M}_w$ of the fractions ranged from the high of 1×10^6 to a low of 1×10^4 and the polydispersity of each fraction was 1.2 or less. Despite, in some cases, the difficulty of handling large volumes of solvent, preparative scale GPC is here to stay. Fractionation by size difference is the best technique available for many polymers. It circumvents the problems of interference by crystallinity or change in solubility with composition of copolymers which plague so many fractionation techniques.

V. APPLICATIONS OF GEL PERMEATION CHROMATOGRAPHY

A. Introduction

In the preface to "Analytical Gel Permeation Chromatography," which is a compilation of papers from a recent American Chemical Society symposium on GPC, editors J. F. Johnson and R. S. Porter (132) state that for many years it has been known that the MWD of polymers strongly affects many of their physical properties. However, because of the extreme difficulty in determining complete MWDs, there has been more speculation than experiment on the extent of these effects. Until gel permeation chromatography, the direct elucidation of a MWD required fractionation, isolation, and characterization of numerous fractions from a whole distribution. To reduce the labor involved, indirect methods have been sought, for example, turbidimetric titration; these have often been inaccurate and imprecise. Gel permeation chromatography, on the other hand, is not laborious, is precise, can be accurately calibrated, and does not require the isolation and characterization of fractions. Therefore gel permeation chromatography is being readily and generally applied to characterize the MWD of polymers. Even as early as 1966 Cazes had compiled a list of 42 polymer systems studied by GPC with references (2).

The literature reports many examples of problems that have been uniquely solved by gel permeation chromatography. A few examples will be discussed wherein it was necessary to know the whole MWD rather than certain moments of the distribution, for example, $\overline{M}_w$, $\overline{M}_n$, $\overline{M}_z$, or $\overline{M}_v$.

B. Polymeric Materials

At the Sixth International Gel Permeation Chromatography Seminar, Boyer (133) cited four examples in which he considered it imperative that the entire MWD of the sample be known. In one example workers had measured the spinnability of three polyacrylonitrile samples that had approximately the same $\overline{M}_w$ but varied from poor to fair to excellent in spinnability. It was ascertained from the MWDs of these samples that they consisted of blends of more than one molecular weight and more than one individual MWD even though the samples averaged approximately the same $\overline{M}_w$. Because it is possible to prepare blends of an infinite number of distributions that give the same $\overline{M}_w$, nothing short of the full MWD would be sufficient for analyzing these blends to predict the spinnability of the composite. Thus analysis of blends is an area that requires full characterization of the MWD.

Another example cited by Boyer involved polymers added to lubricating oils to improve their viscosity index by evening out the viscosity/temperature profiles of the oils. Normally in commercial samples the additives are low molecular weight polymers with broad molecular weight distributions. However, the low molecular weight end of the polymer additive contributes nothing to the viscosity and is an expensive inert filler, whereas the high molecular weight end gets broken down rather quickly in the high shear fields of an automotive engine and loses its usefulness rather quickly. Thus a broad MWD polymer is not the most economically desirable additive as an index improver. Rather, a narrow MWD polymer of the right molecular weight is preferred. Again nothing short of knowing the full MWD is sufficient to optimize the properties of the polymer additive.

Purdon and Mate (134) have studied the dependence of processability of elastomers on the MWD. They concluded that certain processing difficulties encountered with styrene, butadiene, and acrylonitrile/butadiene copolymers, as well as stereospecific polybutadiene polymers, were explained quite readily when the MWD data from GPC were added to the usual information. They investigated specific processing characteristics such as carbon black incorporation time, mill stickiness, and cold-flow characteristics. Extension of the investigation revealed that MWD data for masticulated polymers related better to final physical properties of vulcanizates than did the MWD data of original raw polymers.

Harmon (135) has shown by five examples how GPC was applied to studies of various polymer systems in his laboratory. He points out that the calculation of molecular weights from the curves was not even pertinent to the studies. The only requirements were separation on the basis of size and reproducible results. Harmon's work includes: (1) "The Study of the Mastication of Elastomers"; (2) "The Use of GPC to Make Controlled Distributions by Blending"; (3) "The Effect of Molecular Weight Distribution on the Melting Point of Poly-2, 2,4-trimethyl-1,2-dihydroquinoline"; (4) "The Effect of Addition of Hydrogen on the Molecular Weight Distribution of an Experimental Copolymer"; and (5) "Examination of Extended Polymer by GPC." Harmon concludes that the information obtained was of great practical interest to polymer chemists and physicists and that the data in most cases could not have been obtained in any other way.

Law (136-137) has studied the characterization of HB polymer, a low molecular weight liquid composed of polybutadiene, acrylic acid, and acrylonitrile, used as a fuel binder fraction in solid composite propellant. He found the GPC determination of low molecular weight materials in HB polymers corresponded significantly with final propellant physical properties. By analysis of HB fractions prepared in his studies, Law detected the presence of three major types of chemical species in the molecular weight range 100-500. These are the reaction products of butadiene with itself, with DDM, and with acrylic acid. Further work showed that the HB fraction of molecular weight 250-375 is the one that most influences the propellant physical properties and it does so either directly or by interaction with other propellant ingredients.

C. Small Molecules

More and more attention is being given to the GPC separation of small molecules. Much of the work to date has been involved with theory, the mechanism of separation, controlling factors in the separation, predicting the order of elution, and so on (24, 138-141). Included in the above references, however, are many tables that list molar volumes and elution orders for a large number of small molecules. Such tables should be of considerable value for application problems dealing with small molecules. Bombaugh et al. (142) indicate that one of the important advantages to using GPC for small molecules (assuming that a liquid chromatographic procedure is needed) is that the porous gels require no stationary liquid phase. Therefore there are no bleeding problems and long column life is achieved without a saturated carrier. While Bombaugh et al. demonstrated the separation of complex organic mixtures conventionally and by brute force, up to 40 4-ft columns in series, Heitz and Ullner (143) have shown that recycling chromatography is very effective in the same molecular size range. Excellent, nearly complete separations were obtained for the first 10 repeat units (oligomers) of styrene, butyl methacrylate, and ethylene oxide. Again in a straightforward manner, Neddermeyer and Rogers (144) separated various polyphosphates and concluded that the elution order was as expected from a steric exclusion mechanism. Still other separations and another interesting recycling chromatograph have been described by Porath and Bennich (110).

Only a few examples of applications to polymers and small molecules have been mentioned. Literally hundreds of others could have been chosen to indicate the usefulness of GPC. In fact utility of the GPC technique at this time seems to be limited only by the imagination of the user. Other examples have been noted in Section II-A.

ACKNOWLEDGMENTS

The author gratefully acknowledges the editorial and technical suggestions made by F. W. Billmeyer, Jr., J. Cazes, A. E. Hamielec, C. P. Malone, R. J. Schilit, and T. D. Swartz. He also wishes to thank C. P. Malone, H. L. Suchan, and W. W. Yau for their many hours of timely discussions and valuable suggestions.

NOTES

[1]The author is well aware of the tremendous amount of work and numbers of papers devoted to separations of large, natural-product, molecules in (buffered) aqueous solutions on water-swellable gels. The technique known as gel filtration is generally attributed to J. Porath and P. Flodin (9) and has been extensively reviewed (10-14). In 1960 Vaughan reported successful fractionation of polystyrenes in a nonaqueous system (15). Then to some extent molecular size separations using synthetic polymers and nonaqueous solvents were suggested by others before Moore's presentation in 1964 (16-18). However, the gels and solvents employed and results obtained were of marginal utility at best for general work in the nonaqueous area. Distinction is made between aqueous "gel filtration" and organic (solvent) "gel permeation" chromatography. It is this author's opinion that the latter technique did not exist per se in terms of general utility for organic solvents and synthetic polymers until J. C. Moore's work was published and the gels became commercially available. The problems of inventorship and complex nomenclature have been discussed (see references 3, 4, and 4a) and for the purposes of this chapter they are best omitted.

[2]Actually four requirements were fulfilled in demonstrating this principle. (1) The polymer standards had Gaussian, normal, bell-shaped distributions. (2) The distributions were continuous and of one chemical composition for the refractometer detector. (3) Efficient columns were used, $R_S \geq 1.0$. (4) The GPC curve fell within the respective linear log region where W/d is a constant.

[3]See footnote 1.

[4]Reprinted from Ref. 107, p. 122, by courtesy of the American Chemical Society.

GLOSSARY OF TERMS[a]

GPC Parameter	Symbol	Units	Definitions
Curve baseline width	W	cm^3	The distance between the base line intercepts of lines drawn tangent to the points of inflection of the trace
Theoretical plates for system	N	—	$16\,(V_R/W)^2$
Effective plates	n	—	$16\left[(V_R - V_o)/W\right]^2$
Resolution	$R_{1,2}$	—	$R_{1,2} = 2\left[(V_{R_1} - V_{R_2})/(W_1 + W_2)\right]$
Specific resolution	R_s	—	$\dfrac{2(V_2 - V_1)}{(W_1/d_1) + (W_2/d_2)} \cdot \log(\overline{M}W_2/\overline{M}W_1$
Resolution index	R_I	—	$R_I = S/W_{monodisperse}$
Separation factor	S	—	$\left[\dfrac{\log(V_{h_2}/V_{h_1})}{V_{R_1} - V_{R_2}}\right] = \dfrac{d\log V_h}{dV_R}$
Total liquid volume	V	cm^3	$V = V_i + V_o + V_{ext}$

GPC Parameter	Symbol	Units	Definitions
Hydrodynamic volume	V_h	cm^3/mole	A polymer molecular property proportional to $[\eta] \cdot M$.
Exclusion limit of support	$V_{h,max}$	cm^3/mole	Max V_h that entered into pore
Solvent velocity	U	cm/sec	$U = V/60\ A_c$
Corrected solvent flow rate	v	cm^3/min	Solvent volume flow at column temperature
Distribution coefficient	K		$K = (V_R - V_o)/V_i$
Capacity ratio	k'	—	$k' = (V_R - V_o)/V_o = KV_i/V_o$
Q factor	Q	g/mole/nm	Molecular weight of repeat unit divided by contour length of same; not physically realistic
Skewness	Sk	—	Sk is W_b/W_a where b and a are right and left node of the curve per sketch

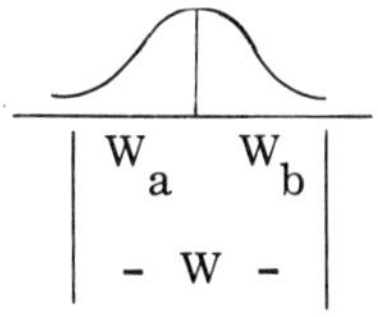

W_a W_b

- W -

GPC Parameter	Symbol	Units	Definitions
Temperature	temp	C°	For C°, temp is used
Peak width at half height	$W_{1/2}$	cm^3	Peak width measured parallel to base line at 50% of its height
Zone width	W_o	cm^3	Width due to injection time
Width due to injector	W_I	cm^3	Contribution made to W by volume and geometry of injector
Width due to column	W_c	cm^3	Contribution made to W by volume and geometry of columns; for column = $W_c = (W^2 - W_{ext}^2)^{1/2}$
Width due to detection system	W_d	cm^3	Contribution made to W by volume and geometry of the detection system
External width	W_{ext}	cm^3	Contributions to W by all system components external to columns
Theoretical plates for columns	N_c	—	$N_c = 16\left[(V_R - V_{ext})/W_c\right]^2$
Retention volume	V_R	cm^3	Volume of liquid that has passed through system from middle of sample injection period to peak maxmum

GPC Parameter	Symbol	Units	Definitions
Volume of mobile phase (interstitial volume or void volume)	V_o	cm^3	Volume of mobile phase within the column
External volume	V_{ext}	cm^3	Contributions to V by all system components external to column
Stationary liquid volume	V_i	cm^3	Total carrier volume contained within the support
Column inlet pressure	Pi	atm or psig	
Height equivalent to theoretical plate: System	H	mm	$H = L/N$
Height equivalent to the theoretical plate (columns)	H_c	mm	$H_c = L/N_c$
Reduced HETP	h	—	$h = H/dp$
Molar volume	V_m	cm^3/mole	
Calibration curve	—	—	The relationship between V_h and V_R
Peak area	A	cm^2	$A = \int^{\infty} \text{height} \cdot dV_R$
Standard deviation	σ	cm^3	$\sigma = W/4 = W_{1/2}/(2 \ln 2)^{1/2}$ for gaussian curve

GPC Parameter	Symbol	Units	Definitions
Pore radius	r_p	cm	(Determined by physical measurement)
Average particle diameter	d_p	cm or μm	Number average particle diameter
Number average molecular weight	$\bar{M}_n$	g/mole	First moment of a polymer distribution
Weight average molecular weight	$\bar{M}_w$	g/mole	Second moment of a polymer distribution
Molecular weight distribution	MWD	—	Weight (or number) fractions as a function of molecular weight
Integral molecular weight distribution	$\int$MWD	—	Sum of weight fractions as a function of molecular weight
Differential molecular weight distribution	d(MWD)	—	Relative abundance of a fraction as a function of molecular weight
Dispersity	d	—	$(M_i + 1)/M_i$ where i is any moment of the distribution
Intrinsic viscosity	η	dl/g	—

[a] Taken in part from the preliminary ASTM D-20.70.04 list.

REFERENCES

Note: The international seminars referenced are sponsored by Water Associated, Inc. See Ref. (48). Preprints and abstracts to some may still be available. Attendees all received copies.

(1) The ASTM D-20.70.04 1968 bibliography, prepared for section members, is cumulative and lists 400 references. A revised addition is under way.

(2) J. Cazes, J. Chem. Ed., 43, A567-582 (1966).

(3) J. F. Johnson, R. S. Porter, and M. J. R. Cantow, Reviews in Macromolecular Chemistry, 1 (2), 393-434 (1966).

(4) K. H. Altgelt and J. C. Moore, in Polymer Fractionation (M. J. R. Cantow, ed.), Academic Press, New York, 1967.

(4a) D. D. Bly, Science, 168, 527-533 (1970).

(5) K. H. Altgelt, "Theory and Mechanics of GPC," in Advances in Chromatography, 1, Dekker, New York, 1968.

(6) See, for example, Du Pont Products Book, E. I. Du Pont, de Nemours and Company, Inc., 1968 ed.

(7) Papers presented at International Symposium on Microchemical Techniques, 1965, The Pennsylvania State University, University Park, Pennsylvania, Aug. 22-27, 1965.

(8) Technical Documentary Report No. ASD-TDR-62-879, December 1962, Armed Services Technical Information Agency, Arlington, Virginia. Cataloged by ASTIA as A.D. No. 294355.

(9) J. Porath and P. Flodin, Nature, 183, 1657-1659 (June 13, 1959).

(10) H. Determann, Angew. Chem., 76, 635-644 (July 1964).

(11) H. Determann, Angew. Chem. Intern. Ed. Engl., 3, 608-617, (Sept., 1964).

(12) A. Tiselius, J. Porath, and P. A. Albertsson, Science, 141, 13-20 (July 5, 1963).

(13) P. Flodin, "Dextran Gels and Their Applications in Gel Filtration," Dissertation, University of Uppsala, Sweden, 1962; available from A. B. Pharmacia, Uppsala, Sweden.

(14) J. Porath, Laboratory Practice, 16, 838 (July, 1967).

(15) M. F. Vaughan, Nature, 188, 55 (Oct. 1, 1960).

(16) P. I. Brewer, Nature, 188, 934-935 (Dec. 10, 1960).

(17) B. Cortis-Jones, Nature, 191, 272-273 (July 15, 1961).

(18) P. I. Brewer, J. Inst. Petroleum, 48, 277-282 (Sept., 1962).

(19) R. M. Wheaton and W. C. Bauman, Ind. Eng. Chem., 45, 228-233 (1953).

(20) W. C. Bauman, R. M. Wheaton, and D. W. Simpson, "Ion Exclusion" in Ion Exchange Technology (F. C. Nachod and J. Schubert ed.), Academic Press, New York, 1956.

(21) Chem. and Eng. News (Dec. 17, page 43, 1962).

(22) J. C. Moore, J. Polymer Sci., A, 2, 835-843 (Feb., 1964).

(23) L. E. Maley, J. Polymer Sci., C, 8, 253-268 (1965).

(24) J. Cazes and D. R. Gaskill, Separation Sci., 2(4), 421-430 (1967).

(25) T. Edstrom and B. A. Petro, J. Polymer Sci., C, 21, 171-182 (1968). See also ACS Division of Polymer Chemistry Polymer Preprints, 8(2), 1227-1240 (Sept. 1967).

(25a) D. D. Bly, K. A. Boni, M. J. R. Cantow, J. Cazes, D. J. Harmon, J. N. Little, and E. D. Weir, Polymer Letters, 9, 401-411 (1971).

(26) D. D. Bly, J. Polymer Sci., C, 21, 13-21 (1968). See also ACS Division of Polymer Chem., Polymer Preprints, 8(2), 1234-1240 (Sept. 1967).

(27) D. D. Bly, Anal. Chem., 41, 477-480 (1969).

(28) D. D. Bly, J. Polymer Sci., A-1, 6, 2085-2089 (1968).

(29) J. Kwok, L. R. Snyder, and J. C. Sternberg, Anal. Chem., 40, 118-122 (1968).

(30) D. J. Harmon, Second International Seminar on Gel Permeation Chromatography, Boston, September 1965.

(31) M. Hess and R. F. Kratz, J. Polymer Sci., A-2, 4, 731-734 (1966).

(32) M. Bohdanecky, P. Kratohvil, and K. Solc, J. Polymer Sci., (A) 3, 4153-4157 (1965).

(33) DeBell and Richardson, Inc., Hazardville, Conn. 06036.

(34) ARRO Laboratories, 1107 West Jefferson Street, Joliet, Illinois 60435.

(35) D. J. Harmon, J. Polymer Sci., C, 8, 243-251 (1965).

(36) L. W. Gamble, ACS Rubber Division Meeting, Miami, May 3-7, 1965.

(37) E. Drott, Fourth International Gel Permeation Chromatography Seminar, Miami Beach, May 1967.

(38) K. A. Boni, F. A. Sliemers, and P. B. Stickney, ibid.

(39) G. Meyerhoff, Ber. Bunsenges. Physik Chem., 69, 866-874 (1965).

(40) H. Benoit, Z. Grubisic, P. Rempp, D. Decker, and J. G. Zilliox, Third International Gel Permeation Chromatography Seminar, Geneva, 1966. Also J. Chim. Phys., 63, 1507-1514 (Nov.-Dec. 1966).

(41) Z. Grubisic, P. Rempp, and H. Benoit, J. Polymer Sci., B, 5, 753 (1967).

(42) J. V. Dawkins, J. Macromol Sci-Phys, B2 (4), 623-639 (1968).

(43) F. Rodriguez and O. K. Clark, I. & E. C. Product Research and Development, 5, 118-121 (June 1966).

(44) W. W. Yau, H. L. Suchan, C. P. Malone, and S. W. Fleming, Fifth International Gel Permeation Chromatography Seminar, London, 1968.

(45) H. E. Pickett, M. J. R. Cantow, and J. F. Johnson, J. Appl. Polymer Sci., 10, 917-924 (1966).

(46) H. E. Pickett, M. J. R. Cantow, and J. F. Johnson, J. Polymer Sci., C, 21, 67-81 (1968).

(47) J. B. Carmichael, J. Macromol. Sci.-Chem., A-2, 1411-1414 (1968).

(48) Waters Associates, Inc., 61 Fountain Street, Framingham, Massachusetts 01701.

(49) J. G. Hendrickson, J. Polymer Sci., A-2, 1903-1917 (1968).

(50) A. E. Hamielec, McMaster University Symposium on Gel Permeation Chromatography, Winter, 1966-1967.

(51) L. H. Tung, J. Appl. Polymer Sci., 10, 375-385 (1966).

(52) L. H. Tung, J. C. Moore, and G. W. Knight, J. Appl. Polymer Sci., 10, 1261-1270 (1966).

(53) L. H. Tung, J. Appl. Polymer Sci., 10, 1271-1283 (1966).

(54) H. L. Berger and A. R. Shultz, J. Polymer Sci., A, 2, 3643-3648 (1965).

(55) R. A. Potter, private communication.

(56) S. T. Goforth, private communication.

(57) S. T. Balke and A. E. Hamielec, J. Appl. Polymer Sci., I. & E. C. Product Research and Development, 8, 54-57 (1969). Also presented in part at the Sixth International Gel Permeation Chromatography Seminar, Miami Beach, Oct. 1968.

(58) S. T. Balke and A. E. Hamielec, J. Appl. Polymer Sci., I. & E.C. Product Research and Development, 8, 54-57 (1969).

(59) R. L. Pecsock and D. Saunders, Separation Sci., 1(5), 613-628 (1966).

(60) W. W. Yau, C. P. Malone, and S. W. Fleming, J. Polymer Sci., B, 6, 803-807 (1968).

(61) J. Porath, Pure Appl. Chem., 6(3), 233-244 (1963).

(62) P. G. Squire, Arch. Biochem. Biophys., 107, 471-478 (1964).

(63) T. C. Laurent and J. Killander, J. Chromatog., 14, 317-330 (May 1964).

(64) G. K. Ackers, Biochem., 3, 5, 723-730 (May 1964).

(65) W. W. Yau and C. P. Malone, J. Polymer Sci., B, 5, 663-669 (1967).

(66) A. J. deVries, Preprint #139, International Symposium on Macromolecular Chemistry, IUPAC meeting, Prague, 1965.

(67) A. J. deVries, M. LePage, R. Beau, and C. L. Guillemin, Anal. Chem., 39, 935-939 (1967).

(68) E. F. Casassa, "Characterization of Macromolecular Structure," Publication 1573, National Academy of Science, Washington, D.C., 1968, p. 285.

(69) E. F. Casassa, J. Polymer Sci., B, 5, 773-778 (Sept. 1967).

(69a) J. C. Moore and M. C. Arrington, Preprint VI-107, International Symposium on Macromolecular Chemistry, Tokyo, Kyoto, 1966.

(70) J. B. Carmichael, Macromolecules, 1, 526-529 (1968).

(71) M. J. R. Cantow and J. F. Johnson, J. Polymer Sci., A-1, 5, 2835-2841 (1967).

(72) M. J. R. Cantow, R. S. Porter, and J. F. Johnson, J. Polymer Sci., A-1, 5, 987-991 (1967).

(73) M. J. R. Cantow and J. F. Johnson, Polymer, 8, 487-488 (Sept. 1967).

(74) J. J. Hermans, J. Polymer Sci., A-2, 6, 1217-1226 (1968).

(75) T. L. Chang, Anal. Chim. Acta., 42, 51-57 (1968).

(76) W. W. Yau, J. Polymer Sci., A-2, 7, 483 (1969).

(77) J. C. Giddings, Anal. Chem., 39, 8, 1027-1028 (July 1967).

(78) G. Meyerhoff, J. Polymer Sci., C, 21, 31-41 (1968).

(79) M. LePage, R. Beau, and A. J. deVries, J. Polymer Sci., C, 21, 119-130 (1968).

(80) H. W. Osterhoudt and L. N. Ray, Jr., J. Polymer Sci., A-2, 5, 569-581 (1967).

(81) G. A. Feldman and W. V. Smith, Sixth International Seminar on Gel Permeation Chromatography, Miami, Oct. 7-9, 1968.

(82) J. H. Purnell, J. Chem. Soc., 1268-1274 (1960).

(83) D. H. Desty, Gas Chromatography, Academic Press, New York (1958), p. xi.

(84) J. C. Giddings and K. L. Mallik, Anal. Chem., 38, 997-1000 (July 1966).

(85) J. Cazes and D. R. Gaskill, Fourth International Seminar on Gel Permeation Chromatography, Miami, May 1967.

(86) L. W. Gamble, E. A. McCracken, and J. T. Wayde, Third International Seminar on Gel Permeation Chromatography, Geneva, May 1966.

(87) C. C. Maitland, Laboratory Practice, 16, 847-850 (July 1967).

(88) W. W. Yau, H. L. Suchan, and C. P. Malone, J. Polymer Sci., A-2, 6, 1349-1355 (1968).

(89) K. A. Boni, F. A. Sliemers, and P. B. Stickney, Fourth International Seminar on Gel Permeation Chromatography, Miami, May 1967.

(90) J. G. Hendrickson, Fourth International Seminar on Gel Permeation Chromatography, Miami, May 1967.

(91) D. F. Alliet and J. M. Pacco, J. Polymer Sci., C, 21, 199-213 (1968). See also ACS Division of Polymer Chem., Polymer Preprints, 8, (2), 1288-1294 (Sept. 1967).

(92) J. C. Moore, Fourth International Seminar on Gel Permeation Chromatography, Miami, May 1967.

(93) D. J. Pollock and R. F. Kratz, Fourth International Seminar on Gel Permeation Chromatography, Miami, May 1967.

(94) J. G. Hendrickson and J. C. Moore, Second International Seminar on Gel Permeation Chromatography, Boston, Oct. 1965.

(95) M. J. R. Cantow, R. S. Porter, and J. F. Johnson, J. Polymer Sci., B, 4, 707-711 (1966).

(96) J. A. Thoma, Anal. Chem., 37, 500-508 (April 1965).

(97) D. J. Winzor and L. W. Nichol, Biochem. Biophys. Acta, 104, 1-10 (1965).

(98) W. B. Smith and A. Kollmansberger, J. Phys. Chem., 69, 4157-4161 (Dec. 1965).

(99) L. R. Snyder, Anal. Chem., 39, 698-704 (June 1967).

(100) C. G. Horvath, B. A. Preiss, and S. R. Lipsky, Anal. Chem., 39, 1422-1428 (Oct. 1967).

(101) D. S. Horne, J. H. Knox, and L. McLaren, Separation Sci., 1 (5), 531-554 (1966).

(102) F. W. Billmeyer, Jr., G. W. Johnson, and R. N. Kelley, J. Chromatog., 34, 3, 316-321 (1968).

(103) F. W. Billmeyer, Jr., and R. N. Kelley, J. Chromatog., 34, 3, 322-331 (1968).

(104) E. M. Barrall II, M. J. R. Cantow, and J. F. Johnson, J. Appl. Polymer Sci., 12, 6, 1373-1377 (1968).

(105) W. Heitz and J. Coupek, Sixth International Seminar on Gel Permeation Chromatography, Miami, October 1968.

(106) D. MacCallum, Die Makromolekulare Chemie, 100, 117-125 (1967).

(107) F. Rodriguez, R. A. Kulakowski, and O. K. Clark, I. & E. C. Product Research and Development, 5, (2), p. 121-125 (June 1966).

(108) J. H. Ross and M. E. Casto, J. Polymer Sci., C, 21, 143-152 (1968). See also ACS Division of Poly. Chem. Polymer Preprints, 8 (2), 1278-1281 (Sept. 1967).

(109) S. L. Terry and F. Rodriguez, J. Polymer Sci., C, 21, 191-197 (1968). See also ACS Division of Poly. Chem. Polymer Preprints, 8 (2), 1270-1277 (Sept. 1967).

(110) J. Porath and H. Bennich, Arch. Biochem. Biophys., Suppl., 1, 152-156 (1962).

(111) 410 Precision Photometer, Instrument Products Division, Photo Products Department, E. I. du Pont de Nemours and Company, Inc., Glasgow, Delaware.

(112) K. P. Hupe and E. Bayer, J. Gas Chromatog., 5, 197-201 (April 1967).

(113) T. Cotgreave, Chem. and Ind., 689 (April 1966).

(114) A. Karmen, Anal. Chem., 38, 286-290 (Feb. 1966).

(115) H. Coll, H. W. Johnson, Jr., A. G. Polgar, E. E. Seibert, and F. H. Stross, J. Chromatog. Sci., 7, 30-35 (1969).

(116) Milton Roy Company, 1300 East Mermaid Lane, Philadelphia, Penn. 19118.

(117) R. E. Jentoft and T. H. Gouw, Anal. Chem., 38 (7), 949-950 (June 1966).

(118) J. F. Johnson and M. J. R. Cantow, J. Chromatog., 28, 128-130 (1967).

(119) Chromatronix Incorporated, 2743 Eighth Street, Berkeley, California 94710.

(120) R. C. Williams, private communication.

(121) R. Aris, Proc. Roy. Soc. (London), 259A, 370 (1960).

(122) C. G. Horvath, B. A. Preiss, and S. R. Lipsky, Anal. Chem., 39, 1422-1428 (1960).

(123) L. R. Synder, Anal. Chem., 39, 698-704 (June 1967).

(124) W. Haller, Nature, 206, 693-696 (1965).

(125) M. LePage, R. Beau, and A. J. deVries, J. Polymer Sci., C, 21, 119-130 (1968). See also ACS Division of Polymer Chemistry, Polymer Preprints, 8 (2), 1211-1219 (Sept 1967).

(126) M. J. R. Cantow and J. F. Johnson, J. Appl. Polymer Sci., 11, 1851-1854 (1967).

(127) Pharmacia Fine Chemicals, Inc., 800 Centennial Ave., Piscataway, New Jersey. Technical Data Sheet No. 13.

(128) Corning Glass Works, Corning, New York, 14830.

(129) J. J. Kirkland, J. Chromatog. Sci., 7, 7-12 (1969).

(130) K. J. Bombaugh, W. A. Dark, and R. N. King, Fourth International Seminar on Gel Permeation Chromatography, Miami, May 1967.

(131) L. E. Maley, W. B. Richman, and K. J. Bombaugh, ACS Division of Polymer Chem. Polymer Preprints, 8 (2), 1250-1258, Sept. 1967, Chicago.

(132) J. F. Johnson and R. S. Porter, J. Polymer Sci., C, 21, iii (1968).

(133) R. F. Boyer, Sixth International Seminar on Gel Permeation Chromatography, Miami, Oct. 1968.

(134) J. R. Purden and R. D. Mate, Third International Seminar on Gel Permeation Chromatography, May 1966 (presented in part at a meeting of the Division of Rubber Chemistry, ACS, San Francisco, May 1966).

(135) D. J. Harmon, J. Appl. Polymer Sci., 11, 1333-1343 (1967).

(136) R. D. Law, J. Polymer Sci., C, 21, 225-251 (1968).

(137) R. D. Law, Sixth International Seminar on Gel Permeation Chromatography, Miami, Oct. 1968.

(138) J. G. Hendrickson, Anal. Chem., 40, 49-53 (1968).

(139) J. G. Hendrickson, J. Chromatog., 32, 543-558 (1968).

(140) G. D. Edwards, ACS Division of Polymer Chemistry, Polymer Preprints, 1326-1336 (Sept. 1967).

(141) W. B. Smith and A. Kollmansberger, J. Phys. Chem., 69, 4157-4161 (1965).

(142) K. J. Bombaugh, W. A. Dark, and R. F. Levangie, Separation Sci., 3, 375-392 (1968).

(143) W. Heitz and H. Ullner, Die Makromolekulare Chemie, 120, 58-67 (1968).

(144) P. A. Neddermeyer and L. B. Rogers, Anal. Chem., 41, 94-102 (1969).

ADDENDUM - GPC

Many significant papers appeared in the literature during the preparation of the preceding manuscript. This addendum presents recent advances made in the theory of the mechanism of GPC, especially the flow-through concept, and advances in calibration, calculation, and application to branched polymer systems. In addition, a few miscellaneous papers are mentioned.

A. Mechanism

DiMarzio and Guttman have discussed the possibility of separating large molecules by flow alone (1). The idea is a simple one. The authors view a separation column made up of many fine capillary tubes down through which a polymer solution is flowing. The molecules are presumed to flow down the tubes but not to be adsorbed on the walls. Because the center of mass of a large molecule is closer to the center of the flow stream than a small molecule, the large molecules flow through the capillary more rapidly and emerge first. It intrigued DiMarzio and Guttman that the conclusions reached from this type of flow model are similar to those obtained experimentally in GPC. Thus separation by flow may be a contributing factor to the GPC mechanism. (This is especially true in light of Benoit's proposal that the separation in GPC depends on the hydrodynamic volume of the dissolved molecules.)

DiMarzio and Guttman expand on the separation by flow theory in their second paper (2) where they seek to place the ideas on a more quantitative basis. The paper is in several parts and complete synopsis is not warranted in this addendum. The main conclusion of Section 2A is that the velocity of the center of mass of a molecule in a general velocity field is approximately that which the fluid would have at the location occupied by the center of mass with the polymer molecule not there. In another section the authors state further that if one is willing to wait long enough, either by using a long tube or equivalently by recycling through one tube, it is always possible to separate two materials that may even differ in size by only 10%.

Whether separation in GPC occurs by the flow mechanism or by actual physical excluding of macromolecules becomes a rather difficult distinction to make. Yau et al. (3) have shown experimental evidence against the flow model. Using glass beads similar to the porous glass beads used in GPC (but lacking the pores), they showed that styrene and polystyrene of 860,000 molecular weight elute at nearly the same retention volume. This experiment demonstrates that the velocity profile in the interstitial spaces does not provide separation capability. Yau et al. have also discussed vacancy permeation chromatography (3, 4) wherein the circulating liquid is the polymer solution and the sample introduced is a block of pure solvent. The resulting chromatogram is characteristic of the polymer and not of the solvent (Fig. 1). From these experiments, Yau et al. conclude that the porous nature of the GPC packing is the essential element of the separation capability. This appears to be in direct contradiction to the capillary, flow-through model.

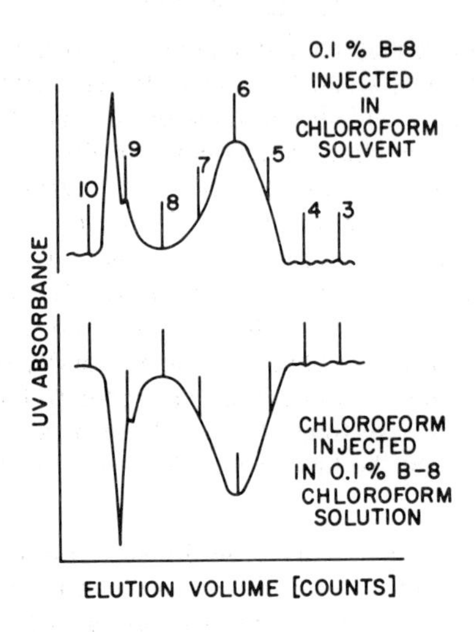

FIG. 1. Polystyrene curves for vacancy and conventional gel permeation chromatography.

It seems that DiMarzio and Guttman are aware of the conclusions drawn by Yau, Malone, and Suchan because in their third paper (5) entitled "Separation by Flow II: Application to GPC," they view the column packing as being constructed of beads of many fine pores in which both flow and diffusion take place. Thus polymer molecules in the beads can diffuse out to the mobile phase or they can be flushed out by the flow of carrier liquid. This leads the authors to the conclusion that in the equilibrium condition one can never distinguish (from measurement of average retention volume) whether the pores are accessible by virtue of a diffusion process or because of flow through them. The same retention volume will be observed for porous beads through which there is flow as for a system in which these pores are blocked, that is, where their spaces are available via diffusion but not via flow. In terms of V_R there seems to be no way to distinguish between the two mechanisms provided the ratio of the radii is large and the flow rate is small.

The separation of polymer molecules by flow alone through an appropriately small capillary or by flow through a column packed with solid glass spheres small enough to yield 2 μ size channels needs to be demonstrated if the flow-through theory is to become significant.

Concurrently, but independently, Verhoff and Sylvester also proposed a flow-through mechanism for GPC; that is, they proposed that flow occurs both through and around the gel phase (6). The theory was developed specifically for glass beads of equal pore size with equal sized connecting spaces, and it predicts the retention volume versus molecular weight curve. Although this theory is not as complete as DiMarzio and Guttman's, the paper is easier to read, especially for the mathematically unsophisticated. After the quantitative development, Verhoff and Sylvester compare predicted to experimental curves for porous glasses obtained by three different workers. The comparisons are in remarkably good agreement as shown here by Fig. 2 (7). The authors feel the good agreement between theory and experiment may occur because the properties of the glass packings correspond most closely to the assumptions in their theory. With crosslinked polymer packings, where a large distribution of pore sizes exists, the theory does not fit well. However, it does predict the general shape of the curve. The authors are continuing their work to determine dispersion effects, methods of dispersion correction, and to obtain a general theory for columns containing a distribution of pore sizes. Yau has worked on the dispersion (variance) problem (8) and presently feels that the flow-through model cannot be very significant in GPC. His calculations of the variance of GPC elution curves, based on the above mechanism, were at least two orders of magnitude too large when compared to experiment.

B. Calibration

Considerable effort has been spent attempting to define a universal calibration method in GPC. Most workers have tried to eliminate the dependence of retention volume on polymer type by determining empirically some polymer parameter universally related to retention volume. Meyerhoff, for example (9), characterized prepared fractions of three different polymers but was unable to correlate retention volume with intrinsic viscosity, diffusion coefficient, or root mean square radius of gyration, R_G. He did find, however, that $M^{1/2} \times R_G$ cor-

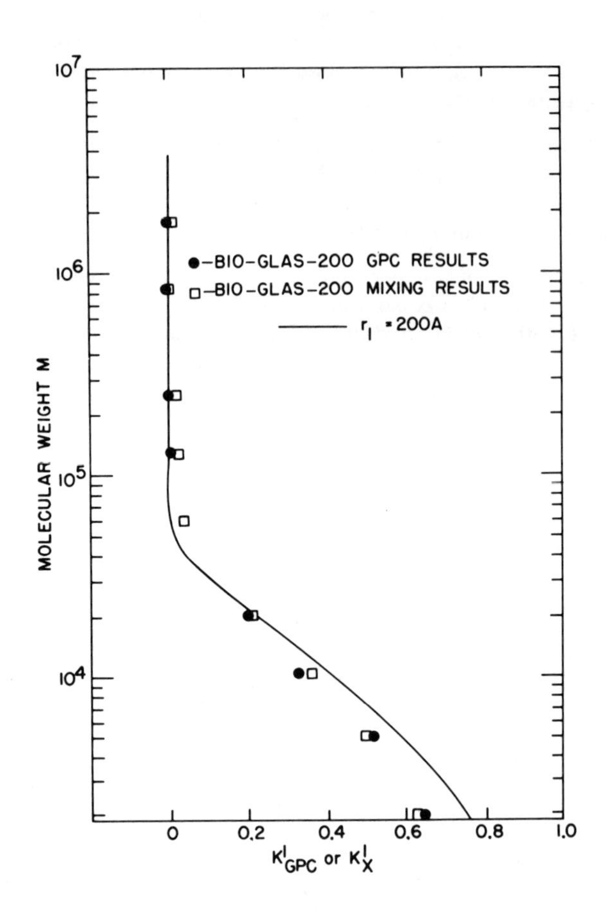

FIG. 2. Retention volume versus molecular weight for polystyrene on Bio-Glas. ———, calculated d; ●, □ , observed.

related with retention volume. As mentioned in the body of the text, Benoit proposed a quantity proportional to the hydrodynamic volume, $[\eta]$ X M, as a parameter that could be related to V_R (10). Much recent work has been done testing Benoit's proposition. Boni et al. (11) pointed out that a single Mark-Houwink relationship will not cover the whole range of molecular weight-intrinsic viscosity data. At low molecular weights polymer molecules behave as if they were in the solvent where $(\alpha) = 0.5$. Generally, the transition region to this value of

α occurs in the 30,000 to 50,000 molecular weight range and therefore linear extension to the low molecular weight range of the log $[\eta]$ M versus V_R curve could introduce serious error. Whitehouse (12) has discussed the effect of polydispersity on the intrinsic viscosity measurement itself. According to Whitehouse in the typical Mark-Houwink expression, $[\eta] = KM^{\alpha}$, the apparent values of K and α will change with the degree of polydispersity. This form of the equation is not the most convenient way to treat polydisperse systems. A better alternative is to use K and α for monodisperse polymers and then correct for polydispersity according to Eq. (1)

$$[\eta] = K\overline{M}_w^{\alpha}(1-\Delta) \tag{1}$$

where Δ is a function of polydispersity $\overline{M}_w/\overline{M}_n$. Whitehouse defines Δ quantitatively and then shows how the intrinsic viscosity is affected by polydispersity. By way of example, for an α of 0.7 and a polydispersity of 5, Δ is approximately 0.15 or 15%, whereas for a polydispersity of 2, Δ still has a value of 7.5%.

Goedhart and Opschoor (13) have also sought to eliminate the dependence of intrinsic viscosity on the polymer molecular weight distribution. They use the GPC fractionation process itself as a preparative tool for narrowly disperse fractions, and measure the viscosity of each fraction (syphon dump) as it elutes from the chromatograph with an automatic capillary tube viscometer. Calibration is achieved with available monodisperse polymers, while computation is made by summing the product of intrinsic viscosity and molecular weight for each fraction of the unknown polymer. It appears that Goedhart and Opschoor's method will be generally applicable to many polymer systems. There are, however, two experimental difficulties that must be resolved. One is the measurement of the concentration of each fraction. Because of zone broadening, it cannot be assumed that the curve height at any given position is a linear function of concentration. Second, the dilute solution viscosity as measured may not be an accurate representation of the intrinsic viscosity. The latter will depend on the line slope relating $\eta_{sp/c}$ to c. The former has been considered by Tung (14) and Meyerhoff (15).

Several other universal calibration methods have appeared recently. Weiss (16) has described a further application of the $[\eta]$ M concept and together with Cohn-Ginsberg (17) describes a method similar to that of Hamielec for specific calibration of gel permeation columns. The method requires standard polymers whose molecular weight distributions are known. The distribution is approximated by a generalized, two parameter, single peak distribution function like that of Schulz and Zimm (18).

Using known $\overline{M}_w$ and $\overline{M}_n$, or $\overline{M}_v$ and $\overline{M}_n$ values, the whole distribution function describing the polymer is used to compute the calibration curve and to calculate the unknowns. Williams and Ward (19) have shown yet another variation of the $[\eta]$ M equation and several other authors have reviewed the theoretical justification for using $[\eta]$ M or similar quantities such as "unperturbed dimension," or both. Dawkins (20), Meyerhoff (21), Coll and Gilding (22), and Boni et al. (11) have all studied the development of the universal calibration for GPC. Each discusses various aspects of the $[\eta]$ M or other -parameter calibrations and show personal viewpoints to the solution of the problem using justifying applications. It should be noted that Meyerhoff also discusses the use of automatic viscosity measurements on GPC fractions and discusses the concentration problem.

In a recent paper Coll (23) developed a different method for calibration using the Mark-Houwink equation and the expressions of Ptitsyn and Eizner (24). Coll's equation has the final form of Eq. (2):

$$\log M_2 = \frac{1}{1 + A_2} \log \frac{k_1 f(\epsilon_1)}{k_2 f(\epsilon_2)} + \frac{1 + A_1}{1 + A_2} \log M_1 \tag{2}$$

where k and A are the parameters of the respective Mark-Houwink equations,

$$\epsilon = (2A - 1)/3$$

$$f(\epsilon) = 1 - 2.63\epsilon + 2.89\epsilon^2$$

In Eq. (2) the subscripts refer to polymers 1 and 2, respectively, one being the standard and the other the unknown. The various molecular averages for polymer 2 are computed from the chromatogram in terms of the polymer 1 values. After the dummy molecular weight values are obtained, they are converted to the true values for 1 using the coefficients of the equation. Or, if the log M_1 versus V_R relationship is known, the log M_2 relationship can be computed from Eq. (2) and a calibration curve for polymer 2 generated. The transformation of the primary calibration curve may have to be carried out by segments if more than one set of Mark-Houwink parameters is required. Certainly more testing of the method is needed at this time; but the idea looks promising. Coupling Coll's method to the automatic viscometric measurement of GPC fractions, one should be able to generate sets of Mark-Houwink expressions for each neighboring set of fractions, provided the concentrations can be propertly normalized. Herein lie some R and D notions.

Finally, in a somewhat philosophically different approach to calibration, Tung and Runyon (25) have discussed the calibration of instrumental spreading for GPC. Significantly they found that instrumental spreading characteristics depended on retention volume but not on the nature of the polymer. Therefore the calibration results from polystyrene standards, for example, can be used to treat the chromatograms for other polymers. However, the appropriate function for universal calibration to define molecular weight was not discussed in this paper.

C. Branching

Several authors have indicated that branching in polymers may be studied by GPC. Most of the detailed work has been done on long-chain branching in polyethylene. Wild and Guliana (26) characterized branched fractions of polyethylene by viscosity and molecular weight measurements and found that the size of the polyethylene molecules varies in a way different from simple linear growth. In addition they found that a single curve was adequate for relating retention volume to molecular weight; however, the curve was different from that for linear polyethylene. The replacement of the molecular weight M by the parameter $[\eta]$ M led to a single relationship for both linear and branched polyethylenes. This further indicates that GPC separation takes place according to the hydrodynamic volumes of the polymer molecules.

Drott and Mendelson (27, 28) also assumed applicability of the universal calibration, that is, $[\eta]$ M can be related to retention volume for long-chain and linear polyethylenes, and put the branching determination on a quantitative basis. In their first paper Drott and Mendelson (27) rigorously developed theory for using viscosity and GPC to determine polymer branching quantitatively, and in their second paper they give examples of applications (28). Fundamental to their theory is the concept that the key parameter governing separation in GPC is a function of the hydrodynamic volume of the polymer molecule.

In order to establish quantitative relations between molecular parameters, the effect of molecular weight and the degree of branching on hydrodynamic volume and the effect of branching on the viscosity must be described in measurable quantities. These relate to the nature of the branch points, trifunctional or tetrafunctional, the length of the branches, long or short, and the number of branch points in each molecule. For whole polymers a branching distribution function must also be defined. Drott and Mendelson considered each of these points in quantitative terms in their theory. For details the reader is referred to the original. It is concluded from Drott's paper that long-chain branching is significant and may be measured by GPC and viscosity, whereas short-chain branching in general leads to an insignificant altering of the data. Although the method is straightforward for determining branching, one must always be aware of the assumptions made (concerning the universality of the calibration and the branching function that one chooses in order to determine the branching information).

D. Miscellaneous

Several miscellaneous papers deserve mention here. Cooper et al. (29) showed that it is possible to interrupt flow, collect fractions, and monitor each fraction by infrared detection without changing the shape of the gel permeation chromatogram. In a rather remarkable experiment a polystyrene sample was chromatographed on a polystyrene gel column using toluene solvent and no difference was found between a normal curve and that interrupted for 17 days. Bombaugh et al. (30) showed the advantages of recycle by comparing a bulk system of 180,000 theoretical plates made by stacking columns to the recycling technique. Waters demonstrated a unique method for determining the width of a very narrow molecular weight distribution by GPC involving the recycling technique and extrapolation to infinite resolution (31). Waters' extrapolation method should be very useful for narrow molecular weight distributions and for testing some of the various computational methods for correcting peak broadening in GPC. Several reviews have appeared on peak resolution, separation, and correction for broadening (32-34). Ackers (35) wrote a review entitled "Analytical Gel Chromatography of Proteins" and Cazes an article entitled "Current Trends in GPC - 1970" (36-37). This author featured an article in Science (38) and is chairman of an ASTM Committee, D-20.70.04 which is writing a standard method for GPC.

REFERENCES

(1) E. A. DiMarzio and C. M. Guttman, Polymer Letters, 7, 267-272 (1969).

(2) E. A. DiMarzio and C. M. Guttman, Macromolecules, 3, 131-146 (1970).

(3) W. W. Yau, C. P. Malone, and H. L. Suchan, Separation Sci., 5, 259-271 (1970).

(4) C. P. Malone, H. L. Suchan, and W. W. Yau, Polymer Letters, 7, 781-784 (1969).

(5) C. M. Guttman and E. A. DiMarzio, Macromolecules, 3(5), 681-691 (1970).

(6) H. F. Verhoff and N. D. Sylvester, Macromolecular Sci.-Chem., A4 (4), 979-1001 (1970).

(7) W. W. Yau, J. Polymer Sci., A-2, 7, 483 (1969).

(8) Private communication.

(9) G. Meyerhoff, Macromol. Chem., 89, 282 (1965).

(10) H. Benoit, Z. Grubisic, P. Rempp, D. Kecker, and J. G. Zilliox, J. Chem. Phys., 63, 1507-1514 (1966). See also J. Polymer Sci., B, 5, 753 (1967).

(11) K. A. Boni, S. A. Sliemers, and P. B. Stickney, J. Polymer Sci., A-2, 6, 1579-1591 (1968).

(12) B. A. Whitehouse, Macromolecules, 4, 463 (1971).

(13) D. Goedhart and A. Opschoor, J. Polymer Sci., A-2, 8, 1227-1233 (1970).

(14) L. H. Tung, Separation Sci., 5, 339-347 (1970).

(15) G. Meyerhoff, paper presented at the Houston ACS Meeting, Feb. 1970.

(16) A. R. Weiss, Polymer Letters, 7, 379-381 (1969).

(17) A. R. Weiss and E. Cohn-Ginsberg, J. Polymer Sci., A-2, 8, 148-152 (1970).

(18) G. B. Schulz, Z. Physik.-Chim., B-45, 25 (1939).

(19) T. Williams and I. M. Ward, Polymer Letters, 6, 621-624 (1968).

(20) J. V. Dawkins, J. Macromol. Sci.-Phys., B-2, 623-639 (1968).

(21) G. Meyerhoff, Pure Appl. Chem., 20, 309-327 (1969).

(22) H. Coll and D. K. Gilding, J. Polymer Sci., A-2, 8, 89-103 (1970).

(23) H. Coll, Separation Sci., 5, 273-282 (1970).

(24) O. B. Ptitsyn and Y. E. Eizner, Soviet J. Tech. Phys. (Eng. Trans.), 4, 1020 (1960).

(25) L. H. Tung and J. R. Runyon, J. Appl. Polymer Sci., 13, 2397-2409 (1969).

(26) L. Wild and R. Guliana, J. Polymer Sci., A-2, 5, 1087-1101 (1967).

(27) E. E. Drott and R. A. Mendelson, J. Polymer Sci., A-2, 8, 1361-1371 (1970).

(28) Ibid, 1373-1385 (1970).

(29) A. R. Cooper, A. R. Bruzzone, and J. F. Johnson, J. Appl. Polymer Sci., 13, 2029-2030 (1969).

(30) K. Bombaugh, W. A. Dark, and R. F. Levangie, J. Chromatog. Sci., 7, 42-47 (1969).

(31) J. L. Waters, J. Polymer Sci., A-2, 8, 411-415 (1970).

(32) D. J. Harmon, Separation Sci., 5, 283-289 (1970).

(33) J. H. Duerksen, Separation Sci., 5, 317-338 (1970).

(34) R. N. Kelley and F. W. Billmeyer, Jr., Separation Sci., 5, 291-316 (1970).

(35) G. K. Ackers, Adv. in Protein Chem., 24, 343-446 (1970).

(36) J. Cazes, J. Chem. Ed., 47(7), A461-A471 (1970).

(37) J. Cazes, J. Chem. Ed., 47(8), A505-A514 (1970).

(38) D. D. Bly, Science, 168, 527-533 (1970).

Chapter 2

INTERACTIONS OF POLYMERS WITH SMALL IONS AND MOLECULES

D. J. R. Laurence

Chester Beatty Research Institute
Fulham Road
London, England

I. INTRODUCTION

The interaction of small molecules and ions with polymers is of considerable theoretical and practical interest. This field of investigation includes a wide variety of the events occurring in living systems and also the formation of complexes of importance in the chemical industry.

There are few general reviews of the methods available. Klotz's review (1) published in 1954 contains most of the current techniques. There is a brief but comprehensive review of optical methods by Carroll (2). A monograph edited by Pullman (3) contains recent (1968) accounts of current work and theory in the field of biopolymers.

In the present chapter the first two sections describe the binding equations and forces involved in the interactions. The equations discussed represent some of the simpler situations to which the binding phenomena might be expected to conform. The extension of these equations to more complex situations is indicated very briefly. The forces discussed are most appropriate to adsorption from aqueous solutions and include many of the observations found in the literature. The relative role of solvent and other forces is indicated. The methods are discussed according to the type of apparatus that is employed. Examples of catalytic change in the adsorbate due to binding by the polymer or reactive irreversible binding are not considered. Finally some brief reviews are given of the application of binding studies to a number of classes of biopolymers and synthetic polymers.

Throughout the chapter the words "ligand" and "adsorbate" are used synonymously for "ion" or "small molecule."

Some General Ideas

1. Partition and Saturation

A polymer in solution may be considered as a highly dispersed second phase. At low concentrations the adsorbate partitions into this phase usually with a constant ratio of adsorbed molecules to adsorbate concentration. As the adsorbate concentration is increased the partition in favor of polymer becomes less pronounced until the amount bound reaches a constant value. At higher concentrations the fraction of adsorbate bound is inversely proportional to the adsorbate concentration.

The saturation effect is demonstrated if a tracer quantity of a radioactively labeled adsorbate is partitioned between the polymer solution and an "external" homogeneous phase. Addition of cold (unlabeled) adsorbate will change the partition in favor of the external phase (4) as the limit to the binding capacity of the polymer is approached or exceeded.

2. Reversibility

The interactions may be made to reverse by placing the polymer-adsorbate complex in contact with a large volume of external phase (dialysis). Alternatively a third substance is added which competes at the polymer sites or complexes with the unbound ligand.

3. Time Effects

Polymer-small molecule complexes differ greatly in their rate of dissociation. This varies from a fraction of a second (5) to hours (6) or even weeks (7). The complex formed initially may be metastable and "relax" to a more stable configuration. The relaxation process can involve changes in either the polymer (8) or the adsorbate (9).

4. Conformational Changes

These are the rule rather than the exception. Adsorption is very often a mutual accommodation between ligand and polymer. Conformational changes play an important part in the metabolic control that occurs in living systems.

5. Site and Diffuse Binding

Interactions may be divided into long-range diffuse interactions and shorter-range binding at polymer sites. Long-range interactions occur by electrostatic forces and do not affect the solvation layer or spectroscopic ground state of a ligand. Closer interactions involve "more than electrostatic" bonding. The ground state and solvation layer are changed and there is a decrease in rotational diffusion of the ligand or its immediate environment.

6. Simple and Complex Sites

The adsorbate may bind at a single independent atom or group in the polymer. Binding may involve a number of polymer residues dependent on a specific secondary and/or tertiary structure in the polymer.

II. ADSORPTION ISOTHERMS

A. Activity and Concentration

Binding of ligand by polymer reduces the activity of a ligand. The extent of adsorption can be measured without further analysis as an activity coefficient which is the ratio of ligand activity to total ligand concentration. Changes in the ligand activity coefficient are an indication of changes in polymer configuration (10).

When a distinction is made between bound and free ligand, it is convenient to introduce activity coefficients to allow for lack of proportionality between activity and concentration of the free ligand. It is customary to apply these corrections when working with ionic adsorbates. Interactions between the ionic components in solution that cannot be represented by conventional equilibrium constants are thereby eliminated. A more radical use of activity coefficients is to choose an environment different from the solvent as the standard state. Organic ligands may be referred to an organic solvent or to the pure organic substance itself as the standard state. In the equations given in this section no distinction will be made between activity and concentration and it is assumed that the side effects are taken care of in an appropriate manner.

B. Detailed Balance

Interaction equilibria usually result in complexes that can be formed in a number of different ways. A complex formed by adsorption at two sites a and b, for example, can be formed by adsorption at a followed by adsorption at b or vice versa. The two alternative paths make a circuit. It will be assumed that such circuits are not subject to net circulation and that the individual steps are each in true equilibrium. A sufficient condition for a detailed balance appears to be that there is no coupling with an external supply of energy. Equilibrium conditions will not necessarily be obtained when the polymer acts as a catalyst or is subjected to a photochemical process.

C. Polymer Concentrations

Polymer concentrations are commonly taken on a molar basis. Where the repeated sequence of residues results in a corresponding sequence of adsorption sites, it is more convenient to take the equivalent concentration with the monomer or some small multiple of it as the unit of concentration (11).

D. Structure of the Equations

In most cases the polymer is multivalent with respect to ligand. If the molecular weight of polymer is unaffected by the binding equilibrium with the activity coefficients of polymer and its complex in constant ratio, it is convenient to express the interaction in terms of the free ligand concentration or activity. This condition may be tested experimentally by diluting the complex with a solution containing the equilibrium concentration of free ligand.

E. Association Constants

The development of our knowledge of multisite binding is shown by references (12-20). The affinity of an adsorbate for a site will depend on the site and on the occupancy of all other sites in the polymer. If m sites out of the total n on the polymer are occupied there are

$$s = \binom{n}{m}$$

different complexes of this type. For a single complex,

$$C_{mj} = l_{mj} A^m P \tag{1}$$

where C_{mj} is the concentration of the jth configuration of the complex with m molecules of adsorbate bound. A is the free ligand concentration, P is the concentration of unbound polymer, and 1_{mj} is a constant. The total concentration of polymer with m molecules bound (C_m) is

$$C_m = L_m A^m P \tag{2}$$

where

$$L_m = \sum_{j=1}^{s} 1_{mj}$$

and the total polymer concentration (T) is given by

$$T = PQ \tag{3}$$

where

$$Q = 1 + \sum_{1}^{n} L_m A^m$$

Similarly the total adsorbate bound (A_B) is seen to be

$$A_B = PA\, dQ/dA \tag{4}$$

Combining Eqs. (3) and (4) the average number of molecules bound per molecule of polymer ($r = A_B/T$) is

$$r = d \ln Q/d \ln A \tag{5}$$

It is possible to derive Eq. (1) by considering the reaction of the complex $C_{m-1,x}$ with adsorbate to give complex C_{my}

$$C_{my} = k_{myx} C_{m-1,x} A \tag{6}$$

y and x refer to the configurations of complexes C_m and C_{m-1}, respectively. Although there are numerous possible values of k, these are not independent. On forming Eq. (1) by iteration, 1_{mj} will be uniquely determined by m and j. This may be represented as

$$1_{mj} = \left(\prod_{i=1}^{m} k_i\right)_j \tag{7}$$

where the brackets refer to any chain of k's compatible with the final configuration j. It is also possible to define a K_i to include all complexes C_i formed from C_{i-1} and

$$L_m = \prod_{i=1}^{m} K_i \tag{8}$$

F. Frequency Distributions

The terms in Q can be used to obtain the fraction of polymer F_i in form of complex C_i as

$$F_i = L_i A^i / Q \tag{9}$$

Thus Q is an unnormalized frequency generating function. From statistical theory this may be converted to a moment generating function by replacing A^i by exp $(i\lambda)$ where $\lambda = \ln A$. Equation (5) is a consequence of this relationship, r being the first moment of the frequency distribution.

From Eqs. (5) and (9) (21)

$$d \ln F_i / d\lambda = i - r \tag{10}$$

With increase in ligand concentration, the frequency distribution moves toward higher values of i. From Eq. (10) the hinge point about which the distribution is turning instantaneously is $i = r$. All F_i are decreasing below $i = r$ and increasing above as A is increased. Putting $P = C_o$,

$$d \ln P / d\lambda = -r \tag{11}$$

This relation, which is helpful in studies of metal chelates, may be applied to polymers if the uncomplexed polymer has some special characteristic. Polymer fluorescence quenched by ligand binding $(i \gg 1)$ would be an example.

G. The Bjerrum Plot (22)

The dependence of r on λ is a convenient representation of the binding data when the total binding, but not binding to individual sites, is known. From this function it would be possible in principle to obtain Q by integration of Eq. (5). The significance of the "Bjerrum" r versus λ plot was appreciated by Sorensen who plotted titration data on acids and bases in this way.

The slope of the plot of r against λ gives the second moment σ^2 of the frequency distribution about the mean (18).

$$\begin{aligned} dr/d\lambda &= d^2 \ln Q/(d\lambda)^2 \\ &= d^2Q/Q(d\lambda)^2 = (d \ln Q/d\lambda)^2 \\ &= \sum m^2 F_m - (\sum mF_m)^2 \\ &= \sigma^2 \end{aligned} \tag{12}$$

H. Homogeneous Binding

From Eqs. (2) and (7) assuming that all k's are equal and so independent of both m and j,

$$\begin{aligned} Q &= 1 + \sum_{m=1}^{n} \binom{n}{m} k^m \Lambda^m \\ &= (1 + kA)^n \end{aligned} \tag{13}$$

From Eqs. (5) and 13)

$$r = nkA/(1 + kA) \tag{14}$$

Equation (14) is a typical Langmuir isotherm. At half saturation $k = A^{-1}$. If the isotherm is normalized by putting $r/n = \overline{p}$, the plot of r against A has a characteristic shape subject to a scaling factor k.

Alternatively, according to Bjerrum's plot,

$$\overline{p} = \frac{\exp(\upsilon + \lambda)}{1 + \exp(\upsilon + \lambda)} \tag{15}$$

where $\nu = \ln k$ and $\overline{p}$ versus λ curves may be brought into coincidence by translation along the λ axis by a distance $-\nu$. The slope of Bjerrum plot may be obtained from Eq. (15) or by use of Eq. (12), remembering that Eq. (13) is the frequency function of a binomial distribution,

$$dr/d\lambda = n\bar{p}(1 - \bar{p}) \tag{16}$$

Consequently $dp/d\lambda$ has a maximum slope of one-quarter at half saturation.

I. Heterogeneous Binding

If the sites do not interact, the distribution is a Poisson modification of the binomial, viz.,

$$Q = \prod_{i=1}^{n} (1 + k_i A) \tag{17}$$

with k_i the association constant at the i'th site. The Bjerrum function breaks down into the sum of a set of Langmuir terms, one for each site,

$$r = \sum_{i=1}^{n} k_i A/(1 + k_i A) \tag{18}$$

The slope of the Bjerrum plot is always less than that for homogeneous binding just as the variance of a Poissonian is always less than that of a Bernoullian distribution. In fact (20),

$$\begin{aligned} d\bar{p}/d\lambda &= \sum p_i(1 - p_i)/n \\ &= p(1 - \bar{p}) - \sum(\Delta p_i)^2/n \end{aligned} \tag{19}$$

Here p_i is the fraction of sites i that are occupied and $p_i = \bar{p} + \Delta p_i$.

Rearrangement and integration of Eq. (19) gives

$$\int_0^{\infty} \bar{p}(1 - \bar{p})\, d\lambda = 1 + \int_0^{\infty} (\Delta p_i)^2\, d\lambda/n \tag{20}$$

The first integral of Eq. (20) consequently has a minimum value of unity when all k's are equal.

For two sites, association constants k and $k + \Delta k$, the fractional decrease in slope of Eq. (19) at $\bar{p} = 1/2$ is $(\Delta k/4k)^2$ to first approximation. The increase in integral, Eq. (20), is $(\Delta k/k)^2/6$. In either case the change in value would be small unless Δk were comparable with k. It is therefore difficult to detect small extents of heterogeneity from observations of the binding isotherms.

J. Cooperation Between Sites

If all sites are intrinsically identical in k but there is interaction, we may utilize Eqs. (2) and (7) dropping the j suffix,

$$Q = 1 + \sum_{m=1}^{n} \binom{n}{m} \prod_{i=1}^{m} k_i \, A^m \tag{21}$$

For two sites (17)

$$Q = 1 + 2k_1 A + k_1 k_2 A^2 \tag{22}$$

The relationship of r with A divides into two Langmuir terms only if Q can be factorized. From Eq. (22) this is only possible for two sites (with real coefficients) if $k_1 \gg k_2$. If $k_1 < k_2$, the binding of a second molecule is favored by the presence of a first molecule on the polymer, a situation known as "cooperative binding." Thus cooperative binding cannot be represented by a sum of Langmuir terms even in the simplest case. The restrictions on the differential and integral of Eqs. (19) and (20) no longer apply to cooperative binding.

K. The Hill Equation (23)

Increase of association constants k_m with m tends to favor certain C_m and in an extreme case it could be imagined that $F_m = 0; i \neq 0$ or j. Even when binding is not as strongly cooperative as this, it is convenient to treat the isotherm as though there were a j such that

$$\bar{p} = LA^j/(1 + LA^j) \tag{23}$$

L being a constant.

For cooperative binding $1 < j < n$ but a similar equation has also been used (24) for heterogeneous binding for which $j < 1$. On a Bjerrum plot λ is subject to a scaling factor j and the slope will be uniformly multiplied by a constant factor j. The integral of Eq. (20) will be $1/j$. An analysis of cooperative binding (25) uses a plot of $dr/d\lambda$ against λ, the differential Bjerrum plot.

The Hill and Sips equations are conveniently plotted as

$$\ln \bar{p}/(1 - \bar{p}) = j\lambda + \ln L \tag{24}$$

L. Group Interaction

The interactions of a set of sites uniformly distributed on a lattice may be treated statistically (Bragg-Williams lattice gas). When the interaction energy is small compared with kT, the probability of interaction is proportional to m^2 (26) (m being the subscript of C_m). A similar proportionality exists (27) for a set of electric charges "smeared out" over a sphere. Taking the interaction constant in the form $l_m = k^m \exp(wm^2)$,

$$Q = 1 + \sum_{m=1}^{n} \exp X_m$$

where $X_m = \ln\binom{n}{m} + m \ln kA + wm^2$; w being the constant relating the interaction energy in units of RT to m^2. X_m has the form of a free energy of the C_m complex. Differentiating with respect to m in order to obtain a partial free energy and equating the result to zero

$$kA \exp(2wm^*) \doteqdot m^*/(n - m^*)$$

When $w = 0$, comparison with Eq. (14) shows that $m^* = r$. Thus in the absence of interaction the mean r and mode m^* are the same. According to Eq. (10) the distribution is pivoting about the maximum value.

If w is not too great and the distribution not too asymmetrical, m^* may be replaced by r in Eq. (24) and

$$r = \frac{kA \exp(2wr)}{1 + kA \exp(2wr)} \tag{25}$$

By the Debye-Huckel theory a uniform spherical distribution of charged ligands gives an interaction constant (18, 27, 28),

$$w = \frac{-z^2 N_o e^2 [\,1/b - K/(1 + Ka)]}{2DRT} \tag{26}$$

where z is the charge on the ligand; e, the electronic charge; b, the radius of the polymer (protein) taken as a charged sphere; a, the distance of closest approach of ligand and polymer measured from the centers; K, the reciprocal radius of Debye; D, the dielectric constant; and N_o, Avogadro's number. It is possible to choose any initial charge since replacement of wm^2 by $w(\pm m_o \pm m)^2$ adds a constant factor to k that may be absorbed into it by a change of reference state.

M. Klotz and Scatchard Plots

According to Klotz (29) r^{-1} is plotted as a function of A^{-1}. Equation (14) then becomes

$$r^{-1} = n^{-1} + (nk)^{-1} A^{-1} \tag{27}$$

Alternatively according to Scatchard (27), rA^{-1} is plotted against r since

$$rA^{-1} = (n - r)k \tag{28}$$

There are two special cases that deserve attention. When k is large compared with A^{-1}, the Klotz plot produces a horizontal and the Scatchard plot a vertical straight line. This is stoichiometric binding with $A^{-1} \gg 1$ unless $r = n$. Alternatively n may be very large but nk comparable with A^{-1}. This is the condition in which the adsorbate "partitions" into the polymer, with r proportional to A. The Scatchard plot then gives a horizontal line but the Klotz plot a line inclined to the axes passing close to the origin (see Fig. 1).

When there is heterogeneity (Eq. 18), the plots are curved; the Klotz plot is concave downward and the Scatchard plot is concave upward. At low values of A the asymptotic plot is linear. The factor k is then replaced in the equations by $\bar{k}_a$, the arithmetic mean, and n is replaced by $n/(1 + V_k^2)$ where V_k is the coefficient of variation of k

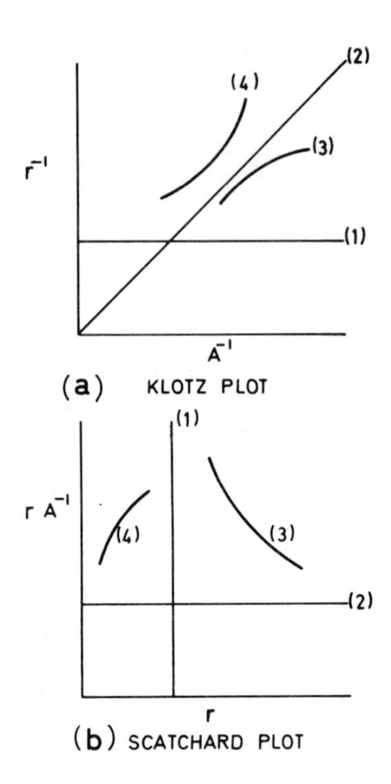

Fig. 1. Examples of Klotz and Scatchard plots: Curve (1), stoichiometric binding; Curve (2), partition; Curve (3), statistical or heterogeneous binding; and Curve (4), (positive) cooperative binding.

(that is, the RMS scatter about the mean divided by the mean). At high values of A there is a different asymptotic straight line with k replaced in the equations by $\bar{k}_h$, the harmonic mean (but n is not replaced in this case). Cooperative interaction curves the plots in the opposite direction (that is, concave up for the Klotz plot and concave down (30) for the Scatchard plot) (see Fig. 1).

If w of Eq. (25) is known, it is enough to replace A by A exp (2wr) in Eqs. (27) and (28). If w is not known, it is then necessary to evaluate the three constants n, k, and w from the experimental data. Usually n is estimated first and k and w are evaluated by plotting $\ln r/(n - r)A$ against r following Eq. (25).

N. The Freundlich Equation (31, 32)

As mentioned in Sec. I.I no evidence of heterogeneity is likely from the isotherms unless there are substantial differences among the association constants. When there is considerable heterogeneity, it is possible to resolve the separate "waves" on a Bjerrum plot and to estimate the number of groups and their binding constants by inspection. The number of groups is the "span" between plateaus of r, and the log k values are the points of maximum slope half way between the plateaus. This procedure is satisfactory if the k's occur in groups differing by several orders of magnitude, for example, the titration curves of proteins. Alternatively an estimate of the value of n_1 and k_1 is made for a dilute solution and the experiments are then repeated with concentrations increased by at least an order of magnitude. In plotting the values obtained for the more concentrated solution, the values n_1 and r_1 are subtracted from n and r to give a new group of constants n_2 and k_2.

An alternative equation which facilitates representation of the results in some cases is the Freundlich isotherm,

$$r = \text{constant } A^t \tag{29}$$

where t is a constant less than unity. This formula is empirical and must be "wrong" in the sense that no saturation effect is included.

O. Stockell's and Jardetsky's Equations (33, 34)

Assuming that the system conforms with the simple Langmuir equation, Eq. (14), it is sometimes convenient to rewrite the equation in terms of the total adsorbate (A_T) and total polymer (T) concentrations. The extent of interaction is given by either the fraction of adsorbate molecules (f) or of polymer sites ($\bar{p}$) that are bound. For convenience of plotting these equations may be written

$$A_T/\bar{p}\,T = k/(1 - \bar{p})T + n \tag{30}$$

and

$$A_T = nT/f - k/(1 - f) \tag{31}$$

P. Nearest Neighbor Interaction (35)

For adsorption to a linear polymer it is not difficult to take condensation effects into account. Consider a linear polymer, total concentration of sites S, and of sufficiently high degree of polymerization so that end effects may be neglected. To preserve symmetry it is considered that there are two adsorbates in the solution A and B and that the polymer sites are filled with either A or B. Finally, it is possible to replace B by solvent and absorb B and its interaction constant k_A into k_B. As before capital letters are used for the free ligand concentrations but small letters (a and b) will be used for the concentration of bound ligand. It is convenient to write the concentrations of nearest neighbors (Fig. 2) aa, bb, and ab as α , β, and 2γ. These can be considered as the concentration of quasichemical "bonds" between neighbors. Then

$$S = a + b$$

$$a = \alpha + \gamma \tag{32}$$

$$b = \beta + \gamma$$

For the quasichemical reaction aa + bb $\rightleftharpoons$ 2ab,

$$\alpha\beta = K_c \gamma^2 \tag{33}$$

In order to relate the concentrations of occupied sites to the free ligand concentrations, consider the reaction involving three adjacent sites, viz.,

$$aab + B \rightleftharpoons abb + A$$

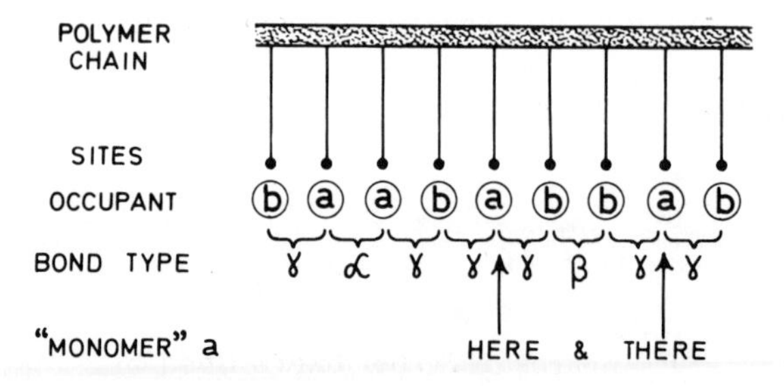

Fig. 2. Nearest neighbor interaction: This diagram illustrates some of the concepts introduced in this section and should be used with reference to the text.

This involves no change in the number of γ bonds on the polymer. The concentrations of aab and abb are $2\alpha\gamma/a$ and $2\beta\gamma/b$, respectively, and so

$$\alpha b/\beta a = k_A A/k_B B \tag{34}$$

From Eqs. (32), (33), and (34), solving for a and taking B as solvent as suggested above,

$$\begin{aligned} \bar{p} &= a/S \\ &= 1/2\left\{1 + (k_A A - 1)/\left[(k_A A - 1)^2 + 4k_c^{-1} k_A A\right]^{1/2}\right\} \end{aligned} \tag{35}$$

If there is no attraction or repulsion between adjacent ligands, that is, $k_c = 1$, this reduces to the Langmuir form. The value of k_c may be obtained by plotting $\log \bar{p}/(1 - \bar{p})$ as a function of λ. The slope at $\bar{p} = 1/2$ is $k_c^{1/2}$ (35). Alternatively the "monomer" configuration bab may be identified spectroscopically (36). The ratio Y of monomer to total adsorbate bound is γ^2/a^2 and from Eqs. (32) and (33)

$$\bar{p}^{-1} = \frac{(k_c - 1)\,Y + 1}{1 - Y^{1/2}} \tag{36}$$

Q. Interaction via Conformation Changes (37, 38)

A positively cooperative interaction is obtained if there is a polymer configuration rare in absence of ligand, which has a higher affinity for ligand than the majority of the unperturbed polymer. Addition of ligand would increase the proportion of this polymer configuration and so favor further interaction. In general there could be many possible such configurations but it will be assumed here that there is only one P' in addition to the "ground state polymer" P.

Let the concentration of the polymer P' be M times that of P in absence of ligand ($M \ll 1$) and binding occur at n noninteracting identical sites on P or P' with constants k and k' ($k' \gg k$). On writing T in terms of P and P'

$$(1 + M)Q = (1 + kA)^n + M(1 + k'A)^n \tag{37}$$

From this r can be obtained by differentation in the usual way.

At low A values, binding is mostly by P and its complexes whereas at high A it is almost exclusively by P' and its complexes. A plot of $k_r = r/(n - r)A$ as a function of A gives a changeover from $(k + Mk')/(1 + M)$ at low A rising to $(k^n + Mk'^n)/(k^{n-1} + Mk'^{n-1})$ at high A. The function is invariant only if n = 1.

The concentration of P and its complexes relative to P and its complexes is M at A = 0 and $M(k'/k)^n$ at saturation with A. On taking logarithms it is seen that the free energy of adsorption tends to reverse the sign of the conformational change free energy in favor of P'. Some change of this sort occurs even when n = 1.

R. Competition (39)

With two adsorbates A and B in the solution each combining with sites on the polymer, an argument similar to that used to develop Eq. (3) gives

$$Q = \sum_{i=1}^{m} \sum_{j=1}^{m} L_{ij} A^i B^j \tag{38}$$

From Q, r_A, and r_B are obtained by partial differentiation with respect to the appropriate variable, for example,

$$r_A = (\partial \ln Q / \partial \ln A)_B \tag{39}$$

On repetition of the differentiation with respect to the other ligand and reversing the order of differentation,

$$(\partial r_A / \partial \ln B)_A = (\partial r_B / \partial \ln A)_B \tag{40}$$

This ensures that all interactions are mutual. For a polymer with only two sites, one for A and the other for B, intrinsic constants k_A and k_B, suppose the presence of the two ligands together on a polymer molecule changes the association constants to ϵk_A and ϵk_B. Then the association constant of A in presence of B is given by k'_A, where

$$k_A' = k_A (1 + \epsilon k_B B)/(1 + k_B B) \tag{41}$$

When $\epsilon = 1$, $k_A' = k_A$ and binding of A is unaffected by B. Mutually exclusive binding of A and B is given by $\epsilon = 0$. This occurs, for example, if A and B occupy the same polymer site. The resulting competition may be plotted as

$$k_A'^{-1} = k_A^{-1} + B\, k_B k_A^{-1} \tag{42}$$

When A is adsorbed only when B is adsorbed, ϵ is very large.

Taking ϵ into k_A we obtain

$$k_A' = k_A k_B \; B/(1 + k_B B) \tag{43}$$

This sort of relation is found, for example, when adsorption is conditional upon the protonation of a group either on the polymer or the adsorbate. B is the hydrogen ion and k_B the reciprocal of the acid dissociation constant of the group.

Conditional binding also occurs in titration of wool (40, 41). Titration is conditional on taking up of the acid anion into the wool fiber.

S. More Complex Systems

The equations developed in this section are mostly applicable only to the simplest types of interaction. For the treatment of more complex cases the reader is referred to discussion of sequential conformation change (42, 43), matrix methods (44, 45), and the sequence generating function (46). It is hoped that binding data will justify the use of these extremely elegant procedures.

III. TYPES OF BONDING

Adsorption of small molecules by polymers in solution may be represented by a "quasichemical" reaction (47) in which the formation of the complex reduces the number of contacts between the solvent and the interacting components. It is necessary to take into account not only the interaction of the adsorbate with the polymer and the interactions of both polymer and adsorbate with solvent but also the energy of interaction of the "liberated" solvent molecules with each other.

A. Coulombic Interactions

The strength of electrostatic interactions is determined by the charge density on the polymer and adsorbate, the distance of closest approach, and the local dielectric constant.

In water the interaction of uniunivalent electrolytes is weak compared with the Boltzmann thermal energy and these ions remain unassociated to a large extent at all but the highest salt concentrations. The association of a bivalent salt on the other hand (cf. magnesium sulphate) is demonstrable at 0.1 M concentration in water (48).

Polyelectrolytes with their high charge density associate with counterions and the association appears to be resistant to dilution over a wide range of concentration (49).

Binding due to coulombic forces is a long-range interaction. It is difficult to choose a distance from the ionic centers within which an ion of opposite charge must be considered to be bound (48). The choice is easier in the case of strongly charged ions of cylindrical symmetry such as polyelectrolytes than it is for salts of microions (49). As shown in Sec. IV.A the Donnan corrected binding includes associations resulting from any fluctuations in electrostatic potential in the polymer solution.

For a given polymer, binding of a series of ions of the same charge is related to the ionic radius of the counterion. When the alkali metals are arranged in order of their hydrated radii, the order is the reverse of that obtained from the crystallographic radii (50). The observed order (51, 52) may therefore give an indication of penetration of the hydration layer of the counterion by the charged groups of the polymer. A displacement of the order in favor of some ion in the middle of the series may also be observed (53). The primary structure or folding of the polymer may favor binding of this ion for steric reasons.

The importance of solvent interactions was discussed by Gurney (54). This author found that small ions such as lithium and hydrogen replaced the normal water structure by a new ionocentric arrangement. Larger ions such as cesium or sulphate with a charge spread out over the ionic surface destroyed the existing structure of the water without replacing it by a new one. Interaction was greatest between ions with a similar effect on the water structure. For example, the carboxyl group is order retaining and interacts strongly with the hydrogen ion but sulfate is order destroying and interacts weakly. These relationships are probably not exclusive of the ideas on hydration mentioned in the previous paragraph.

The electrostatic potential that can be realized is limited by the difference between the pK of the charged groups and the pH of the ambient medium. When the charged groups are on nonadjacent chain elements, changes of the polymer secondary and tertiary structure may also limit the potential that can be obtained.

Differences in dielectric constant between the polymer sites and the surrounding medium influence the electrostatic field and hence the interaction energy and charge specificity (55, 56).

Coulombic forces act alone or in combination with shorter-range interactions. Nuclear magnetic resonance evidence suggests that polysulfonic acids unlike polycarboxylic acids are completely dissociated (57). The binding observed with these substances is therefore electrostatic. Competition between organic ions and alkali metals confirms that binding to sulfated polysaccharides contains little that is attributable to the organic part of the molecule (58). Even when there is a substantial noncoulombic component to the reaction, competition of inorganic salts at the charge centers may be observed. In this case the concentration of salt needed to displace the organic ion is higher than the ligand concentration. The reaction of DNA with basic dyes in competition with neutral salts is of this type (58).

B. Chelation and Hydrophobic Bonding

The small molecule is often considerably larger than a solvent molecule and a number of solvent molecules are displaced into the bulk solution during interaction.

It is necessary to distinguish two types of solvent interaction, viz., with polar and charged groups on the one hand and with relatively nonpolar groups on the other.

It is known from the studies of metal chelates (59-61) that reactions releasing more than one solvent molecule per ligand are assisted by a cratic free energy term of entropic origin. It would be expected that a similar term would occur in interactions between small molecules and polymers in which more than one dipolar or charged center on the ligand was able to find a complementary partner on the polymer. A number of specific biological reactions including the well-known three-point contact of Ogston (62) might be of this type. Pairing of the polar groups is also important in determining interaction of nucleotides with nucleic acids.

If the insolubility of cellulose and uric acid may be taken as an example of interaction (in this case self-interaction) by chelation, it can be seen that the interaction is only broken up by stronger complexing agents (such as hydroxyl or copper ion).

Disruption of the hydrogen bonded structure by partial alkylation increases solubility (cf. caffeine or methyl ethyl cellulose). The alkyl groups would have been expected to make the substances less soluble on account of hydrophobic bonding.

Introduction of less polar groups, for example, paraffin groups, into water leads to a situation where the solvent-solvent interaction energy is larger than any other of the four energy terms in the quasichemical reaction between solvent, adsorbate, and polymer. Interaction may then be attributed to rejection from the water structure and this type of bonding is known as "hydrophobic" or "apolar" bonding. The interaction between nonpolar structures and water follows the van Laar scheme in the sense that nonpolar-water interactions are more energetic than nonpolar-nonpolar interactions. No justification can therefore be given for the term "hydrophobic" unless it is applied to the system as a whole.

There is a direct proportionality between the energy of macroscopic nonpolar-water interface and the energy of interaction of nonpolar groups on a molecular scale in terms of the relative surface areas (63, 64, 56). The water in contact with nonpolar surfaces is able to take up a structure in which the energy required for destruction of hydrogen bonds due to contact with nonpolar material is balanced by the heat derived from changes in local degrees of freedom. It will be remembered that "native" water has an unusually high specific heat. The interaction therefore turns out to be entropic even though hydrogen bonds are probably broken or at least bent (64).

Material held together by hydrophobic bonding is split if the water is replaced by an organic solvent. A polar organic liquid will reduce the interaction energy considerably and the water structure may possess properties apart from its polarity that render cavities in the structure unusually energetic (65). One of these could be the three-dimensional nature of the hydrogen-bonded structure. Other polar liquids are less than tetravalently hydrogen bonded. Aqueous urea solutions, for example, show anomalous viscosity effects that suggest a linear or perhaps sheetlike structure of the hydrogen bonding (66). It would be much easier to introduce nonpolar material into such a lattice without breaking hydrogen bonding.

The difference between chelational and hydrophobic bonding is expressed at the molecular level as a difference in the orientation of the solvent dipoles with respect to the solute molecules.

For chelational bonding the dipoles are directed radially toward polar groupings in the solute molecules. The strength of the bond depends on the number of solvent molecules that are rigidly bound in this way and on differences in dipole-dipole or dipole-ion electrostatic interactions and covalent bonding in the bound and free states. Electrostatic interactions would be stronger in regions of low dielectric constant which occur within certain polymer coils.

For hydrophobic bonding the solvent dipoles tend to lie tangential to the nonpolar surfaces and to associate with each other. Rise in temperature causes the solvent molecules to adopt a more random configuration and the energy of the contact region would then tend to increase. A similar effect would be expected from salts which interfere with the preferred orientation of the water dipoles at the interface.

Salts that interact strongly with the water would produce regions between the hydration layers having a repulsive configuration of the solvent dipoles. The entry of nonionic groupings into these regions would be energetically more favorable. The solvent action of lithium salts on nonionic material could be accounted for in this way.

The possibility of a bias of uncharged nonpolar regions in polymers toward anions was discussed by Schellman (55). The greater hydration of cations would prevent the inclusion of these groups in a phase from which water is excluded.

C. Dispersion and Charge-Transfer Forces

For adsorption of nonpolar molecules from vacuum onto a nonpolar polymer surface these forces would be the predominant ones. In solutions, analysis is made more difficult by the solvent release mechanisms described in the previous paragraph. With the larger atoms (of atomic weight greater than 20) and with conjugated systems of light atoms there is no doubt that a substantial contribution comes from these more conventional sources. An example is the reaction of mercury with sulfur groups in the polymer (67). The association constant is greater than 10^{10} and cannot possibly come from any solvent release mechanism. The reaction of silver with aromatic rings is another example (68). For other less weighty atomic centers the task of disentangling the four interaction energies has only just been started.

D. Lock and Key Relationships

Mention was made in the discussion of chelational bonding of the importance of a good fit between adsorbate and polymer groups. Williams has observed (69) that a pair of charged groups brought together in a ligand by covalent bonding will have an advantage over two independent charges at a binding site. The repulsion energy of the groups in a multivalent ligand is included in the energy of formation of the compound and therefore does not enter into the interaction as a negative term. Similarly, a rigid ligand with just the right spacing of charged groups would have an advantage over a flexible ligand (70) for which the correct spacing is one of a large number of configurations that are possible in solution.

IV. DISTRIBUTION METHODS

A. Equilibrium Dialysis

The polymer solution is separated from polymer free medium by a membrane of regenerated cellulose or collodion. The membrane is chosen to be impermeable to polymer but freely permeable to ligand. The distribution of ligand across the membrane is observed.

When the polymer is charged, electrical neutrality requires that an excess of counterions be present in the polymer phase sufficient to neutralize the polymer charge. This results in an excess of counterions and a deficiency of coions in the polymer phase. A Donnan correction is applied (71) to give the value for true binding by the polymer. According to the Donnan concept there is a concentration ratio D_z across the membrane for an ion of charge z which is $(D_{+1})^z$. D_{+1} is the ratio for the distribution of a positive univalent ion across the membrane. When it is known that a certain ion is not bound, the distribution of this ion may be taken as defining the Donnan ratio. Radioactive sodium ion which is bound weakly, if at all, by serum proteins has been employed in this way (32). The Donnan ratio may also be determined (71) by measurement of the membrane potential $e_m = (RT/F) \ln D_{+1}$. For strongly bound ions it may be considered acceptable to add a swamping excess of a more weakly bound electrolyte to a low polymer concentration in order to reduce D_{+1} to near unity.

The Donnan correction is less satisfactory if the polymer solution cannot be treated as a phase with uniform D_{+1} or e_m. Polyelectrolyte solutions contain significant volumes from which the coions are effectively excluded. This is an extreme example of a general effect, viz., that the average D_{+1} is $\sum v_i (D_{+1})_i$, v_i being the volume fraction of the solution with ratio $(D_{+1})_i$. This is not the reciprocal of D_{-1}, which is $\sum v_i (D_{+1})_i^{-1}$. From the relationship between the arithmetic and harmonic mean it appears that "true binding" after removal of the Donnan correction includes a contribution attributable to the variance of the electrostatic potential in the polymer solution. The electrochemical methods might also contain this contribution whereas the spectroscopic and magnetic resonance methods would not. Ropars and Viovy observed that a copper DNA complex which was nondialyzable nevertheless contained a large fraction of copper with an electron spin resonance signal indistinguishable from that of free copper ion (72).

Rapid equilibrium is achieved by stirring the solutions or by increasing the surface to volume ratio in the compartments (73). Regenerated cellulose contains sulfur compounds that are a trap for metallic ions. These may be removed by controlled heating in large volumes of water or by dilute nitric acid treatment (74, 75).

A steady state method with the dialysis system of the automatic analyzer (Technicon, Chauncey, N.Y.) has been developed. This allows a very rapid run through of samples (76).

B. Ultrafiltration

The ultrafiltration method is analogous to equilibrium dialysis but the polymer free phase is obtained by expressing some of the medium through the membrane under pressure. In order to test for binding of ligand in the membrane it is advisable to analyze several small consecutive samples from the same polymer solution (77). New filters are now available with a more rapid filtration rate than have the conventional collodion or cellophane (from Amicon, Cambridge, Mass.). An alternative to filtration in a filter assembly is to place the sample in a semipermeable sac on a centrifuge (78, 79). The medium is expressed under the gravitational force of centrifugation.

C. Gel Filtration

Crossed linked dextran (Pharmacia, Sweden) and acrylamide gel (Biorad, Richmond, Calif.) are permeable to small molecules but not to molecules of high molecular weight. Passage through a Sephadex or Biogel column is thus a simple way of analyzing certain interactions (80, 81). If the complex is preformed and stable, the bound and free ligand may be separated completely. Alternatively, the column is pre-equilibrated with a ligand solution (82). The polymer is washed down the column, and equilibrium between polymer and ligand then occurs. The peak due to the complex is followed by a trough or negative peak of equal magnitude. The resolution is adequate when the base lines before and behind the complex are flat and equal.

Sephadex is not indifferent to certain ligands or polymers and this limits the simplicity of the method in these cases (83, 84).

D. Liquid-Liquid Partition

The ligand is partitioned between two liquid phases only one of which dissolves the polymer and inorganic salts. The application of this method to polymer interaction studies is due to Karush (85). The method has been applied both with proteins (85, 86) and with nucleic acids (87).

E. Gas-Liquid Partition

This may be done by vacuum-line techniques and radioactive counting (88). Another method is to measure the saturation concentration in a gas-liquid chromatography apparatus after addition of an internal standard (89).

F. Increased Solubility (90, 91)

Increased solubility may result from complex formation. It is important to ascertain whether the complex that is formed is not polydisperse consisting of blocks of ligand molecules coated with polymer. Slightly soluble materials are easily adsorbed by filters and some surfaces even when the solution is monodisperse. One should not therefore reject the possibility of a monodispersed solution because the ligand is retained by a filter (90).

G. Decreased Solubility

Decreased solubility of the complex may also provide evidence of interaction. The precipitation of a multivalent complex is not evidence that a similar complex exists as the majority component of the solution.

It is necessary to show that the ligand is distributed in favor of the precipitate. A precipitation might occur if the precipitant interacts with the solvent but if this happens, the precipitated polymer will contain less ligand than does the solution. The stability of the complex can be tested by washing with more ligand solutions (92). It is found that precipitates of charged polymers are frequently isoelectric and the composition gives a measure of the acid or base bound of the polymer (93, 94, 95).

H. Surface Methods

Reaction of a polymer with a ligand at a surface was observed by Brigando who measured the spreading of carcinogens on DNA solutions (96, 97). The reaction of lauryl sulfate with proteins was studied by Bull (98) by surface pressure measurements. Sher and Sobotka (99) observed the binding of stearic acid to proteins by the change in interference colors of multilayers.

I. Complexing Reactions

Binding of ligand to an ion exchange resin is a convenient way of feeding the polymer a large quantity of ligand at a controlled concentration (100, 101). The transfer of ligand to the polymer solution in excess of the equilibrium concentration is taken as bound ligand. Alternatively, the ligand may be supplied as a soluble chelated complex.

This is necessary if the association constant between ligand and polymer is very high and it is required to avoid very large volumes of ligand solution or extremely dilute polymer solutions. Chromatographic methods have been employed to facilitate the analysis of the partition equilibrium or the total ligand bound (102, 103, 104). Certain dyestuffs aggregate readily in solution and the adsorbed complex may be less sensitive to changes in salt concentrations and other factors than is the colloidal dye (105).

V. ELECTROCHEMICAL METHODS

Electrochemical methods are of special value in studies of inorganic ion binding and there are also some applications to organic ions.

A. Potentiometric Methods

According to the Nernst equation

$$E = (RT/zF) \ln a/a_o$$

where a and a_o are ligand activities in the solution under test and in the standard solution or state adjacent to it, E is the electromotive force observed between the test solution and standard, z is the ionic charge, and F is the Faraday constant. The standard together with its limiting membrane comprises the electrode. There are various forms of electrode. An electrode of the first kind consists of an element in contact with the solution, for example, silver wire. This electrode has recently been used in a study of reaction between DNA and silver ions (106). Another example is the hydrogen gas electrode. This is taken as a standard to determine drift in the pH glass electrode (107).

An electrode of the second kind consists of a metal in contact with an insoluble salt of the same metal (108). The best known examples are of silver with silver halides or thiocyanate used by Scatchard and coworkers in studies of binding by plasma albumin (109, 110). Botré and coworkers (111) have developed a silver-silver dodecyl sulfate electrode for investigation of detergent complexes. The calomel electrode which serves as a reference potential is another example.

1. Glass Electrodes

Glass electrodes were first developed for determination of hydrogen ion concentration (112). Some forms of pH glass electrode showed an "alkali error" at high pH and sodium ion concentration. At first all efforts were made to reduce this error by change in the glass composition. More recently it has been found possible to obtain electrodes specially sensitive to certain alkali and alkaline earth metals by systematic attempts to make the alkali error worse. This is done by replacing part of the silicon oxide in the glass by oxides of boron, aluminum, or phosphorus, a development pioneered by Eisenman (113, 114). Sodium electrodes have also found application to cells with transference by Ise and Okubo (115) in an investigation of the average activity of polymer together with counterions. These authors consider, following Guggenheim (116), that it is theoretically unsound to attempt to define a single ion activity in a polyelectrolyte solution. With the average activity it is possible to test the Gibbs-Duhem relationship by a comparison with the osmotic coefficient.

Glass electrodes have fairly good specificity toward the cation namesake. There are some strong cross reactions between silver and a sodium electrode or zinc and a calcium electrode, for example.

2. Membrane Electrodes (117, 118)

Membrane electrodes are sheets of an ion exchange material, solutions of polyelectrolytes in collodion (119), or suspensions of insoluble salts in a paraffin base (120). Only the latter have a specificity comparable with the glass electrode. According to Botré et al. (121) a complete electrochemical circuit without salt bridges may be constructed by placing the sample between two half cells identical except for the charge density of the permselective membranes. A dissymmetry potential develops which may be taken as a measure of the ionic activity in the sample. Glass electrodes are sensitive only to cations; membrane electrodes also detect anions if the membrane contains fixed positive charges.

A "membrane" consisting of a liquid ion exchanger is suitable for alkaline earth and organic ions that form complexes soluble in an organic medium (122, 123).

3. Connecting Bridges

Completion of the electrochemical circuit with a connecting bridge presents special difficulties with polymer solutions as the high concentrations of salt required to avoid a variable junction potential can precipitate some polymers and ligands (121). The salts, potassium chloride and ammonium nitrate, are often used because of a relatively high solubility and similarity in mobility of anion and cation. There are also some theoretical difficulties in the use of connecting bridges (115).

B. Polarographic Methods

For many metals there is no suitable reversible electrode. The polarographic method (124) records the reaction of ions with an electrode surface under non-equilibrium conditions. The voltage region in which variations of current are obtained is characteristic of the metal. Under suitable conditions the span of the polarographic wave depends on the concentration and average diffusion constant of the metal. Complexing of metal by polymer can affect the polarographic result in two ways. Certain complexes no longer react with the electrode (125, 126). As the extent of binding increases the polarographic wave falls to zero. In other cases the complex is sufficiently reversible to react with the electrode but attachment to the polymer reduces the rate of diffusion and hence also the current which is diffusion limited. The relative rates of diffusion for polymer bound and free ligand may be estimated by plotting a saturation curve from zero to high polymer concentrations, provided that there are no large changes in viscosity over the range of polymer concentration necessary to achieve complete binding. In earlier work it was customary to assume that the complex and free ligand approached the electrode independently (126). More recently dissociation and reassociation of the complex during its path to the electrode are considered (127). It is desirable that the polymer and ligand should have opposite charges. If the charges are the same, the polymer can interfere with deposition of the ligand on the electrode surface (128).

C. Conductivity and Transport

There are two simple methods of carrying out conductivity experiments, viz., by conductivity titration or extrapolation to infinite dilution. In the titration procedure (129) the ligand is added to a polymer solution and the increments of conductivity are compared with those obtained when the addition is to polymer free medium under otherwise identical conditions. A plot of ligand added against conductivity may give a straight-line relationship with a different slope between polymer solution and control. Saturation of the polymer sites is indicated by a change of slope of the polymer curve toward that of the control. Although the simplicity of this method is attractive, it may be difficult to correct for changes in viscosity brought about by interaction.

Extrapolation to infinite dilution has been used to show that polyelectrolytes do not become completely dissociated on dilution over the practicable range of the method (130).

Conductivity measurements are simpler to evaluate if the counterion is hydrogen. The high mobility of this ion reduces the contribution of the polymer to the conductivity. For polyelectrolyte salts it is necessary to make a correction for the polymer mobility. This may be done (131, 132) by simultaneously analyzing the transport of material into the end compartments by Hittorf's method.

Transport measurements have also been used to demonstrate the electrophoresis of radioactive counterions in the "wrong" direction due to polymer binding (133, 134, 135). It is possible to determine the extent of association by this method. The polyelectrolyte appears to be less associated by transport methods than by osmotic determinations. The difference has been explained by a mobility of the counterions along the long axis of polyelectrolyte molecule (49).

D. Electrophoresis

Variations in mobility of proteins depending on the salt content of the buffered medium drew attention to salt binding by proteins and its specificity. Methods of estimating binding in the classical moving boundary cell have been suggested (136, 137). Constituent mobilities are calculated by averaging the mobility of a component, free and in all its complexes, using the fraction of component in the various mobility states as weighting factors. A simplification of the equation results if it is assumed that there is a linear relationship between the mobility and the number of molecules bound. The serum albumin-methyl orange interaction was analyzed in this way (136, 137).

Electrophoresis is also informative if there is a strong preference for certain multiligand complexes. Putnam and Neurath (138) analyzed the interaction of detergents with proteins by the splitting of the protein pattern into complexes of characteristic mobility.

Association of small molecules with proteins can lead to a double-peaking artifact even if the interaction is rapidly reversible. The partial differential equations describing this effect were investigaged recently by Cann (139).

For polyelectrolytes, mobility is found to be dependent on intrinsic properties of the coil, viz., the ratio of charge to frictional coefficient of a coil element (140). This simple relationship allows an estimate of binding to be made by the change in mobility. Strauss and Woodside (141) have taken advantage of this situation to study binding of alkali metals to polyphosphate.

Binding of adsorbates to blood plasma proteins has been studied by zone electrophoresis on various supporting media. The phenomenon is most readily observed if the ligand is colored, radioactively labeled, or located by a colorimetric reaction. Westphal and Devenuto (142) have identified the proteins that bind radioactively labeled steroids in the plasma. Many investigations have been made with labeled thyroxine and analogues. The result has been shown to depend on the choice of electrophoretic buffer and supporting medium (143).

Electrophoresis on a hanging curtain through which there is a buffer flow is another useful technique (144). Polymer and ligand are applied at the top of the curtain. At some point the trajectories of the components cross and deviation or accumulation at the crossing point is taken as evidence of an interaction. Interaction of certain amines with serum albumin has been detected in this way by Bickel and Bovet (145).

VI. OPTICAL METHODS

Optical methods include absorption spectroscopy, fluorescence, and measurements of optical rotation. There have been considerable developments in these fields in the last 20 years especially in the application to polymer small molecule interactions. Combined with rapid reaction techniques, a resolution of kinetics on a time scale of 10^{-7} sec or less is possible (146). In contrast with distribution and electrochemical methods, observations of optical changes give most weight to short-range interactions and site binding. There is therefore a similarity with magnetic resonance which may be regarded as complementary with a certain area of overlap.

A. Absorption Spectroscopy

The changes in color of hemoglobin with ligand binding have been a subject of investigation for many years. A recent study has shown that the absorption changes are closely proportional to the extent of binding (147).

The basic assumption that is made in the analysis of mixtures by absorption spectroscopy is that absorption by molecular species in the mixture is additive. Other simplifying features, which cannot, however, be taken for granted, are correspondence with Beer's law for free ligand and equivalence of the adsorption sites with respect to effect on the spectrum of the ligand. Failure of Beer's law, like deviations of activity coefficients from "unity," is an indication that something interesting is happening to the ligand in the solution.

The development of commercial spectrophotometers since 1942 has given new scope to analysis of interactions by optical density changes. This was followed by publication of Klotz's papers from 1946 onward on the interaction between plasma albumin and other proteins with azodyes (148).

1. Ligand Spectra

Binding of ligand may result in change in the wavelength of maximum absorption (λ_{max}), in the absorption at the maximum (ϵ_{max}), and in splitting and sharpening of spectrum bands and the appearance of new bands. Some of the parameters commonly used to specify the effect of absorption (such as λ_{max} and ϵ_{max}) have only a weak relationship with the parameters usually regarded as of importance by spectroscopists (oscillator strength, transition moment, 0-0 transition, etc.). There have been appeals, for example, by Stone (149), for a more meaningful recording of the optical data.

When there are only two relevant absorption coefficients, viz., of bound and free ligand, the difference in optical density at a given wavelength is linearly proportional to the fraction of dye that is bound as the polymer concentration is varied. If a wavelength, usually on the red edge of the absorption band, is found for which ϵ for the free dye is negligible, experiments with variable dye concentration give a linear scale of occupation of polymer sites (33). Small changes in absorption coefficient may be used with a sensitive differential spectrophotometer such as that designed by Chance (150) which permits measurement of a change of 10^{-3} in optical density.

The heterogeneity of absorption sites can be investigated by the method of isosbestic points. A strictly defined isosbestic point is only likely if there are not more than two absorbing components in the solution.

The trypan blue-plasma albumin system shows evidence of heterogeneity of absorption spectrum from which the occupancy of certain specified sites on the polymer may be determined (151). More than one spectrum is also observed when basic dyes combine with a flexible polyelectrolyte such as polyadenylic acid. The additional spectra result in this case from dye-dye interaction (152) on the polymer. The spectrum obtained at low levels of occupancy resembles the monomer spectrum of the free dye whereas the spectrum at high levels of occupancy resembles that of the aggregated dye. The equilibrium between monomer and aggregate is affected by changes in flexibility and coiling of the polymer.

Changes in ligand spectrum on adsorption can also occur as a result of preferential adsorption of one of a conjugate pair of acid and base. As a general rule, preference is given to the member of the pair with charge opposite to the charge on the polymer. Plasma albumin is an exception (153). This polymer associates more strongly with anions even when the net charge on the polymer is negative (56).

When aggregation and pH indicator effects are excluded, there remain some notable changes in ligand spectrum which may throw light on the nature of the binding sites. In a study of binding of fluorescent dyes by plasma albumin (56) it was noticed that changes of absorption spectrum and also of fluorescence were similar to those produced by transfer of the dyes to a nonpolar medium. Such ligands can serve as probes of the polarity at the adsorption sites. The absorption spectrum and fluorescence changes were not due to ionic interactions as similar effects were observed with sulfonic acids and with neutral sulfonamides of naphthylamine derivatives. Hydrogen bonding of the polymer with the amine group is also discounted as the N-methylated naphthylamine also gave similar results even in aprotic solvents such as dioxan. The most likely explanation for the spectrum changes is by interactions of the solvent with the conjugated n and electrons of the dye.

Azo dyes, as observed by Klotz and coworkers, are sensitive to species differences and conformational changes that could not be detected by optical rotation (154). These dyes contain an elongated conjugated system and the spectrum may be influenced by axial electrostatic fields. Carroll was able to detect binding of large numbers of azo dye molecules to proteins in acid media by spectrum changes (155).

The spectra of transition metals bound to polymers also gives information on the nature of the complex (156, 157). The weak visible absorption bands are due to "ligand field" transitions between d orbitals. The more symmetrical octahedral complexes have ϵ_{max} = 1-10, whereas tetrahedral complexes are more strongly absorbing, ϵ max = 10^2. In addition to these weak bands there are stronger bands (ϵ_{max} = 10^3 - 10^4) in the near ultraviolet. In model compounds the intensity and frequency of these ultraviolet bands depend on the ease of oxidation or reduction of the ligand. These intense bands are therefore ascribed to charge transfer transitions between ligand and metal. In tetrahedral complexes there is mixing between ligand field and charge transfer bands. It is also appropriate to mention in this section the analysis of binding by the use of colored complex between ligand and a spot reagent (158, 159), for example, calcium and murexide. Binding of the calcium by polymer decreases the amount of murexide-calcium complex and the reaction may be followed by absorption measurements.

2. Polymer Spectra

Apart from changes in absorption spectrum of the ligand, conformational changes in the polymer result in changes in the polymer absorption spectrum and these may be caused or prevented by ligand binding.

An increase in absorption of nucleic acids at 260 mμ results from thermal denaturation. The normal optical density temperature curve is sigmoid and known as a "melting profile." The midpoint of the curve is called the T_m or melting temperature. At "normal" salt concentrations T_m is about 80 degrees and varies with the base composition of the nucleic acid. A polymer rich in guanine and cytosine (G-C) has a higher T_m than an adenine-thymine (A-T) rich polymer. The T_m increases linearly with the logarithm of the alkali salt concentration. At less than 10^{-3} M NaCl, DNA denatures even at room temperature. Addition of ligands to the nucleic acid results in changes of melting profile. Many substances cause a rise in T_m, for example, actinomycin which binds most strongly to the native form of DNA. Copper binds weakly to the native structure but a stronger binding and denaturation of the DNA occurs if the solution is heated (160). The effect of mercury and silver depends on the ratio of metal to nucleotide. At low ratios the metal comes between the base pairs and prevents denaturation. At higher ratios the metal competes with base pairing and favors loss of organized structure. The normal order of thermal stability (G-C $>$ A-T) may be reversed if the metal has a special affinity for the G-C component (161).

The spectrum of DNA may be changed not only by denaturation but also by interaction of the metal with the bases. This is most clearly shown in complexes with colorless metals such as silver and mercury (162, 163). The melting profile is not only shifted up or down the temperature axis, it also pivots around the midpoint. Temperature then has no further effect on the absorption spectrum which resembles neither the native nor the denatured spectrum of the uncomplexed polymer.

Changes in absorption spectrum also occur when proteins combine with ligands. The aromatic residues in proteins are sensitive to polarity of the environment in a manner similar to the fluorescent dyes described in Sec. VI.A.1 (164, 165).

B. Fluorescence

The application of fluorescence to studies of interaction of small molecules with polymers has been discussed in a previous chapter in this series (Vol. 1, Chap. 5). Intensity, polarization, lifetime, polarization spectrum, and energy transfer may all be used to characterize the interaction.

The changes in fluorescence intensity on adsorption are variable. Some fluorescent compounds become quenched whereas others are only fluorescent when adsorbed (56). This gives a useful scale to measure either the free or the bound ligand. The polarization increases on adsorption as the molecular tumbling is slowed down. The polarization is not a linear function of the extent of binding unless the intensity is unaltered on adsorption. A convenient linear combination of polarization and intensity measurements has been described (56).

Both ligand and polymer fluorescence may be changed by the interaction (166, 167, 168). Nonradiative transfer of excitation energy between polymer and ligand may also occur. This can produce an isoemissive point, similar to the isosbestic point of absorption spectroscopy, which is a test of site homogeneity (169).

Certain fluorescent dyes are very sensitive to the polarity (170) of the polymer sites. They act as a molecular transponder beacons with emitted frequency and intensity typifying the local environment. From analogy with space technology these substances are known as molecular probes (see Fig. 3).

C. Optical Rotatory Dispersion

Gross changes in optical rotation of polymers due to reaction with small molecules have been known for some years (171, 172). More recently the analysis of polymer conformation by circular dichroism or optical rotatory dispersion methods has been developed. The helical structure of some polymers is believed to give rise to a pair of closely spaced transitions. These can be analyzed by the Moffit-Yang (173) or Schechter-Blout (174) methods to give an estimate of the amount of helical structure.

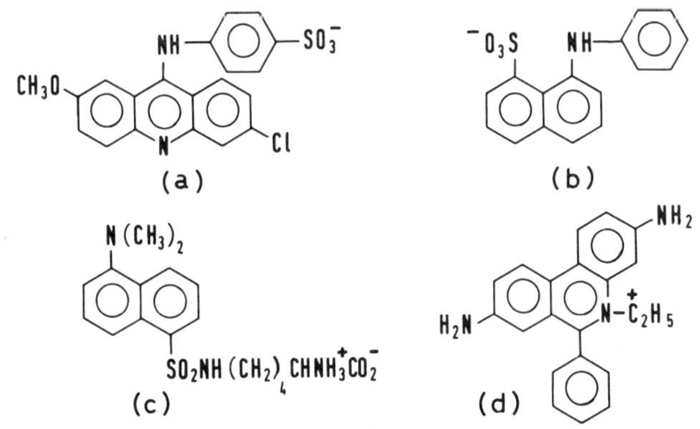

Fig. 3. Hydrophobic probes: These compounds share the property of being very weakly fluorescent in aqueous solution but brightly fluorescent when adsorbed on to a nonpolar region of a polymer. (a) and (b) are mentioned in ref. 170, (c) in refs. 259 and 260, and (d) in ref. 294.

Changes in optical rotation of the polymer due to ligand binding have been described for albumin and azodyes (175, 176), sulphonamides (177), and anesthetic gases (178).

Induced activity in ligands was noticed on binding of dyes to helical poly-L-glutamic acid and DNA (179, 180). At first it was thought that a helical configuration was necessary in order to induce optical activity in the dye. Recent work has shown that an extended regular configuration of the polymer is not necessary (181) and that activity is also found for isolated monomer-bound dye (182). It appears therefore that vicinal effects are responsible for some of the induced activity. Most biopolymers contain many optical active centers that are an intrinsic part of their primary structure. Interaction of bilirubin with albumin gives rise to a very large optical activity centered on the ligand transition. This suggests that the bilirubin molecule may become "twisted" when it is bound to the protein (183).

Optical activity has been detected in ligand-field bands following reaction of an external small molecule with the metal in the polymer (184).

VII. MAGNETIC METHODS

A. Susceptibility

Magnetic susceptibility measurements were used in binding studies of the heme proteins by Coryell et al. (185). Later Theorell and Ehrenberg (186) utilized a similar technique to investigate the cobalt analogue of the enzyme carbonic anhydrase. The naturally occurring zinc enzyme was taken as a measure of the diamagnetic background.

B. Magnetic Resonance

Magnetic resonance methods (187, 188) have recently found wide application in investigations of electron spin and nuclear paramagnetism. With resonance techniques the diamagnetic background is largely eliminated. Fine detail in the behavior of the atomic and nuclear magnets due to environmental influences becomes apparent. In the basic resonance apparatus there is a steady field H_1 and an alternating field H_2 at right angles to it. Transitions occur when the frequency of ν_0 of H_2 corresponds with resonance of the magnetic dipoles in the field H_1. The resonance phenomenon is observed by absorption of energy from the alternating field or by the generation of an alternating component at right angles to H_1 and H_2. The spectrum is plotted by varying H_1 or ν_0. The frequency ν_0 gives a time $\tau_0 = (2\pi\nu_0)^{-1}$ which serves as a criterion for sorting environmental effects. For proton magnetic resonance (pmr) ν_0 is about 10^{-8} sec and for electron spin resonance (esr) about 10^{-11} sec. The instability of the magnetic dipoles in absence of the alternating field H_2 is determined by fluctuation of magnetic fields in the environment that have a component near the resonance frequency. This magnetic noise spectrum is determined by a correlation time τ_c. When τ_c is less than τ_0, the spread in resonance frequencies depends on the frequency of transitions between parallel and antiparallel in the orientation of the dipoles with respect to H_1.

1. Relaxation Time T_1

The stability of the dipoles in direction of H_1 is characterized by a longitudinal relaxation time T_1 which is of the order of 1 sec for protons in water. Addition of paramagnetic ions to the water causes a fall in T_1 (189). This fall is greater when the paramagnetic ion is attached to an outer site of the molecular framework of a polymer, as τ_c then more nearly approaches τ_0. Inclusion of the metal in an "interior" site of the polymer, on the other hand, removes the ion from contact with the water protons and an increase of T_1 occurs relative to the polymer free control. An analysis of polymer-metal interactions has been made by Eisinger, Schulman, and others (190-193) using determinations of T_1 for protons.

2. Relaxation Time T_2

When τ_c is greater than τ_0, the persistence of inhomogeneities of the environment adds to H_1 and causes further spread of the resonance frequencies that is not reflected in a change of T_1. A second "transverse" relaxation time T_2 is required. This is a measure of the cohesiveness of a pulse of magnetic dipoles released in synchrony to precess about H_1 as axis. T_2 may be estimated from the width of the resonance line with which there is an inverse relationship $T_2 = (\pi\Delta_L \nu_0)^{-1}$, $\Delta_L \nu_0$ being the line width at half maximum peak height.

A drastic fall in proton T_2 (10^3-fold) was observed by Jardetzsky and coworkers (34, 194) when certain antibacterial drugs were bound to plasma albumin. The system was analyzed with the drugs in D_2O as solvent and in great excess relative to the concentration of polymer sites.

3. Chemical Shift

The diamagnetism of the ligand and its environment appears as a constant factor of unity or less which multiplies H_1. In proton magnetic resonance this is termed the "chemical shift" which can be utilized to determine the nature of the bonding to the proton. If reversible chemical reaction occurs involving the proton, the signal is split when the rate of constant chemical exchange is small compared with $(2\pi\Delta_0 \nu_0)$; $\Delta_0 \nu_0$ being the difference in resonance frequencies between the two states. As the rate of exchange increases, the spectrum lines spread and fuse together into a single peak. At high rates of exchange this signal then sharpens again at an average frequency determined by the extent of binding. In suitable cases it is possible to determine both extent of binding and binding kinetics. This method was utilized by Kotin and Nagasawa (57) to show that polystyrene sulfonic acid is completely dissociated.

4. Anisotropy Effects

The chemical shift and other environmental effects with influence on the spectrum are anisotropic. With electron spin resonance the anisotropy is such that $(2\pi\Delta_a \nu_0)^{-1}$ for differences in orientations of the molecular framework falls between the relaxation time of the bound and free ligand. The spectrum of the free ligand is sharpened by averaging over various orientations but the spectrum of the bound ligand may be spread by resolution into its spatial components. In disorientated solutions all the various spatial components show simultaneously. With DNA as polymer and chlorpromazine as ligand it was possible to make the spatial asymmetry explicit by orientating the complex in a flow gradient (195).

A similar spatial anisotropy is obtained in crystals. Griffiths (196) has utilized this effect to determine the orientation of heme with respect to the axes of hemoglobin crystals.

The anisotropy with respect to ligand molecular axes provides a useful index of rotational diffusion of the ligand. The method may be calibrated according to Stryer and Griffiths (197) by a comparison with fluorescence polarization. McConnell (198) obtained an internal standardization by the splitting of resonance due to hyperfine interaction. The electron spin resonance method may be applied in a manner similar to the fluorescence method to measure the extent of binding. Alternatively, the free radical may be attached firmly to the polymer by covalent bonding and then serve as an index of rotational relaxation processes in the polymer. Unlike the fluorescence method, the electron spin resonance method is applicable to colored polymers. It has been used to study conformational changes during ligand binding by hemoglobin (199).

5. Other Uses of Resonance Methods

Magnetic resonance methods may be used empirically to observe binding by the spreading of the spectral lines. Often the unbound ligand has a sharp spectrum whereas the bound ligand spectrum is so diffuse that is is difficult to distinguish from the background (200).

The extent of delocalization of the electron spin due to partial covalent bonding between ligand and substrate may be assessed by measurements of hyperfine splitting (201, 202).

With improved techniques it is now possible to observe the spectrum of those protons in polymers that are closely associated with the binding site for certain ligands (203).

VIII. DYNAMICAL METHODS

In this section the classical trio of sedimentation, viscosity, and diffusion are considered.

A. Ultracentrifuge Experiments

Ligand bound to polymer is usually more rapidly sedimented than is free ligand. Analysis of the sedimentation is facilitated by the monochromator optical systems now available for ultracentrifuges. By suitable selection of wavelengths, ligand and polymer sedimentation may be observed during the same centrifuge run. Calculation of the extent of binding is done from the constituent sedimentation constants (204) in a manner analogous to that used for boundary electrophoresis experiments (Sec. V.D).

Sedimentation equilibrium experiments may be analyzed by calculation of the effective molecular weight of the ligand $\overline{M}_A$. According to Schachman (204) the number of molecules bound is given by $r = (A_T/T)(\overline{M}_A/M_P)$, where M_P is the polymer molecular weight. This equation includes the assumption that polymer and complexes all have an effective molecular weight M_P. It is sometimes possible to sediment the polymer and complexes completely (205) from the top section of the centrifuge tube. This gives the free ligand concentration directly. Ultracentrifuge measurements of binding are done at extremely high pressures (up to 100 atm) and the results may differ from those found at normal pressures.

The ultracentrifuge, like the electrophoretic method, is valuable in helping to identify the adsorbing species in a mixture of polymers (206). Resolution of a polymer mixture may be improved by binding. A component of albumin (mercaptalbumin) is resolved on addition of mercury which dimerizes this fraction (207). Denatured and satellite DNA may be separated from the usual native polymer by complexing with mercury (208). Circular DNA is separated from linear DNA by reaction with ethidium bromide (209). The DNA separations were carried out in a strong salt gradient and depend on changes in buoyant density rather than on molecular weight (isopycnic separations).

B. Viscosity

The viscosity method has been much used in work with polyelectrolytes. According to Cox (210) a plot of log (viscosity) against log (salt concentration) is a straight line with inflections indicating conformational changes due to ligand binding. With the increase in viscosity of DNA on binding of acridine orange and certain other dyes, Lerman (211) showed that the ligand causes a lengthening of the molecule by partial unrolling of the DNA helix.

C. Diffusion

Observations on the diffusion of dyes into gels was one of the oldest methods of studying binding to proteins (212). According to Bennhold, dyes migrate at different rates into gels. Monodisperse dyes migrate rapidly and polydisperse dyes very slowly, if at all. Addition of certain proteins to the dye solution (for example, plasma albumin) caused many dyes to diffuse at the same rate, viz., that of the protein carrier. More recently, diffusion of isotopically labeled metal ions in polyelectrolyte solutions has been used to measure binding (213).

Combination of sedimentation and diffusion methods has confirmed that the changes in properties of polyelectrolytes with salt concentrations are due to changes in shape and not of molecular weight (214).

IX. X-RAY DIFFRACTION

X-ray diffraction can, in suitable cases, give a detailed account of the orientation and configuration of the ligand in the macromolecular binding sites.

An early application was to the relatively simple structures of cellulose-dye (215, 216) and starch-iodine complexes (217). The starch structure was the first of the biological helices to be recognized.

Solution of the more complex spatial relationships in protein crystals was made possible by the introduction of metal ions into the protein structure by the method of isomorphous replacement (218, 219). The heavy metal allowed the correlation of phases between X-ray intensities to be made. Once the structure of the protein has been elucidated a relatively simple analysis of the intensity changes resulting from the introduction of a ligand into the crystal can be made (220). Cyanide binding by methemoglobin (220) and binding in the non-polar region of the hemoglobin molecule (221-223) have been characterized in this way. Dye binding by DNA (224) and substrate and inhibitor binding by the enzymes, lyzozyme and chymotrypsin, have also been investigated (225, 226). Differences in crystal structure between polymer and complex throw light on the conformational changes resulting in the polymer due to binding (227, 228). The binding equilibrium may be observed in the crystalline state. The conformational change occurs more slowly in the crystal (8) as there are steric restraints on conformational changes that are not present in solution.

X. APPLICATIONS

A. Proteins

In living systems it is at the level of organization of the proteins that the information content of the organism becomes explicit. Many of the functions of the proteins are expressed in terms of interaction with other molecules. Therefore binding studies with proteins have a special interest when they are interpreted with reference to the physiology of the organism.

In this section the interactions of proteins are illustrated by a number of examples that have been the subject of detailed investigation. With one exception the proteins discussed are components of mammalian blood, for example, hemoglobin, plasma albumin, and so on. The exception is histone which is of general occurrence in the cell nuclei of both plants and animals.

1. Plasma Albumin

It has been recognized for some 50 years that the albumin fraction of the blood plasma is mainly responsible for binding dyes introduced into the body as a test of kidney function. The regulation of the concentration of certain potentially toxic products of metabolism, for example, bile pigments and free fatty acids, is also assisted by reversible binding to albumin.

At physiological pH albumin carries a net negative charge, the isoelectric point being at about pH 5.4. Lepper and Martin (153) were the first to notice that the indicator error in albumin solutions was anomalous. The albumin bound negatively charged or neutral forms of an indicator in preference to the positively charged forms. The adsorption site therefore acts as a region of charge opposite in sign to the average charge of the protein. The cationic nature of the albumin binding site has been confirmed in later work with acridines (56, 229) and azodyes (230). Cations at higher concentrations (about 10^{-3} M) induce conformational changes in albumin (10). Other conformational changes, perhaps not as extensive, are produced by anesthetic gases (nonionic) and azodyes (anionic) (175-178, 231, 232).

Spectrophotometric observations on azodyes bound to albumin (154, 232, 233) have shown differences between the proteins, obtained from various animal species (for example, man and the ox). These appear to arise from variations in the stability of the protein with respect to pH and from differences in the distribution of groups in the binding sites.

Albumin binding studies have also played a part in demonstrating the importance of hydrophobic bonding in proteins. The changes in fluorescence and absorption of dyes resulting from albumin binding were similar to the changes on replacement of the solvent water by a nonpolar organic liquid (56). It was concluded that the albumin sites are regions of low polarity. Analysis of the nuclear magnetic resonance spectra of penicillin and sulphonamide derivatives subsequently showed that protons in the nonpolar parts of the ligand are those most strongly perturbed by binding to albumin (34, 194).

Before 1952 it was customary to ignore the contribution of water-water interactions to the energy of protein binding of organic groupings from aqueous medium. This simplified view was abandoned as a result of albumin binding studies (56).

Comparing the affinity of the natural L isomer of tryptophan with that of the D isomer shows the binding to be stereochemically specific (234). Only one site is involved in this interaction. Another ligand that binds strongly at one site is Cu^{II}. This chelates with several amino acids at the N terminal end of the albumin molecule (125, 235, 236).

More than one site is found on an albumin molecule for most ligands and there may be several groups of sites different both in number and affinity constant (86). The sites of lower affinity are more numerous and also less specific with respect to ligand structure (110, 118). The fluorescence method which gives weight to the most rigidly bound ligand rarely detects more than five binding sites (56).

2. Hemoglobin (39, 237-240)

The binding of oxygen by the red pigment of the blood corpuscles exhibits a characteristic sigmoid binding isotherm. This phenomenon has attracted the attention of physiologists for many years, being functionally adapted to transfer of the oxygen from blood to tissues. In muscle there is another red heme pigment myoglobin which also reacts reversibly with oxygen but which has the typically hyperbolic Langmuir isotherm of oxygen uptake. At the low oxygen tension in the blood around the muscles the myoglobin may be fully saturated with oxygen when the hemoglobin is only partially saturated.

Normal adult hemoglobin consists of four peptide chains which are two sets of identical twins. These are known as the α and β sets and each peptide chain binds one heme group. The overall configuration may be written $\alpha_2{}^H \beta_2{}^H$, the H superscript referring to the presence of the heme component. It is possible to dissociate the heme reversibly from the peptide chains by alterations of pH and solvent.

Myoglobin consists of only one peptide chain and heme group and is about one quarter of the molecular weight of hemoglobin. Adair (12) was the first to attribute the sigmoid dissociation curve of hemoglobin to interaction among the four oxygen binding sites.

The nature of the interaction has been clarified by observations on certain genetic mutants of hemoglobin which occur infrequently in the human population, by chemical modification of the normal protein and by X-ray and peptide analysis (240, 241).

The hemoglobin-oxygen isotherm may be represented by the Hill equation which was developed (23) specifically with this system in mind. The exponent j is found to be 2.8 and the value for myoglobin is unity. When hemoglobin is in urea solutions, the j index falls toward unity as the subunits dissociate. A similar fall in j occurs when the subunits are prevented from recombination by chemical modifications, for example, succinylation.

The heme groups on the separated α and β chains do not differ intrinsically in their dissociation constants. Any difference would lead to a j value less than unity. The departure of the isotherm from the Langmuir equation must be due to some interaction among the binding sites. The presence of four hemes together in a hemoglobin molecule is not a sufficient condition for a sigmoid curve. Hemoglobin H, a mutant with configuration $\beta_4{}^H$, has a curve similar to that of myoglobin. The important interaction is that between α and β chains (240, 241).

Gibson (237) observed differences in rates of combination between a mixture of hemoglobin and oxygen obtained by photodissociation on the one hand and a mixture made in a rapid mixing apparatus on the other. Hemoglobin that has recently combined with oxygen will recombine faster than a hemoglobin that has been out of contact with the adsorbate for some time. A memory for the adsorbate with a lifetime of some milliseconds persists in the polymer as a structural change, metastable in absence of oxygen.

The importance of the α - β interaction was clearly shown by the X-ray studies of Perutz and coworkers who compared the structure of hemoglobin and its oxygen complex (242). When oxygen is adsorbed there is a movement of the two units relative to each other. The mode of transmission of information between the hemes and the α - β junctions is not known at present see however Addendum Section XII E.

Quite apart from the O_2 interaction there is the problem of reversible combination of the hemes with the peptide chains. This has been studied by Beychok (243) and by Winterhalter (244, 245). In the absence of the heme the hemoglobin tends to dissociate toward the configuration $\alpha\beta$ with a halving of molecular weight. The combination of heme at the α and β chains may be distinguished by ORD measurements and it appears that the α chains are occupied preferentially. It is possible to separate the $\alpha_2{}^H\beta_2$ isomer electrophoretically. This isomer has a Langmuir isotherm similar to that of hemoglobin H and myoglobin.

From an analysis of O_2 binding at various pH values a titratable group interacting with the heme has been identified as a histidine residue of the protein. The pH dependence is known as the "Bohr effect." The histidine has been located by X-ray analysis.

Some hemoglobin mutants (hemoglobin M) have a histidine replaced by tyrosine. A slow oxidation of the heme iron to the ferric state occurs and is stabilized by complex formation with the phenolic group of the tyrosine. The mutant hemoglobin M is therefore associated with a clinical state of methemoglobinemia in which the hemes on the mutant chains become permanently oxidized in the blood. Methemoglobin M has a characteristic spectrum different from that of normal methemoglobin (238, 246, 247).

A well-defined nonpolar cavity is present in the center of the hemoglobin subunits and has been characterized by X-ray analysis. This cavity may interact with certain dyes and nonpolar compounds (221-223).

3. Immunoglobulins (248-250)

These proteins have a remarkable specificity toward their determinant groupings. Their study by small molecule interactions was initiated by Landsteiner (248). The small molecules are known in this context as "haptens." The classical work of Marrack (249) led to the hypothesis that antibodies are typically bivalent. Eisen and Karush (251) confirmed this hypothesis by an equilibrium dialysis study using haptens. Valentine and Green (252) photographed bivalent hapten-antibody complexes through the electron microscope (see Fig. 4).

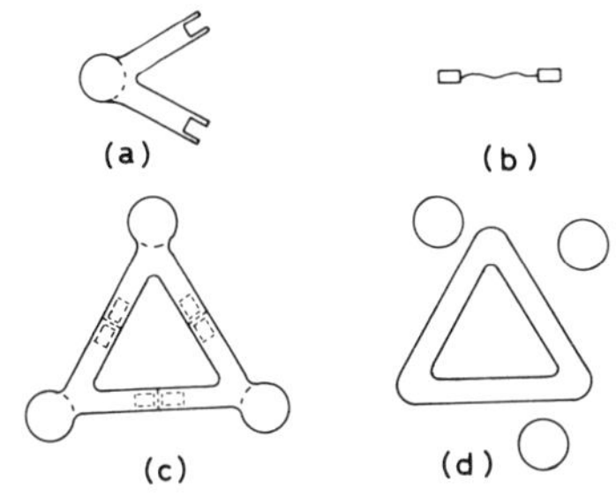

Fig. 4. Formation of closed ring complexes by reaction of immune γ -globulin with a bivalent hapten (a) suggested structure for antibody molecule with two combining sites (b) bivalent hapten, (c) complex of antibody with hapten, and (d) effect of proteolysis which removes the structures at the corners. After Valentine and Green, ref. 252.

Certain antibodies, termed "incomplete," combine with but do not precipitate the conjugated antigens. It was supposed that these might be monovalent but recent work has confirmed that n = 2, as for a precipitating antibody. Failure to precipitate must therefore be attributed to other factors (253).

Binding studies with antibodies require the isolation of specific immune protein. The part of the antibody directed against the carrier protein is removed by precipitation with unconjugated carrier. The determinant-specific proteins are then precipitated by determinant groupings that have been conjugated to a different unrelated carrier protein. The precipitate is resolved by addition of an excess of the determinant as haptene and then separated on a chromatographic column (254).

The determinant may be either positively or negatively charged, or it may be a neutral polar grouping such as glycosido or nitro. Binding studies and analysis of antibody show that the γ globulin directed toward a charged group contains an excess of the oppositely charged amino acid residues at the binding sites (254, 255). Antibodies against nitrogroups appear to contain some amino acid, possibly tryptophan, which is able to form charge-transfer complexes with the nitrogroup (256). Although the binding sites have a considerable specificity (249, 250), binding is also aided by nonspecific interactions with organic groups in the "tail" of the hapten (257).

Antibodies are a heterogeneous population with association constants that vary by as much as 10^3-fold. The mean association constant increases during a course of immunization. The situation is analyzed in terms of Sips' equation (Eq. 24).

Early studies on hapten interactions were done by the equilibrium dialysis method. Recently there has been an increasing use of fluorescence methods in this field.

Velick and coworkers (168), and Eisen (258) observed the quenching of tryptophan fluorescence in the antibody during combination with colored determinants (for example, nitrocompounds). The fluorescence is quenched by nonradiative transfer of the excitation energy between the tryptophans and the haptens and also perhaps by formation of charge transfer complexes (256). The method is extremely satisfactory in view of the symmetry that exists between the two halves of the antibody molecule. Both binding sites have equal quenching potential.

Hydrophobic probes (170) have also found a place in hapten studies. These are anilino compounds of naphthylamine and acridine which are poorly fluorescent except when bound to sites of low polarity or dissolved in certain nonpolar solvents. The polarity of the adsorption sites may be determined in suitable cases by these reagents. Parker and others (259, 260) have employed the ϵ-amino lysine amide of dimethylaminonaphthalene sulphonic acid as a probe. Albumin contamination of antibody to the probes may be tested as the albumin will react with many probes, whereas the antibody reacts only with those containing the correct determinant grouping (261).

Fluorescence polarization (262) and electron spin resonance relaxation methods (263, 264) have also been of value in observing the binding of small molecules to immunoglobulins.

4. Other Carrier Proteins in Blood

In addition to the proteins in Secs. X.1-3 there are some minority protein constituents of blood that have high affinity and specificity toward small molecules of biological importance.

Cortisol, a steroid hormone, is bound to the protein transcortin which constitutes about three parts in 10^4 of the normal blood plasma proteins. There is one site for cortisol on each transcortin molecule (265). A number of minor constituents of plasma bind the hormone thyroxine (266, 267). One of these in the "prealbumin" group occurs at three parts in 10^5 of the total protein.

Carrier proteins for iron and copper (transferrin and ceruloplasmin) also occur in plasma (268). Copper added to plasma is first bound to albumin but migrates to ceruloplasmin in the course of a few hours.

Plasma albumin serves as a second store for some of these substances either temporarily or when the specific carrier becomes saturated. Conversely, certain other proteins share with albumin the ability to bind acidic dyestuffs (269).

5. Histones (270)

Histones are proteins that occur in the nucleus of all the higher organisms. There is little or no difference between the histones of plant or animal origin or between the proteins derived from animals such as trout and calf. The material is isolated as a complex with DNA (deoxyribonucleic acid) and must have some essential role to play in the structure of chromosomes in view of the extreme conservatism of its structure.

Histone may be fractionated into about five main fractions which have an appearance of homogeneity. The fractionation procedures utilize differential extraction and precipitation with organic solvents. The amino acid compositions of the fractions have one main feature in common, viz., that the positively charged residues comprise very nearly a quarter of the total amino acid residues. The negatively charged residues are about one twelfth of the total or less.

The relative solubilities of the fractions in organic solvents are reflected in their amino acid composition. Fractions F2a and F3 are more readily soluble in organic solvents and contain more of the large nonpolar side chains (of leucine and valine) and less of the polar side chains (of serine and threonine). These fractions also contain regions of the peptide chain with a higher nonpolar content and with an absence of the charge imbalance that is typical of the molecule as a whole. These regions become precipitated as a "core" when the fractions are digested with trypsin. Fractions 1 and 2b have a higher polar and lower nonpolar residue content. Their organic solvent solubility is low and there is no core. Fraction 1 is distinguished by a high proline content and an extremely low content of negatively charged residues (about 3% of the total).

The very hydrophilic F1 with its scarcity of potential salt linkages and "kinkiness" introduced by the prolines takes the form of an expanded coil which is difficult to condense even at high salt concentrations. This fraction is the slowest to run in cross-linked acrylamide gel although its molecular weight is somewhat higher than that of other fractions. The configurations of other fractions, viz., F2a, F2b, and F3, are all condensed as the salt concentration is increased as may be shown by ORD measurements.

In spite of the dissimilarities among the fractions, dye binding studies have usually found striking similarities. Alfert and Geschwind (271) utilized the association of dyes with histones at high pH to detect these polymers in the presence of more acidic proteins. The association is mainly coulombic in nature and the specificity obtained is due to the high isoelectric point of histones. This is a property common to all the histone fractions. Loeb (272) has measured binding of eosin in histone solutions spectrophotometrically. There were no evident differences between the histone fractions. In this case charge transfer and polarization forces as well as coulombic interactions may predominate. Winkleman (273) tested a sulfonated porphyrin and this also reacted similarly with all the histone fractions. There was evidence that charge neutralization together with strong dye-dye interaction (stacking) were responsible for the binding.

In contrast with these investigations, the binding of the hydrophobic probe anilinonaphthalene sulfonate (ANS) proved to be very sensitive to differences between the histone fractions and to their physical state in solution (274). The "lipophilic" histones F2a and F3 bound ANS more strongly than did the F1-F2b group which are lipophobic. As the salt concentration was increased there was a sharp cooperative change in F2a and F3 to an even more strongly associating state. This was attributed to the formation of a nonpolar microphase through association of nonpolar residues of the core. The subfraction F2a1 has the sharpest transition with increasing salt concentration and this fraction probably has the most highly connected core region.

The fraction F2b was weakly activated by salt in a noncooperative manner. The binding of ANS by F2b was itself cooperative as though the dye molecules themselves provided the nonpolar region lacking in the protein. Binding of F2b by DNA also raised the fluorescence of its ANS complex to a level similar to that obtained with F2a and F3. Thus F2b may be held together by DNA in a configuration that is not probable in aqueous salt solutions.

Fraction F1 binds ANS weakly and its affinity is not increased when the protein is adsorbed by DNA. This fraction is known to bind to DNA in an extended configuration contrasting with the other fractions.

A similar relationship between binding and the histone core has been described by Sluyser (275) for carcinogen-histone interaction.

B. Nucleic Acids (276-278)

These polymers occur in living systems as two distinct forms, viz., ribonucleic acid (RNA) and deoxyribonucleic acid (DNA). The linear sequence of bases on the polymer chain acts as a code that determines the sequences of amino acids of the proteins. The bases that form the code are guanine, cytosine, adenine, and thymine, or uracil (G, C, A, and T, or U). DNA contains T but not U whereas RNA contains U but not T.

The symbols specifying the amino acids that are to be placed in the peptide sequences are composed of ordered groups of three bases taken from the four possible alternatives. The phenylalanine residue may be specified by the sequence U_3. Thus poly U will induce the formation of polyphenylalanine in a system containing the necessary catalysts and cofactors. Most nucleic acids occurring in nature are composed of all four of the bases which are present in significant amounts. The polypeptides that are synthesized contain most or all of the possible amino acid residues.

Many living systems store and reproduce their genetic information as DNA. The typical structure of DNA consists of a pair of polynucleotide chains. The chain containing the information is paired with a complementary chain. The complementary relationship is determined by a one-to-one pairing of the bases, A in one chain is opposite T in the other and G is opposite to C. The paired chains form a ladder structure that is twisted into a double spiral (278a). (See Fig. 5.) The phosphate and sugar residues are on the outside of the

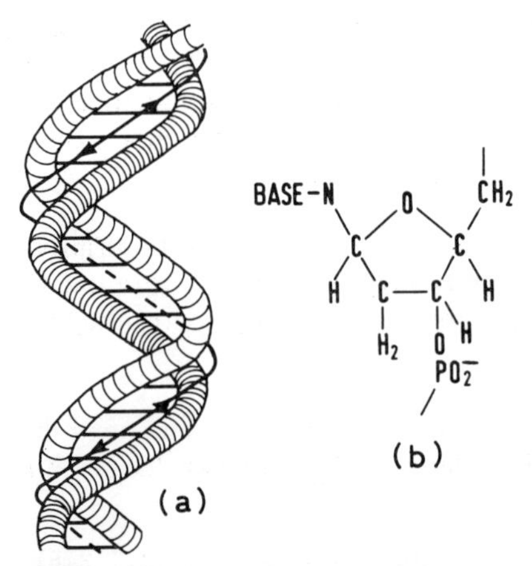

Fig. 5. The double helical structure of deoxyribonucleic acid (a) part of the double helix. The wormlike structures represent the polymer backbone and the horizontal bars the base pairs. The line with arrows indicates the position of the minor grove. (b) chemical structure of part of the backbone with attached base. Fig. 5a. After Watson and Crick ref. 278a.

spiral. Solvent and small molecules can gain access to the bases within grooves of the double spiral. One of these grooves is deeper than the other and they are known as the "major" and "minor" grooves. Two stable configurations of the DNA spiral (A and B) are observed in the crystal. Strong interactions of small molecules with the bases result in a partial unwinding of the helix in order to increase the exposure of the bases.

Some viruses have RNA as a genetic store and replicating unit. These have a base-paired double spiral with AU and GC pairs. The B configuration is not possible for helical RNA.

In addition to this exceptional genetic function, RNA has the function in all living organisms of transmitting information from the genetic store and catalysis of the genetically specified protein synthesis. The RNA used for this purpose is single stranded, but some base-pairing occurs by reflexing or pleating of the chain.

Both DNA and RNA synthesis are catalyzed by a polymerase which requires a template of the genetic nucleic acid (usually DNA) as cofactor. The RNA so formed associates in the region of protein synthesis with amino acids which are then joined together to form proteins. There are specialized RNA's for binding amino acids (tRNA) and sequence determination (mRNA). There is also a large amount of rather inert RNA (rRNA) which occurs at the side of protein synthesis (the ribosome).

Binding of amino acids to the ribosome requires activation and covalent bonding to an amino acid specific tRNA. This complex will then bind to the ribosome if an appropriate mRNA base triplet is present.

1. Interactions of Nucleic Acids with Nucleosides and Amino Acids

Experiments with small molecules have reproduced some of the effects observed with the polymer-polymer interactions. A base-paired spiral is obtained when a single-stranded polynucleotide interacts with the nucleosides of bases complementary to the bases in the polymer (279). The interaction has been observed by equilibrium dialysis and also by the changes in ORD that occur when a spiral structure is formed.

Nirenberg and Leder (280) obtained a combination of amino-acid tRNA complexes with ribosomes using various trinucleoside diphosphates as a replacement for mRNA. The correlations between the amino acid and the trinucleotide gave evidence of the nature of the genetic code.

There are some evident relationships between the base triplets and their corresponding amino acids. Pelc and Welton (281) have used atomic models to account for these correlations in terms of molecular forces. According to Volkenstein (282) the variation of base coding with amino-acid polarity has a genetic explanation. Natural selection has resulted in a code in which polar-nonpolar replacements in the proteins due to mutation are improbable.

There are interactions between nucleic acids and proteins at a number of points during the readout and translation of genetic messages. There is little direct information about the nature of specificity of these processes. A considerable simplification would result if they could be simulated by small molecule interactions.

2. Interactions with Cofactors, Drugs, Mutagens

A large number of studies have been made with other small molecules as adsorbate. Some of these, for example, Mg^{II} and spermidine, occur naturally in living systems. Others have important biological effects as mutagens, cell poisons, or carcinogens which have their principal action upon the nucleic acids of the organism.

a. Acridines, quinolines, and anthridium compounds. These compounds have bacteriostatic and mutagenic properties. They are also useful colorants for revealing cell structures containing nucleic acids. Binding studies in this field have recently been reviewed by Blake and Peacocke (283). There are numerous acridine dyes with somewhat different modes of binding to nucleic acids.

Binding may occur in at least two different ways. At low acridine/phosphate ratios, binding tends to be monomeric (211). The dye molecules interact with each other only by long-range forces, both coulombic and via deformations in the nucleic acid configuration. A partial unwinding of the DNA helix (284) results from interaction between the conjugated ring system of the dye and the bases of the nucleic acid. Circular DNA, that is, DNA in which the chains are in the form of closed circles, cannot unwind as readily as can linear DNA. This restricts the uptake of the monomeric dye (209).

At high dye/phosphate ratios a second binding form is observed, especially if the polynucleotide chains are not held rigidly together by base pairing (36, 46, 180, 152, 285). This is the polymeric or stacked form of binding in which the associations of the conjugated ring systems are dye-dye, not dye-base as with the monomeric type. In this case the polymerization of the acridine is facilitated by the reduction in coulombic repulsion resulting from charge neutralization on the polymer. The tendency to stack is very pronounced in the case of acridine orange and much less so with proflavine.

Even when stacking does not occur, for example, proflavine with DNA, analysis of the isotherm shows that there are two types of binding. The more strongly combining sites are about one fifth or one tenth of the total anionic centers of the polymer.

The molecules of bound acridine do not have identical fluorescent yields (286). The average fluorescence intensity is related to the relative amounts of G-C and A-T in the DNA (287).

Acridines are cooperatively desorbed when the DNA double helix is destroyed by heating the solution to above the denaturation temperature (284, 288). The affinity of DNA for acridine returns when the denatured solution is cooled although the helical structure is not regained. In contrast the ability of DNA solutions to bind the dye methyl green does not return after denaturation and cooling (289, 290).

Chloroquine, a quinoline derivative, interacts with DNA and the binding is assisted by contributions from the aliphatic chain and the chloro group (291).

Acridine interactions are weakened by competition with neutral salts but 6 M urea is not effective in preventing interaction (292, 293).

The phenanthridium compound ethidium bromide binds to DNA and competes with acridines at the same binding sites. The compound becomes brightly fluorescent when bound to double helices and is a useful measure of the extent of these structures (294).

b. Actinomycins. An important group of antibiotics that bind strongly to helical DNA are the dicarboxamides of dimethyl2-aminophenoxazone. The amide groups are derived from certain cyclic peptides. Binding causes a characteristic change in the absorption spectrum of the ligand and the DNA helix is stabilized against thermal denaturation (295).

Deoxyguanine appears to be essential as no interaction occurs with the poly dAT (d = deoxy) double helix. Strong binding is found at about 1 in 20 of the phosphate residues. The amount of actinomycin bound by poly dGdC would be smaller than expected if every GC pair were able to serve as a strong binding site. It is therefore likely that steric hindrance prevents occupation of closely adjacent sites. The presence of A-T groups may assist binding at neighboring G-C sites by permitting a rearrangement of the DNA chains.

Muller and Crothers (9) have made kinetic studies of binding to DNA by actinomycin and some related compounds. The cyclic peptide groups reduce the rates of both association and dissociation of the ligand but the latter rate is reduced to a much greater extent. Consequently the peptide groups increase the association constant by a factor of about 10^4. The time course of the reaction of actinomycin and DNA is complex. The dissociation contains a time constant of about 10 min which is attributed to the relaxation of the configuration of the cyclic peptides. These can be rotated in order to present a hydrophobic aspect to the polymer.

Actinomycin is poorly soluble in water and is a nonelectrolyte. The interaction with DNA is reversed by replacing the water by isopropanol in which the ligand is more soluble. The complex is stable in strong salt solution but dissociated in 6 M urea (293). The free energy of association is predominantly entropic as would be expected from a hydrophobic and chelational linkage.

RNA will not bind actinomycin even when in the form of a double helix. The binding is believed to require the B form of the polymer which is only possible with DNA (224).

Actinomycin inhibits RNA polymerase at concentrations much lower than required to inhibit DNA polymerase. This makes the compound a useful tool for physiologists. The ligand is considered (295) to block the minor groove of the double helix along which the enzyme is believed to travel during RNA synthesis. The drug is effective as an RNA polymerase inhibitor over the range of concentrations in which binding occurs at the more strongly combining sites.

c. Various basic dyes. Kurnick (289, 290) studied the binding of methyl green by nucleic acids. This dye fades in neutral aqueous solution by a slow hydrolysis reaction. Native DNA (but not heat denatured DNA or RNA) prevents this fading reaction. Kurozami et al. (296) observed that the fading could be speeded up by addition of hydrogen peroxide to the dye solution, whereas DNA-bound methyl green was resistant to the action of the peroxide.

The fluorescence intensities of hydroxystilbamidine (297), auramine (298), and berberine (299) are all increased by nucleic acid binding. Klimek and Hnilica (300) found that the DNA complex (but not the RNA complex) of berberine was stable to electrophoresis on filter paper.

The staining properties of nucleic acids have considerable interest for the histologist. Kurnick (289) found that fairly small changes in dye structure would alter the reaction of the dyes towards DNA. Goldstein (301) has observed that the differential staining of DNA and RNA by basic dyes is related to the dye molecular weight. Dyes of lower molecular weight, for example, pyronine, stain RNA more strongly than DNA. Preferential staining of DNA is obtained by dyes of higher molecular weight, for example, methyl green. If the molecular weight is too high, for example, Alcian blue, neither DNA nor RNA will be stained. These observations resulted in the discovery of a new mixture of dyes (301) which stained DNA and RNA in different colors.

d. Polybases. Polybases are compounds with up to four amino groups in a fairly short aliphatic chain. They are widely distributed among living species and react strongly with nucleic acids. It is possible that the polybases take the place of histone in the bacteria.

Binding of spermidine to DNA prevents the binding of actinomycin (91). It is believed that the configurational requirements of binding for the two substances are incompatible. Unlike actinomycin, spermidine does not inhibit RNA polymerase.

e. Binding by nucleoproteins. DNA occurs in higher organisms as a complex with histone and other proteins. Only one strand of the DNA is used to generate genetic instructions. The binding of actinomycin (302) and azure A (303) to nucleoprotein is half of the binding to nucleic acid under the same conditions.

C. Polysaccharides

This group of polymers includes reserve carbohydrates stored for metabolic purposes (starch and glycogen) and also structural materials that contribute to the physical properties of the living organism (cellulose and hyaluronic acid). Certain polysaccharides have a specialized function; for example, heparin acts as an anticoagulant in the blood of higher animals.

The neutral polysaccharides resemble water in the frequency of hydroxyl dipoles in their primary structure. A polymer secondary structure is formed if the dipoles of the polymer can be matched one with another in a regular way. If no regular structure is possible, the polymer associates with the water to form a typically gummy or mucilaginous solution.

Because of the compatibility between carbohydrate and water and the similarity of their external fields, neutral carbohydrates have a restricted range of interactions in aqueous solution.

The acidic polysaccharides, which contain oxidized sugar (uronic acid) residues and acid sulphate groupings, have some of the binding properties of the neutral polymers. Their charge leads to other interactions that are mainly electrostatic. Interaction among the adsorbate cations by the formation of stacks or micelles also contributes to the interaction energy.

1. Neutral Polysaccharides

An interaction of historical interest is the starch-iodine reaction with production of the characteristic blue color. The reaction is due to unbranched polyglucose fraction of starch known as "amylose." Hanes (304) and Freudenberg et al. (305) considered the blue color to be evidence of inclusion of the iodine in a nonpolar region within the polymer coils. The existence of this region was suggested by the reaction of amylose with alcohols and phenols having bulky hydrocarbon groups such as thymol.

Using X-ray analysis Bear (217) found that the iodines are inside a spiral of glucose residues (the V form of starch) and are arranged in a linear manner. Rundle and collaborators (306, 307) in the early 1940's found the adsorption isotherm to be remarkably steep (306) indicating cooperative adsorption of the iodine molecules. The flow dichroism (307) of the complexes was further evidence for a linear arrangement of the iodines and for the presence of an optical transition parallel to the glucose spiral axis. It was also observed (308) that the spectrum of the adsorbed iodine extended further into the red than in any solution of iodine in a nonpolar solvent. It was concluded that there was a significant interaction extending along the helix is between adjacent molecules of the adsorbate.

Similar complexes of iodine in films of cellulose, polyvinylalcohol, and nylon were studied by West (309) using X-ray crystallography. The bond distances between the iodines were intermediate between the van der Waals and the covalent bond distances.

Carroll and van Dyke (310) obtained a complex between amylose and Congo red with a characteristic spectrum. This occurred with polyglucoses with low degrees of polymerization and was sensitive to chain length from 5 to 14 sugar residues. The starch-iodine complex requires longer chains in order to produce any characteristic color. The Congo red reaction was a useful supplement in following degradation of starch to extreme limits.

The interaction of cyclodextrins with dyes and aliphatic substances was investigated by Broser and Lautsch (311) and by Miller (312). The adsorption site on the polysaccharide is within a central "bowl" which has some measure of nonpolarity.

The interaction of cellulose with azodyes is of considerable importance in the commercial dyeing of plant fibers. The cellulose molecule is linearly extended within the crystallite and the structure of direct cotton dye molecules is also linear (313). Robinson (313) was unable to find a relationship between fast dyeing and the spacing of polar groupings in the dye.

The interaction of the bacterial product dextran was mentioned in the discussion of gel chromatography (Sec. IV. C).

Binding of borate by neutral polysaccharides can be used in separation and preparation of these compounds from complex mixtures using an electrophoretic method (314).

2. Acidic Polysaccharides

The presence of a negative charge on these polymers extends the range of interactions to certain cations that react weakly, if at all, with the neutral polysaccharides.

Carroll and Cheung (315) investigated the interaction between oxidized starch and methylene blue. With starch in excess, the amount of dye bound was proportional to the number of alcohol groups in the starch that had been oxidized to carboxyl groups.

Binding to sulfated polysaccharides has been reviewed by Schubert (316). These polymers are precipitated by multivalent cations. The large tetravalent ion, Alcian blue, has been widely used as a fixative and stain for mucopolysaccharide since its introduction by Stedman (317).

Binding of dyes to these polymers is often accompanied by a dimerization or stacking of the adsorbate. Pal and Schubert (318) found that dye-polysaccharide complexes differed in the stability of their aggregates toward alcohol. Mukerjee and Ghosh (319) found that urea also causes dissociation of the aggregates in conformity with a hydrophobic bonding between the adsorbate molecules.

Mucopolysaccharides are precipitated by cationic detergents and the precipitates are redissolved in more concentrated salt solutions (320). It is likely that interactions between adsorbate molecules contribute to the binding free energy.

Binding of ions with more than one valence would be favored if the mucopolysaccharide coils were condensed from their expanded structure. Woodward and Davidson (321) found that digestion of the protein component of a mucopolysaccharide resulted in liberation of the bound calcium. Association of the polysaccharide with protein would permit a more condensed structure which would be favorable to calcium binding.

Evidence of an ordered structure of the heparin molecule was obtained by Stone and Moss (322) who carried out an ORD investigation of a heparin-dye complex. Conformation changes were detected due to alterations in pH and salt concentration.

D. Synthetic Polymers

There are many studies of the interactions of the synthetic polyelectrolytes with counterions. There are a small number of investigations of interactions involving neutral polymers such as polyvinylpyrrolidone.

1. Synthetic Polyelectrolytes (323-325)

These polymers carry a characteristically high density of charges, usually all of the same sign. The electrostatic problem is best treated by taking a cylindrical symmetry about a coil element (326, 327). According to Katchalski (325) the potential at the surface of a typical polyelectrolyte is 3×10^8 v/cm. This produces very strong binding of the counterions and exclusion of coions in the region close to the coil. Some method of charge neutralization is obligatory in this region of intense electric field. If metallic ions are not available, the field is neutralized by ions of the appropriate charge derived from the dissociation of the water.

Taking an acidic polyelectrolyte as an example, titration to form a salt simultaneously reduces the hydrogen ion concentration and raises the metal ion concentration in the medium surrounding the polymer coils. An ion exchange reaction then occurs in the region of the electric field with replacement of hydrogen ions by metal ions (328). The sum of hydrogen and metal ions in the field is approximately constant. It is not difficult to see that the concentration of the alkali metal outside the field also remains similarly constant. This is the osmotic buffer effect described by Katchalsky (325).

Raising the counterion concentration by addition of neutral salt would tend to increase the amount of counterion bound. Because the coion cannot enter the binding region, this can only occur by ion exchange with hydrogen ion. If the pH is not too low, only a small replacement of hydrogen by metal ion would be necessary to restore the balance in the external medium. Thus the extra binding due to addition of neutral salt is often small (329). There is a reciprocal effect of salt on the titration curve of the polyacid. This is shifted to higher pH as the salt concentration increases. The electric field of the coil is bounded by the cylindrical surface on which the difference in energies of coion and counterion falls to below the thermal energy kT. Outside the surface the potential conforms with the Debye-Huckel approximation. The neutral salt "sees" the polyelectrolyte coil through its partial screen of counterions (330). Interaction between the polymer and external electrolyte is not larger than between a pair of simple electrolytes (331).

Binding of counterions depends on the surface density of the polyelectrolyte coil (330, 332) and on the nature of the counterion (328, 333). The surface presented by the polyelectrolyte to the external medium is strongly hydrophilic. Large alkyl groups around an alkylammonium ion decrease interaction by increasing the minimum distance between the ions. Strauss and Siegel (334) used a tetralkylammonium salt as an indifferent supporting electrolyte when studying metal ion binding by polyphosphate.

Bivalent ions are bound more strongly than monovalent ions (325). As mentioned in Sec. III. A, the binding of a series of alkali metals depends on the radius of the ions and the penetration and compatibility of hydration sheaths around the ions and the polymer (51, 328). Strongly bound ions will profit from any additional forces, covalent and dipole ion, that can be established with the polymer. The liaison between Cu^{II} and a polymethacrylic acid has been studied spectroscopically (335).

Typical changes in viscosity occur when a flexible polyelectrolyte is diluted with water. These were analyzed by Fuoss and Strauss (336) who showed that the specific viscosity increased with dilution. There was little or no change with dilution in the fraction of counterion bound (331, 337).

At lower polymer concentrations there is a radial expansion of the characteristic field of force of the polyelectrolyte. The forces between coil elements increase as the counterion screen expands but the number of ions within the field remains approximately constant. Moderately concentrated polymer solutions have no region free of some electrostatic field and the analysis of the binding under these conditions has been discussed by Oosawa (332).

Hydrophobic bonding by polyelectrolytes may be detected in certain circumstances. Kressman and Kitchener (338) obtained evidence for van der Waals forces in a study of binding to a phenolic ion exchange resin. Barone et al. (339) measured the solubility of alkanes in solutions of polyacrylic and polymethacrylic acid and of their salts. Increased solubility was associated only with the polymethacrylic acid solution and was attributed to an interaction with the methyl groups of the polymer.

Some interesting optical effects are obtained when dyes are bound by synthetic polyelectrolytes, as described by Oster (340).

2. Neutral Polymers

The interaction of polyvinylpyrrolidone solutions with dyes was noticed by Bennhold and Schubert (341). Scholtan (342) and Molyneaux and Frank (343) have observed that binding of basic substances is less than that of acidic substances.

Congo red and detergents will bind with polyvinylalcohol (342, 344). Detergents will solubilize certain insoluble neutral polymers (345). The neutral polymers take on polyelectrolyte character when they have bound ionic adsorbates (342, 345).

XI. CONCLUSIONS

Development of techniques for the study of polymer-small molecule interactions has continued throughout the period covered by the present survey.

The initial stimulus that has provoked an advance in this field has often been an adequate fractionation, characterization, and production in bulk of the polymer. Thus the separation of starch subfractions allowed Bear (217), Rundle (306-308) and others to elucidate the starch-iodine interaction. Bulk production of plasma albumin by the ethanol method and fractionation of histone by acid ethanol extraction have provided the material from which the necessary detailed studies can be made.

In other cases the interest in a new adsorbate, for example, actinomycin, has resulted in a period of intensive study of the phenomenon of binding.

At the same time improvements in theoretical treatment have resulted in a more meaningful treatment of the isotherms. The Bjerrum (22), Klotz (29), and Scatchard (27) equations are examples of advances in the systematization of binding data. These equations are based on an analysis of multisite binding. They replaced earlier ideas derived from surface chemistry in which the polymer surface was treated as an amorphous interface determined by a zeta potential. A notable exception in the earlier work was the interpretation of the hemoglobin-oxygen equilibrium by Adair (12).

Advances in techniques have also contributed to the progress recorded. Optical studies were facilitated by production of commercial spectrophotometers and electronic polarimeters. Fluorescence techniques have undergone considerable development during the last 20 years. Electron and nuclear magnetic resonance techniques have also made a noticeable contribution. Recent advances in nuclear magnetic resonance instrumentation will allow a detailed description of interactions by the resolution of reacting groups in polymer and adsorbate. A similar resolution can be obtained by X-ray diffraction together with analytical studies to the sequence of residues in the polymer.

Among the most recent works perhaps the most significant trend is toward routine analysis of the kinetics of formation and dissociation of the complex by rapid reaction techniques.

The general problem of reliability of isotherms was discussed some years ago by Scatchard (27) and has received a new dimension in terms of information theory by Weber (20).

To the questions posed by Scatchard (27) as a model for binding studies, viz., "How many?" "How tightly?" "Where?" "Why?" was later (16) added, "What of it?" As the techniques for binding studies improve it is also becoming more usual to answer even this last question.

XII. ADDENDUM

The following account summarizes events during the past two years in the field of review of the main chapter.

A. Reviews

A general account of protein interactions was given by Steinhardt and Reynolds (346). Their book contains a detailed description of the binding of hydrogen ions, nonpolar molecules, and long-chain detergents. A book by Cann (347) discusses transport phenomena with a section on interactions with small ions. Griffiths and Waggoner (348) have reviewed spin-labeling methods. Wood and Cooper (349) described the application of gel filtration to protein binding. Mayer and Guttman (350) have given an account of recent work on the binding of drugs by the plasma proteins.

B. Isotherms

Wyman (351) developed the theory of adsorption isotherms by a study of allostery in large arrays of polymer units. A kinetic analysis of the one-dimensional Ising problem was given by Schwarz (352). The Adair (12) equation is now accepted as the most general formulation of multiple binding at a finite number of sites. The evaluation of the constants of this equation using a computer program is described (353).

C. Forces

A complete analysis of the forces and geometry of interaction has not yet been attained even when the actual configuration of the sites and adsorbate is known from X-ray data. Sequence studies and X-ray analysis have shown how a single charged group (or an isolated polar group) of the polymer, if embedded in a nonpolar cavity, can exert a large effect on interaction specificity (354, 355). It is not generally necessary to postulate both multiple charge and nonpolar backing when adsorption is found to be charge specific. For the example of plasma albumin, where these postulates were first made (56), recent evidence (356) shows that they were justified.

On a different level Hansch (357) has obtained good correlation between protein-binding energies and the partition of adsorbates in a simple octanol-water system. In order to include the binding of charged molecules, for example, the naphthylamine disulfonate and trisulfonates it would be necessary to add some lipid-soluble amine to the octanol.

The development of our knowledge of the preferred configurations of acyclic compounds during the last 10 years has already had its effect in this field. Beers and Reich (358) describe the most probable conformations of acetylcholine when it binds to its physiological receptor.

D. Methods

1. Affinity Labeling

Affinity labeling is a new and powerful method of obtaining information on adsorption sites. The polymer is reacted with a compound that can make covalent bonds with some group in the sites. Specific binding at these sites is ensured by giving the reagent a chemical structure known to favor strong binding and by avoiding excess reagent. For example, Thorpe and Singer (359) covalently coupled a nitrobenzene diazonium compound with an antibody known to have a specificity toward aromatic nitrocompounds. The peptide containing the labeled tyrosine from the binding region was then isolated from a partial hydrolysate and analyzed.

Covalently bonded groups at the binding site can also be used to signal information to a monitor such as an optical or magnetic spectrometer. When they are used in this way, the groups are known as "reporter groups."

2. Affinity Chromatography

Affinity chromatography has also been recognized and developed in the last two years. Groups that are related to the groups of the "usual" ligand are attached to a granular solid or gel.

Passage of a mixture containing the binding protein down a column packed with the granules results in trapping the binding component. This component is released subsequently after the impurities have been washed away. Isolation of the thyroxine-binding protein of plasma by absorption by thyroxine-labeled agarose is an example (360). Some adsorbents for specific binding of polymers are now available commercially (from Miles-Seravac, Maidenhead, England).

3. Distribution Methods

Ultrafiltration by the centrifuge method can now be carried out in commercially produced thimbles (Centriflo of Amicon, Cambridge, Mass.). Colowick and Womack (361) have further developed the steady-state dialysis method (76). Each determination requires only a minute to establish the steady state.

4. Optical Methods

Fluorescence probes have received attention to interpret the message relayed from the polymer sites. Turner and Brand (362) found that arylaminonapthalene sulfonates respond mainly to polarity of the medium as defined by Kosower's (363) empirical parameter. Thomson (364) showed that the retinyl polenes respond to local viscosity. Yagi and co-workers (365) have developed a new stilbene derivative that responds both to polarity and viscosity.

5. Magnetic Methods

Magnetic resonance methods have advanced rapidly. Nuclear relaxation dispersion (variation of T_1 with υ_0, Sec. VII. B) can detect exchange of protons between water and polymer (366) or between different regions of the same polymer (367). Conformational changes in myoglobin on oxygenation have been analyzed by proton magnetic resonance (368). Slight movements of the neighboring groups relative to the porphyrin ring can be detected by shifts in the upfield temperature-independent resonances. These are due to protons that come within the field of ring currents of the porphyrin ring. This method can detect shifts of a few tenths of an Angstrom unit in the relative positions of the groups and ring.

E. Applications

Applications of the various methods to specific problems has resulted in emergence of new discoveries and concepts.

1. Proteins

a. Serum albumin. Rosen and Weber (369) found that 1-anilino-8-napthalene sulfonate became dimerized when bound to bovine plasma albumin in the presence of traces of nitrous acid. For the reaction to occur, the molecules must be bound at closely adjacent sites in the protein molecule. Once formed the dimer is tailor made as a reagent for detection and estimation of minute quantities of albumin, its association constant for the first molecule bound being greater than 10^8.

b. Hemoglobin. A series of papers by Perutz (370, 371), based on the X-ray work of his group, has clarified many of the problems associated with the functioning of hemoglobin as an oxygen carrier.

Combination of oxygen results in a displacement of iron with respect to the plane of the porphyrin ring. A tyrosine residue is expelled from the interhelical space of the protein. This in turn causes the C-terminal arginine in the α chain to lose connections with groups in the β chain. Once these connections are broken by the first oxygen bound, subsequent oxygen molecules can react without expending some of their binding energy in breaking interchain bonds.

The Bohr effect is explained by a change in the environment of the basic amino acids, notably the C-terminal histidine in the β chain. This is also displaced as a result of the expulsion of its neighboring tyrosine.

The slow reaction observed by Gibson (237) for hemoglobin is now considered to be at least partly due to dissociation of the subunits of hemoglobin on oxygenation. The dissociation has been well characterized by Goers and Schumaker (372) using an ultracentrifuge.

c. Immunoglobulins. The binding sites of antibodies have been investigated by studying the inhibition of the precipitin reaction by various small molecules. Shechter (373) took an antibody to poly-L-alanine which precipitated the antigen-attached carrier protein. He measured the concentration of various alanyl peptides required to inhibit precipitation. It was found that the lowest concentration (and hence the strongest binding) required a sequence of three or more L-alanyl residues. A free amino at the N-terminal end was necessary for maximum binding. Substitutions at the C-terminal end did not reduce the binding affinity. Replacement of L-alanine by some other smaller or more bulky amino acid at one of the three critical residues resulted in lowered binding affinity. The changes in affinity differed considerably depending on which of the three residues was replaced.

The heterogeneity of antibodies from a single animal reflects the origin of these proteins from a group of antibody-producing cells having a number of independent ancestral cells. When a single antibody-producing cell is isolated and its progeny grown, a much more uniform population of antibody molecules results (374). This technique, known as "cloning," should improve the quality of the antibody used for physical methods.

An electron microscope study of the high molecular weight IgM immunoglobulin (M = Macro) (375) has revealed that this is composed of five of the Y-shaped structures, visualized by Valentine and Green (252), joined together at their stems. This molecule would have a site number (n) of 10, two for each subunit. The chemical specificity of the sites for this immunoglobulin is now known.

d. Histones. Sequencing of histones has proceeded apace during the last two years (376-378). The results confirm the general structure of the histones which was known when the main text of this chapter was prepared. The details of sequence do not add greatly to the interpretation of those binding data now available.

2. Nucleic Acids

In the nucleic acid field there has been a considerable increase in information on the primary, secondary, and tertiary structures of transfer RNA's. Difference spectrometry and intrinsic fluorescence have proved to be valuable tools to investigate the changes in conformation that result from variation of temperature and magnesium ion concentration (379). Five conformational transformations were identified by varying these parameters.

Gabbay (380) has carried out a very detailed study of binding of an aliphatic diamine with a reporter group to DNA and RNA. He found hypochromism and circular dichromism of the reporter group following binding to the polymer. Displacement of the compound by salt suggested that the diamine was bound electrostatically to the nucleic acid phosphates. The circular dichromism of a symmetrical nitroaniline had the same sign both for RNA and for DNA. An unsymmetrical dinitroaniline gave a circular dichroism opposite in sign for RNA and for DNA. Gabbay concluded that the second unsymmetrical nitro group was bound outward by one nucleic acid and inward by the other. In order to decide which nucleic acid bound the reporter group internally, a nitronaphthylamine reporter group was used. This gave circular dichroism with DNA but none with RNA, showing that the nitro group of this larger molecule was outside the range of interactions with the RNA structure. As RNA is occluded in its minor groove by the 2'-hydroxyl of ribose it was concluded that this groove was the site of binding of the reporter group.

Gabbay's conclusion is relevant to other interactions that are DNA specific (Section XB2b).

Bauer and Vinograd (381) have made detailed analyses of the binding of ethidium bromide by circular DNA. Changes in torsion of the main helix due to dye binding cause the macromolecule to adopt a coiled-coil (or superhelical) configuration.

Wells and Larson (382) studied binding of actinomycin by synthetic DNA polymers of known sequence. A DNA of structure poly d(A-T-C) poly d(G-A-T) would not bind the antibiotic even through there was 33% of G-C base pairs in the molecule. Clearly the sequence of base pairs is important in determining the binding affinity. Conversely, one polymer with no guanine residues bound actinomycin as well as a DNA with a 50% G-C base pairs. These findings contradict the model of reference 224 but support the model of Reference 9.

3. Carbohydrates

On the carbohydrate front, Stone (383) has found Cotton effects in the amide bonds of chondroitin sulfate. This suggests a more ordered structure for this polymer than has occasionally been assumed. There were some well-defined hydrophobic molecular grooves with large nonpolar areas in the model proposed.

Chondroitin sulfate forms a tight complex with histone (384) and also with the acridine dye rivanol. The rivanol complex is poorly soluble in aqueous salt solutions but dissolves rapidly in 0.5 or 1 M sodium acetate in isopropanol water (1:1). The requirement for an organic solvent suggests that there are nonpolar forces involved. These may be adsorbate-adsorbate interactions as rivanol salts are often poorly soluble in water.

Schneider et al. (385) have studied cooperative binding of iodine by starch using equilibrium dialysis. Thoma and French (386) obtained binding of I_3^- to oligomers of four or five d-glucose units. These complexes were colorless but the interaction was demonstrated potentiometrically.

ACKNOWLEDGMENTS

I am grateful to Professor K. Yagi for sending his paper prior to publication, and to my wife for considerable help with the manuscript.

REFERENCES

(1) I.M. Klotz, in The Proteins (H. Neurath and K. Bailey, eds.), Vol 1, Part B, Academic Press, New York, 1953.

(2) B. Carroll, Conf. Surface Effects Detection, Washington, D. C. 1965. p 239.

(3) Molecular Associations in Biology (B. Pullman, ed.), Academic Press, New York, 1968.

(4) G.W.G. Sharp, C.L. Komack, and A. Leaf, J. Clin. Invest., 45, 450 (1966).

(5) A. Froese, A.H. Sehon, and M. Eigen, Can. J. Chem., 40, 1786 (1962).

(6) R. Schmid, I. Diamond, L. Hammaker, and C. B. Gundersen, Nature (Lond.), 206, 1041 (1965).

(7) R. Tupper, B. W. E. Watts, and A. Wormall, Biochem. J., 50, 429 (1952).

(8) H. Theorell, A. Ehrenberg, and C. de Zalenski, Biochem. Biophys. Res. Commun., 27, 309 (1967)

(9) W. Muller and D. M. Crothers, J. Mol. Biol., 35, 251 (1968).

(10) C. Botré, M. Marchetti, and S. Borghi, Biochim. Biophys. Acta, 154, 269 (1968).

(11) R. E. Gosselin and E. R. Coghlan, Arch. Biochem Biophys., 45, 301 (1953).

(12) G. S. Adair, J. Biol. Chem., 63, 529 (1925).

(13) H. S. Simms, J. Am. Chem. Soc., 48, 1239, (1926).

(14) A. L. von Muralt, J. Am. Chem. Soc., 52, 3518 (1930).

(15) A. Goldstein, Pharmacol. Rev., 1, 102 (1949).

(16) G. Scatchard, W. L. Hughes, F. R. N. Gurd, and P. E. Wilcox, in *Specificity in Biological Interactions* (F. R. N. Gurd ed.), Academic Press, New York 1954, p. 193.

(17) J. T. Edsall and J. Wyman, *Biophysical Chemistry*, Vol. 1, Academic Press, New York, 1958, Chapter 11.

(18) K. Linderström-Lang and S. O. Nielsen in *Electrophoresis* (M. Bier ed.), Academic Press, New York 1959, Chapt 2.

(19) C. Tanford, *Physical Chemistry of Macromolecules*, Wiley, New York, 1961, Chap. 8.

(20) G. Weber, in *Molecular Biophysics* (B. Pullman and M. Weissbluth, eds.), Academic Press, New York, 1965 p. 369.

(21) J. C. Sullivan and J. C. Hindeman, J. Am. Chem. Soc., 74, 6091 (1952).

(22) J. Bjerrum, Metal Amine Formation in Aqueous Solution, F. Haase, Copenhagen (1941).

(23) A. V. Hill, J. Physiol. (London), 40, iv (1910).

(24) R. Sips, J. Chem. Phys., 16, 490 (1948).

(25) G. E. Perlman, A. Oplatka, and A. Katchalsky, J. Biol. Chem., 242, 5163 (1967).

(26) R. H. Fowler and E. A. Guggenheim, Statistical Thermodynamics, Cambridge Univ. Press, Cambridge, England, 1939, Chap. 10.

(27). G. Scatchard, Ann. N. Y. Acad. Sci., 51, 660 (1949).

(28) R. K. Cannan, A. C. Kibrick, and A. H. Palmer, J. Biol. Chem., 142, 803 (1942).

(29) I. M. Klotz, Arch. Biochem., 9, 109 (1946).

(30) C. Frieden and R. F. Coleman, J. Biol. Chem., 242, 1705 (1966).

(31) A. Grollman, J. Biol. Chem., 146, 85 (1925).

(32) M. Pollay, A. Stevens, and C. Davis, Anal. Biochem., 17, 192 (1966).

(33) A. Stockell, J. Biol. Chem., 234, 1286 (1959).

(34) J. J. Fisher and O. Jardetzsky, J. Am. Chem. Soc., 87, 3237 (1965).

(35) G. N. Ling, Federation Proc., 25, 958 (1966).

(36) D. F. Bradley, Trans. N. Y. Acad., Sci., 24, 64 (1961).

(37) J. Monod, J. Wyman, and J. Changeux, J. Mol. Biol., 12, 88 (1965).

(38) R. J. Watts Tobin, J. Mol. Biol., 23, 305 (1967).

(39) J. Wyman, Adv. Protein Chem., 19, 223 (1964).

(40) G. A. Gilbert and E. K. Rideal, Proc. Roy. Soc. (London), A-182, 335 (1944).

(41) G. A. Gilbert, Proc. Roy. Soc. (London) A-183, 167 (1944).

(42) D. E. Koshland, G. Nemethy, and D. Filmer, Biochem., 5, 365 (1966).

(43) J. E. Haber and D. E. Koshland, Proc. Natl. Acad. Sci., 58, 2087 (1967).

(44) G. Karreman, Bull. Math. Biophys., 27, 91 (1965).

(45) D. M. Crothers, Biopolymers, 6, 575 (1968).

(46) D. F. Bradley and S. Lifson in Molecular Associations in Biology (B. Pullman, ed.), Academic Press, New York, 1968, p. 261.

(47) E. A. Guggenheim, Proc. Roy. Soc. (London), A-182, 167 (1935).

(48) E. A. Guggenheim, Discussions Faraday Soc., 24, 53 (1957).

(49) A. Katchalsky, Z. Alexandrowicz, and O. Kedem, Transactions Symp. Electrolyte Solutions, Electrochem. Soc. Wiley, New York, 1965.

(50) H. Jenny, Kolloid Beihefte, 23, 428 (1927).

(51) G. E. Boyd, J. Schubert, and A.W. Adamson, J. Am. Chem. Soc., 69, 2818 (1947).

(52) V. P. Strauss and Y. P. Leung, J. Am. Chem. Soc., 87, 1476 (1965).

(53) C. Botré, S. Borghi, H. Marchetti, and M. Baumann, Biopolymers, 5, 483 (1967).

(54) R. Gurney, Ionic Processes in Solution, McGraw-Hill, New York, 1953.

(55) J. A. Schellman, J. Phys. Chem., 57, 472 (1953).

(56) D. J. R. Laurence, Biochem. J., 51, 168 (1952).

(57) L. Kotin and M. Nagasawa, J. Am. Chem. Soc., 87, 3237 (1965).

(58) J. E. Scott and I. H. Willet, Nature, (London), 209, 985 (1966).

(59) A. W. Adamson, J. Am. Chem. Soc., 76, 1578 (1954).

(60) H. A. Bent, J. Phys. Chem. 60, 123 (1956).

(61) L. E. Orgel, An Introduction to Transition Metal Chemistry, Methuen, London, 1960, p. 14.

(62) A. G. Ogston, Nature (London), 162, 963 (1948).

(63) I. Langmuir, Colloid Symp. Monogr., 3, 48 (1925).

(64) K. W. Miller and J. H. Hildebrand, J. Am. Chem. Soc., 90, 3001, (1968).

(65) M. Abu-Hamdiyyah, J. Phys. Chem., 69, 2720 (1965).

(66) T. C. Tsao, K. Bailey, and G. S. Adair, Biochem. J., 49, 27 (1951).

(67) Webb, J. L., Enzyme and Metabolic Inhibitors, Vol. II, Academic Press, New York, 1966.

(68) L. J. Andrews, Chem. Rev., 54, 713 (1954).

(69) R. J. P. Williams, J. Phys. Chem., 58, 121 (1954).

(70) S. Ehrenpreis and M. M. Fishman. Biochim. Biophys. Acta, 44, 577 (1960).

(71) G.S. Adair, Proc. Roy. Soc. (London), A-120, 573 (1928).

(72) C. Ropars and R. Viovy, Comptes Rendues, 257, 3499 (1963), 261, 1129 (1965).

(73) J. R. Florini and D. A. Buyske, J. Biol. Chem., 236, 247 (1961).

(74) T. R. Hughes and I. M. Klotz in Methods of Biochemical Analysis (D. Glick, ed.), Vol. 3, Wiley-Interscience, New York, 1956, p. 265.

(75) R. C. Warner and I. Weber, J. Am. Chem. Soc., 75, 5094 (1953).

(76) H. H. Stein, Anal. Biochem., 13, 305 (1965).

(77) L. Rodkey, Arch. Biochem. Biophys., 94, 526 (1961).

(78) R. E. Clegg, Chem. Analyst, 38, 87 (1949).

(79) A. S. Prasad and E. B. Flink, J. Appl. Physiol., 10, 103 (1957).

(80) J. P. Hummel and W. J. Dreyer, Biochim. Biophys. Acta, 63, 530 (1962).

(81) J. Lee and J. R. Debro, J. Chromatog., 10, 68 (1963).

(82) G. Hartmann, V. Coy, and G. Kniese, Z. Physiol. Chem., 330, 227 (1963).

(83) B. J. Gelotte, J. Chromatog., 3, 330 (1960).

(84) D. Eaker and J. Porath, Separation Sci., 2, 507 (1967).

(85) F. Karush, J. Am. Chem. Soc., 73, 1246 (1951).

(86) D. W. S. Goodman, J. Am. Chem. Soc., 80, 3892 (1958).

(87) L. F. Cavalieri, A. Angelos, and M. E. Balise, J. Am. Chem. Soc., 73, 4902 (1951).

(88) A. Wishnia and D. W. Pinder, Biochem., 3, 1377 (1964), 5, 1534 (1966).

(89) K. Mohammadzaheh, R. E. Feeny, R. B. Samuels, and L. M. Smith, Biochim. Biophys. Acta, 147, 583 (1967).

(90) E. Boyland and B. Green, J. Mol. Biol., 9, 589 (1964).

(91) F. Ascoli, M. Savino, and A. M. Liquori, Nature (Lond.), 217, 162, (1968).

(92) I. Y. Shaferstein and A. P. Zinova, Biokhimiya, 17, 7 (1952).

(93) L. M. Chapman, D. M. Greenberg and G. L. A. Schmidt, J. Biol. Chem., 72, 707 (1927).

(94) H. Fraenkel-Conrat and M. Cooper, J. Biol. Chem., 154, 239 (1944).

(95) G. E. Perlmann, J. Biol. Chem., 127, 107 (1941).

(96) J. Brigando, Bull. Soc. Chim. (France), 1797 (1956).

(97) F. Robert, J. Chim. Phys., 60, 684 (1963).

(98) H. B. Bull, J. Am. Chem. Soc., 67, 10 (1945).

(99) I. H. Sher and H. Sobotka, J. Colloid, Sci., 10, 125 (1955).

(100) J. Schubert, J. Phys. Chem., 56, 113 (1952).

(101) F. N. Briggs and M. Fleischmann, J. Gen. Physiol., 49, 131 (1965).

(102) V. Westphal, P. Gedigk, and F. Meyer, Z. Physiol. Chem., 285, 36 (1950).

(103) E. Kallee, Z. Physiol. Chem., 290, 207 (1952).

(104) N. Hayama, Nippon Kagaku Zasshi, 84, 943, 948, 953. (1963)

(105) C. Wunderly, Artzl. Forsch, 4, 29 (1950).

(106) R. H. Jensen and N. Davidson, Biopolymers, 4, 17 (1966).

(107) A. J. Zielen, J. Phys. Chem., 67, 1474 (1963).

(108) D. J. G. Ives and G. J. Janz, Reference Electrodes, Academic Press, New York, 1961.

(109) G. Scatchard, I. H. Scheinberg, and S. H. Armstrong, J. Am. Chem. Soc., 72, 535 (1950).

(110) G. Scatchard and W. T. Yap, J. Am. Chem. Soc., 86, 3434 (1964).

(111) C. Botré, V. Crescenzi, and A. M. Liquori, Proc. 3rd Intern. Congr. Surface Activity, Cologne, Abstract 11. 3. 302.

(112) R. G. Bates, Electrometric pH Determinations, Wiley, New York, 1954.

(113) G. Eisenman, Biophys. J. Suppl. 2, 259 (1962).

(114) G. Eisenman, Adv. Anal. Chem. Instrumentation, 4, 213 (1965).

(115) N. Ise and T. Okubo, J. Phys. Chem., 69, 4102 (1965).

(116) E. A. Guggenheim, J. Phys. Chem., 33, 842 (1929).

(117) N. Lakshmirayanaih, Chem. Rev., 65, 491 (1965).

(118) G. Scatchard, J. S. Coleman, and A. L. Shen, J. Am. Chem. Soc., 79, 12 (1957).

(119) C. W. Carr, Arch. Biochem. Biophys., 40, 286 (1952).

(120) H. C. Tendeloo and A. Krips, Rev. Trav. Chim., 77, 678 (1958).

(121) C. Botré, M. Marchetti, and S. Borghi, Biochem. Biophys. Acta, 154, 269 (1968).

(122) J. H. Johnson and B. C. Pressman, Biochem. Biophys. Acta, 153, 500 (1968).

(123) O. D. Bonner and D. C. Lunny, J. Phys. Chem., 70, 1140 (1966).

(124) I. M. Kolthoff and J. J. Lingane, Polarography (2nd ed.) Wiley-Interscience, New York, 1952.

(125) I. M. Kolthoff and B. R. Willeford, J. Am. Chem. Soc., 80, 5673 (1960).

(126) H. A. Saroff and H. J. Mark, J. Am. Chem. Soc., 75, 1420 (1153).

(127) D. Bach and I. R. Miller, Biopolymers, 5, 161 (1967).

(128) Y. F. Frei and I. R. Miller, J. Phys. Chem., 69, 3018 (1965).

(129) G. Felsenfeld and S. Huang, Biochim. Biophys. Acta, 34, 234 (1959).

(130) J. A. V. Butler, A. B. Robins and K. V. Shooter, Proc. Roy. Soc. (London) A 241, 299 (1957).

(131) R. H. Doremus and P. Johnson, J. Phys. Chem., 62, 203 (1958).

(132) R. L. Darskus, D. O. Jordan, and T. Kuracsev, Trans. Faraday Soc., 62, 2876 (1966).

(133) J. R. Huizenga, P. F. Grieger, and F. T. Wall, J. Am. Chem. Soc., 72, 2656, 4228 (1950).

(134) S. J. Gill and G. V. Ferry, J. Phys. Chem., 66, 995 (1962).

(135) L. A. Noll and S. J. Gill, J. Phys. Chem., 67, 498 (1963).

(136) R. F. Smith and D. R. Briggs, J. Phys. Chem., 54, 33 (1950).

(137) R. A. Alberty, in The Proteins (H. Neurath and K. Bailey, eds.) Vol. 1, Part A, Academic Press, New York, 1953, Chap. 6.

(138) F. Putnam and H. Neurath, J. Biol. Chem., 159, 194 (1945).

(139) J. R. Cann, Biochem., 5, 1108 (1966).

(140) J. T. G. Overbeek and D. Stigter, Rec. Trav. Chim. Pays-Bas, T 75, 543 (1956).

(141) V. P. Strauss and D. Woodside, J. Phys. Chem., 61, 1353 (1957).

(142) V. Westphal and F. Devenuto, Biochem. Biophys. Acta, 115, 187 (1966).

(143) L. E. Braverman and S. H. Ingbar, Endocrinology, 76, 547 (1965).

(144) W. Grassmann, Ciba Foundation Symposium on Paper Electrophoresis (G. E. W. Wolstenholme and E. C. P. Millar, eds.) Churchill, London, 1956 p. 2.

(145) M. H. Bickel and D. Bovet, J. Chromatog., 8, 466 (1962).

(146) M. Eigen and L. de Maeyer, in Technique in Organic Chemistry (A. Weissburger, ed.) 2nd ed., Vol. 8, Part 2, Wiley, New York (1963) p. 895.

(147) S. R. Anderson and E. Antonini, J. Biol. Chem., 243, 2918 (1968).

(148) I. M. Klotz, J. Am. Chem. Soc., 68, 2299 (1946).

(149) A. L. Stone, Biochem. Biophys. Acta, 148, 193 (1967).

(150) B. Chance, Rev. Sci. Instrum., 22, 634 (1951).

(151) J. H. Lang and E. C. Larser, Biochem., 6, 2403 (1967).

(152) D. F. Bradley and M. K. Wolf, Proc. Natl. Acad. Sci., 45, 944 (1959).

(153) E. H. Lepper and C. J. Martin, Biochem. J., 21, 356 (1927).

(154) I. M. Klotz, R. K. Burkhard, and J. M. Urquhart, J. Am. Chem. Soc., 74, 202 (1952).

(155) B. Carroll, J. Am. Chem. Soc., 68, 2763 (1950).

(156) R. J. P. Williams, Biopolymers Symp., 1, 515 (1964).

(157) S. Lindskog and P. O. Nyman, Biochim. Biophys. Acta, 85, 462 (1964).

(158) J. Shack and B. S. Bynum, Nature (London), 184, 635 (1959).

(159) J. Ettori and S. M. Scoggan, Arch. Biochem. Biophys., 91, 27 (1960).

(160) J. H. Coates, D. O. Jordan, and V. K. Srivastava, Biochem. Biophys. Res. Commun., 20, 611 (1965).

(161) T. Yamane and N. Davidson, Biochim. Biophys. Acta, 55, 780 (1961).

(162) M. Daune, C. A. Dekker, and H. K. Schachman, Biopolymers, 4, 51 (1966).

(163) V. Santelli, Biochim. Biophys. Acta, 120, 239 (1966).

(164) S. Yanari and F. A. Bovey, J. Biol. Chem., 235, 2818 (1966).

(165) A. Ray, J. A. Reynolds, H. Polet and J. Steinhardt, Biochem., 5, 2606 (1966).

(166) H. F. Fischer, Proc. Natl. Acad. Sci., 51, 1285 (1964).

(167) S. F. Velick, J. Biol. Chem., 233, 1455 (1958).

(168) S. F. Velick, S. W. Parker, and H. N. Eisen, Proc. Natl. Acad. Sci., 46, 1470 (1960).

(169) S. R. Anderson and G. Weber, Biochem., 4, 1948 (1965).

(170) G. Weber and D. J. R. Laurence, Biochem. J., 56, xxxi (1954).

(171) C. Neuberg and I. S. Roberts, Arch. Biochem. Biophys., 20, 185 (1949).

(172) S. N. Timasheff and F. F. Nord, Arch. Biochem. Biophys., 31, 309 (1951).

(173) W. Moffit and J. T. Yang, Proc. Natl. Acad. Sci., 42, 596 (1956).

(174) E. Schechter and E. R. Blout, Proc. Natl. Acad. Sci., 51, 695 (1964).

(175) G. Markus and F. Karush, J. Am. Chem. Soc., 80, 89 (1958).

(176) M. H. Winkler and G. Markus, J. Am. Chem. Soc., 81, 1873 (1959).

(177) W. Scholtan, Arzneimittel Forsch., 14, 1234 (1964).

(178) D. Balasubramanian and D. B. Wetlaufer, Proc. Natl. Acad. Sci., 55, 762 (1966).

(179) E. R. Blout and L. Stryer, Proc. Natl. Acad. Sci., 45, 1591.

(180) D. M. Neville and D. F. Bradley, Biochim. Biophys. Acta, 50, 397 (1961).

(181) A. Blake and A. R. Peacocke, Biopolymers, 4, 1901 (1966).

(182) K. Yamaoka and R. A. Resnick, J. Phys. Chem., 70, 4051 (1966).

(183) G. Blauer and T. E. King, Biochem. Biophys. Res. Commun., 31, 678 (1968).

(184) J. E. Coleman, Biochemistry, 4, 2644 (1965).

(185) C. D. Coryell, F. Stitt, and L. Pauling, J. Am. Chem. Soc., 59, 633 (1937).

(186) H. Theorell and A. Ehrenberg, Arkiv Fysik, 3, 299 (1951).

(187) J. A. Pople, W. B. Schneider, and H. J. Bernstein, High Resolution Nuclear Magnetic Resonance, McGraw-Hill, New York, 1959.

(188) D. J. E. Ingram, Spectroscopy at Radio and Microwave Frequencies (2nd ed.), Butterworths, London 1961.

(189) N. Davidson and R. Gold, Biochim. Biophys. Acta, 26, 370 (1957).

(190) J. Eisinger, R. G. Schulman, and W. E. Blumberg, Nature (London) 192, 963 (1961).

(191) J. Eisinger, R. G. Schulman, and Szymanski, J. Chem. Phys., 36, 1721 (1962).

(192) W. E. Blumberg, J. Eisinger, P. Aisen, A. Morell, and I. H. Scheinberg, J. Biol. Chem., 238, 1675 (1963).

(193) M. Cohn, Biochem., 2, 623 (1963).

(194) O. Jardetzsky and N. G. Wade-Jardetzsky, Mol. Pharmacol., 1, 214 (1965).

(195) S. Ohnishi and H. M. McConnell, J. Am. Chem. Soc., 87, 2293 (1965).

(196) J. S. Griffiths, Nature (London), 180, 30 (1957).

(197) L. Stryer and O. H. Griffiths, Proc. Natl. Acad. Sci., 54, 1785 (1965).

(198) T. J. Stone, T. Buckman, P. L. Nordio, and H. M. McConnell, Proc. Natl. Acad. Sci., 58, 1010 (1965).

(199) S. Ogawa and H. M. McConnell, Proc. Natl. Acad. Sci., 58, 19 (1967).

(200) O. Jardetzsky and J. E. Wertz, J. Am. Chem. Soc., 82, 318 (1960).

(201) B. G. Malmström and T. Vänngård, J. Mol. Biol., 2, 118 (1960).

(202) B. G. Malmström, R. Cessa, and T. Vänngård, J. Mol. Biol., 5, 301 (1962).

(203) D. H. Meadows, J. L. Markley, J. S. Cohen, and O. Jardetzsky, Proc. Natl. Acad. Sci., 58, 1307 (1967).

(204) J. Z. Steinberg and H. K. Schachman, Biochem. 5, 3728 (1966).

(205) S. F. Velick, J. E. Hayes, and J. Hartnung, J. Biol. Chem., 203, 527 (1953).

(206) J. C. Gerhardt and H. K. Schachman, Biochem., 4, 1054 (1965).

(207) W. L. Hughes, J. Am. Chem. Soc., 69, 1836 (1947).

(208) U. S. Nandi, J. C. Wang, and N. Davidson, Biochem., 4, 1687 (1965).

(209) W. Bauer and J. Vinograd, J. Mol. Biol., 33, 141 (1968).

(210) R. A. Cox, J. Polymer Sci., 47, 441 (1960).

(211) L. S. Lerman, J. Mol. Biol., 3, 18 (1961).

(212) H. Bennhold, Ergeb. inn. Med. u. Kinderheilk., 42, 473 (1932).

(213) J. Steigman, Trans. N. Y. Acad. Sci., 21, 220 (1959).

(214) K. Iso, Nippon Kagaku Zasshi, 84, 95 (1963).

(215) J. M. Preston and Y. F. Su, J. Soc. Dyers. Colorists, 66, 357 (1950).

(216) J. M. Preston and D. C. Tsieu, J. Soc. Dyers. Colorists, 66, 361 (1950).

(217) R. S. Bear, J. Am. Chem. Soc., 64, 1388 (1942).

(218) D. W. Green, V. M. Ingram, and M. F. Perutz, Proc. Roy. Soc. (Lond.), A 223, 287 (1958).

(219) I. M. Bluhm, G. Bodo, H. Dintzis, and J. C. Kendrew, Proc. Roy. Soc. (Lond.), A246, 369 (1958).

(220) L. Stryer, J. C. Kendrew, and H. C. Watson, J. Mol. Biol., 8, 96 (1964).

(221) R. H. Kretsinger, H. C. Watson, and J. C. Kendrew, M. Mol. Biol., 31, 305 (1968).

(222) B. P. Schoenborn, Federation Proc., 27, 888 (1968).

(223) B. P. Schoenborn and C. L. Nobbs, Mol. Pharmacol., 2, 496 (1966).

(224) L. D. Hamilton, W. Fuller, and E. Reich, Nature, 198, 538 (1963).

(225) L. N. Johnson, Proc. Roy. Soc. (Lond.), B167, 439 (1967).

(226) J. Collins, Proc. Roy. Soc. (Lond.), B167, 441 (1967).

(227) F. Haurowitz, Z. Physiol. Chem., 238, 266 (1938).

(228) H. Muirhead and M. F. Perutz, Nature (Lond.), 199, 633 (1963).

(229) F. S. Parker and J. L. Irvin, J. Biol. Chem. 199, 889 (1952).

(230) I. M. Klotz, E. W. Gelewitz, and J. M. Urquhardt, J. Am. Chem. Soc. 74, 209 (1952).

(231) R. J. Goldacre, Ph.D. Thesis, London University (1952).

(232) T. D. Sokoloski, D. T. Witiak, M. W. Whitehouse, and F. Hermann, Fed. Proc. 28, 614 (1969).

(233) J. R. Colvin, Can. J. Chem., 31, 734 (1953).

(234) R. H. McMenamy and J. L. Oncley, J. Biol. Chem., 233, 1436 (1959).

(235) T. Peters, Biochim. Biophys. Acta, 39, 546 (1960).

(236) T. Peters and F. A. Blumenstock, J. Biol. Chem. 242, 1574 (1967).

(237) Q. H. Gibson, Progr. Biophys. Biophys. Chem., 9, 1 (1959).

(238) G. Braunitzer, K. Hilse, V. Rudloff, and N. Rilschman, Adv. Protein Chem. 19, 1 (1964).

(239) J. Wyman, Adv. Protein Chem. 19, 224 (1964).

(240) E. Antonini, Physiol. Rev., 45, 123 (1965).

(241) M. F. Perutz and H. Lehmann, Nature, 219, 902 (1968).

(242) M. F. Perutz, Endeavour, 26, 3 (1967).

(243) K. Jahaverian and S. Beychok, J. Mol. Biol., 37, 1 (1968).

(244) K. Winterhalter, Nature, 211, 932 (1966).

(245) K. M. Winterhalter, G. Amiconi, and E. Antonini, Biochem. 7, 2228 (1968).

(246) P. S. Gerald and P. George, Science, 129, 393 (1959).

(247) P. S. Gerald and M. L. Efron, Proc. Natl. Acad. Sci., 47, 1758 (1961).

(248) K. Landsteiner, The Specificity of Serological Reactions, Harvard Univ. Press, Cambridge, Mass. (1947).

(249) J. R. Marrack, The Chemistry of Antigens and Antibodies, Med. Res. Council, Special Report Series, London, (1938).

(250) F. Karush, Adv. Immunol. 2, 1 (1962).

(251) H. N. Eisen and F. Karush, J. Am. Chem. Soc., 71, 363 (1949)

(252) R. C. Valentine and N. M. Green, J. Mol. Biol. 27, 615 (1967).

(253) N. R. Klinman, J. H. Rockey, and F. Karush, Science, 146, 401 (1964).

(254) M. E. Koshland, F. Englberger, Proc. Natl. Acad. Sci., 50, 61 (1963).

(255) A. L. Grossberg and D. Pressman, Biochem. 7, 272 (1968).

(256) J. R. Little and H. N. Eisen, Biochem. 6, 3119 (1967).

(257) S. Joffe, Mol. Pharmacol., 3, 399 (1967).

(258) H. N. Eisen, Methods Med. Res. 10, 115 (1964).

(259) C. W. Parker, T. J. Yoo, M. C. Johnson, and S. M. Godt, Biochem. 6, 3408 (1967).

(260) C. W. Parker, S. M. Godt, and M. C. Johnson, Biochem., 6 (3417 (1967).

(261) T. J. Yoo, O. A. Roholt, and D. Pressman, Science, 157, 707 (1967).

(262) W. B. Dandliker, H. C. Shapiro, J. W. Meduski, R. Alonso, and J. R. Hamrick, Immunochemistry, 1, 165 (1964).

(263) L. Stryer and O. H. Griffiths, Proc. Natl. Acad. Sci., 54, 1785 (1965).

(264) R. P. Haughland, L. Stryer, T. R. Stengle, and J. D. Baldeschweiler, Biochem., 6, 498 (1967).

(265) W. R. Slaunwhite, S. Scheider, F. C. Wissler, and A. A. Sandberg, Biochem. 5, 3527 (1965).

(266) J. H. Oppenheimer, M. I. Surks, C. Smith, and R. Squef, J. Biol. Chem., 240, 173 (1965).

(267) S. C. Thornton, W. N. Tauxe, and H. F. Taswell, J. Clin. Endocrinol. Metab., 26, 181 (1966).

(268) L. A. Kazal, Progr. Hematol., 3, 294 (1962).

(269) D. Watson, Clin. Chim. Acta, 15, 121 (1967).

(270) J. A. V. Butler, E. W. Johns, and D. M. P. Phillips, Progr. Biophys. Mol. Biol. 18, 209 (1968).

(271) N. Alfert and I. I. Geschwind, Proc. Natl. Acad. Sci., 39, 991 (1953).

(272) J. Loeb, Compt. Rend., 258, 5087 (1964).

(273) J. Winkleman, Biochim. Biophys. Acta, 147, 577 (1967).

(274) D. J. R. Laurence, Biochem. J., 99, 419 (1966).

(275) M. Sluyser, Biochim. Biophys. Acta, 154, 606 (1968).

(276) I. Asimov, The Genetic Code, New American Library, New York (1963).

(277) J. D. Watson, The Molecular Biology of the Gene, Benjamin, New York (1965).

(278) C. R. Woese, The Genetic Code, the Molecular Basis for Genetic Expression, Harper and Row, New York (1967).

(278a) J. D. Watson and F. H. C. Crick, Cold Spring Harbor Symposium, Quant. Biol., 18, 125 (1953).

(279) W. M. Huang and P. O. P. Ts'o, J. Mol. Biol., 16, 523 (1966).

(280) M. Nirenberg and P. Leder, Science, 145, 1399 (1964).

(281) S. R. Pelc and M. G. E. Welton, Nature, 209, 868 (1966).

(282) M. V. Volkenstein, Nature, 207, 294 (1965).

(283) A. Blake and A. R. Peacocke, Biopolymers, 6, 1225 (1968).

(284) J. Chambron, M. Daune, and Ch. Sadron, Biochem. Biophys. Acta, 123, 306, 319 (1966).

(285) L. Michaelis, Cold Spring Harbour Symp. Quant. Biol., 12, 131, (1947).

(286) G. Weill, Biopolymers, 3, 567 (1965).

(287) R. K. Tubbs, W. E. Ditmars, and Q. van Winkle, J. Mol. Biol., 9, 545 (1964).

(288) E. V. Anufrieva, M. V. Volkenstein, and T. V. Shevleva, Biofizika, 7, 554 (1962).

(289) N. B. Kurnick, J. Gen. Physiol., 33, 243 (1950).

(290) N. B. Kurnick, J. Am. Chem. Soc., 76, 417 (1954).

(291) T. Finklestein and I. B. Weinstein, J. Biol. Chem., 242, 3763 (1967).

(292) R. L. O'Brien, J. G. Olenick, and F. E. Hahn, Proc. Natl. Acad. Sci. 55, 1511 (1966).

(293) M. Liersch and G. Hartmann, Biochem., Z. 340, 390 (1964).

(294) J-B. Le Pecq and C. Paolleti, J. Mol. Biol., 27, 87 (1967).

(295) E. Reich, Science, 143, 634 (1964).

(296) T. Kurozami, Y. Kurihara, and K. Shibata, J. Biochem. (Tokyo), 53, 135 (1963).

(297) I. Snapper, B. Scheid, F. Lieban, and E. Greenspan, J. Lab. Clin. Med., 37, 562 (1951).

(298) G. Oster, Compt. Rend., 232, 1708 (1951).

(299) H. Yamagishi, J. Cell. Biol., 15, 589 (1962).

(300) M. Klimek and L. Hnilica, Arch. Biochim. Biophys., 81, 105 (1959).

(301) D. J. Goldstein, Nature, 191, 407 (1961).

(302) L. Jurkowitz, Arch. Biochem. Biophys., 111, 88 (1965).

(303) F. Klein and J. A. Szirmai, Biochim. Biophys. Acta, 72, 48 (1963).

(304) C. S. Hanes, New Phytologist, 36, 189 (1937).

(305) K. Freudenberg, E. Schaaf, G. Dumpert, and T. Ploetz, Naturwiss, 27, 850, (1939).

(306) R. L. Bates, D. French, and R. E. Rundle, J. Am. Chem. Soc., 65, 142 (1943).

(307) R. E. Rundle and R. R. Baldwin, J. Am. Chem. Soc., 65, 554 (1943).

(308) R. E. Rundle, J. F. Foster, and R. R. Baldwin, J. Am. Chem. Soc., 66, 2116 (1944).

(309) C. West, J. Chem. Phys., 15, 689 (1947).

(310) B. Carroll and J. W. van Dyke, J. Am. Chem. Soc., 76, 2506 (1954).

(311) W. Broser and W. Lautsch, Z. Naturforsch., 8b, 711 (1953).

(312) J. N. Miller, Starch Chem. Technol., 1, 309 (1967).

(313) C. Robinson, Disc. Faraday Soc., 16, 125 (1954).

(314) A. B. Foster, P. A. Newton-Hearn, and M. Stacey, J. Chem. Soc., 1956, 30.

(315) B. Carroll and H. C. Cheung, J. Phys. Chem. 66, 2585 (1962).

(316) M. Schubert, Biophys. J., 4 (supplement), 119 (1964).

(317) H. F. Stedman, Quart J. Microscop. Sci., 91, 477 (1950).

(318) M. K. Pal and M. Schubert, J. Am. Chem. Soc., 84, 4384 (1962).

(319) P. Mukerjee and A. K. Ghosh, J. Phys. Chem., 67, 193 (1963).

(320) J. E. Scott, Biochem. J., 84, 270 (1962).

(321) C. Woodward and E. A. Davidson, Proc. Natl. Acad. Sci., 60, 201 (1968).

(322) A. L. Stone and H. Moss, Biochem. Biophys. Acta, 136, 56 (1967).

(323) S. A. Rice and M. Nagasawa, Polyelectrolyte Solutions: a Theoretical Introduction, Academic Press, New York (1961).

(324) H. Morawetz, Macromolecules in Solution, Interscience - Wiley, New York (1966).

(325) A. Katchalsky, Biophys. J., 4, Part. 2, 9 (1964).

(326) T. Alfrey, P. W. Berg, and H. Morawetz, J. Polymer Sci. 7, 543 (1951).

(327) R. M. Fuoss, A. Katchalsky, and S. Lifson, Proc. Natl. Acad. Sci., 37, 579 (1951).

(328) H. P. Gregor and M. Frederick, J. Polymer. Sci., 23, 451 (1957).

(329) F. T. Wall and M. J. Eitel, J. Am. Chem. Soc., 79, 1556 (1957).

(330) L. Kotin and M. Nagasawa, J. Chem. Phys., 36, 873 (1962).

(331) R. A. Mock and C. A. Marshall, J. Polymer Sci., 13, 263 (1954).

(332) F. Oosawa, J. Polymer Sci., 23, 421 (1957).

(333) H. Morawetz, Fortschr Hochpolymer Forsch., 1, 1 (1958).

(334) V. P. Strauss and A. Siegel, J. Phys. Chem., 67, 2683.

(335) M. Mandel and J. C. Leyte, J. Polymer Sci., A2, 2883, 3771 (1964).

(336) R. M. Fuoss and V. P. Strauss, J. Polymer Sci., 3, 246, 602, (1948).

(337) R. A. Mock, C. A. Marshall, and T. E. Slykehouse, J. Phys. Chem., 58, 498 (1954).

(338) T. R. E. Kressman and J. A. Kitchener, J. Chem. Soc., 1949, 1208.

(339) G. Barone, V. Crescenzi, B. Pispisa, and F. Quadrifolio, J. Macromol. Chem., 1, 761. (1966).

(340) G. Oster, J. Polymer Sci., 16, 235 (1955).

(341) H. Bennhold and R. Schubert, Z. Ges. Expt. Med., 113, 722 (1943).

(342) W. Scholtan, Makromol. Chem. 11, 131 (1953).

(343) P. Molyneux and H. P. Frank, J. Am. Chem. Soc., 83, 3168 3175 (1961).

(344) S. Saito, Kolloid Z., 154, 19, (1957).

(345) K. Isemura and A. Imanishi, J. Polymer Sci., 33, 337 (1958).

(346) J. Steinhardt and J. A. Reynolds, Multiple Equilibria in Proteins, Academic Press, London and New York (1970).

(347) J. R. Cann, Interacting Macromolecules, Academic Press, London and New York (1970).

(348) O. H. Griffith and A. S. Waggoner, Accounts Chem. Research, 2, 17 (1969).

(349) G. C. Wood and P. F. Cooper, Chromatog. Rev., 12, 88 (1970).

(350) M. C. Meyer and D. E. Guttman, J. Pharmaceutical Sci., 57, 895 (1968).

(351) J. Wyman, J. Mol. Biol., 39, 523 (1969).

(352) G. Schwarz, Eur. J. Biochem., 12, 442 (1970).

(353) A. J. Cornish-Bowden and D. E. Koshland, Biochem., 9, 3325 (1970).

(354) M. F. Perutz and H. Lehmann, Nature (Lond.), 219, 902 (1969).

(355) D. M. Shotton and B. S. Hartley, Nature, 255, 802 (1970).

(356) J. B. Swaney and I. M. Klotz, Biochem. 9, 2570 (1970).

(357) F. Helmer, K. Kiehs, and C. Hansch, Biochem., 7, 2858 (1968).

(358) W. M. Beers and E. Reich, Nature (Lond.), 228, 917 (1970).

(359) N. O. Thorp and S. J. Singer, Biochem., 8, 4523 (1969).

(360) J. Pensky and J. S. Marshall, Arch. Biochem. Biophys., 135, 304 (1969).

(361) S. P. Colowick and F. C. Womack, J. Biol. Chem., 244, 774 (1969).

(362) D. C. Turner and L. Brand, Biochem., 7, 3381 (1968).

(363) E. M. Kosower, J. Am. Chem. Soc., 80, 3253 (1968).

(364) A. J. Thomson, J. Chem. Phys., 51, 4106 (1969).

(365) A. Kotaki, M. Naio, and K. Yagi (1971) Biochim. Biophys. Acta (in press).

(366) S. M. Koenig and W. E. Schillinger, J. Biol. Chem., 244, 3283 (1969).

(367) R. Kimmich and F. Noack, Z. Naturforsch., 25a, 299 (1970).

(368) R. G. Shulman, K. Wuthrich, T. Yamane, D. J. Patel, and W. E. Blumberg, J. Mol. Biol., 53, 143 (1970).

(369) C. G. Rosen and G. Weber, Biochem., 8, 3915 (1969).

(370) M. F. Perutz, Nature (Lond.) 228, 726 (1970).

(371) M. F. Perutz, Nature (Lond.) 228, 734 (1970).

(372) J. W. Goers and V. N. Schumaker, J. Mol. Biol., 54, 125 (1970).

(373) I. Schechter, Nature, 228, 639 (1970).

(374) B. A. Askonas, A. R. Williamson, and B. E. G. Wright, Proc. Natl. Acad. Sci., 67, 1398 (1970).

(375) A. Feinstein and E. A. Munn, Nature (Lond.) 224, 1307 (1970).

(376) K. Iwai, K. Ishikawa, and H. Hayashi, Nature (Lond.), 224, 1058 (1970).

(377) R. J. De Lang, D. M. Fambrough, E. L. Smith, and J. Bonner, J. Biol. Chem., 244, 319 (1969).

(378) M. Bustin, S. C. Rall, R. M. Stellwagen, and R. D. Cole, Science, 163, 391 (1969).

(379) R. Romer, D. Riesner, and G. Mass, F. E. B. S. Letters, 10, 352 (1970).

(380) E. J. Gabbay, J. Am. Chem. Soc., 91, 5136 (1969).

(381) W. Bauer and J. Vinograd, J. Mol. Biol., 47, 419 (1970).

(382) R. D. Wells and J. E. Larson, J. Mol. Biol., 49, 319 (1970).

(383) A. L. Stone, Biopolymers, 7, 173 (1969).

(384) D. J. R. Laurence and J. M. Higginson, J. Chromatog., 40, 145 (1969).

(385) F. W. Schneider, C. L. Cronan, and S. K. Podder, J. Phys. Chem., 72, 4563 (1968).

(386) J. A. Thoma and D. French, J. Phys. Chem., 65, 1825 (1961).

Chapter 3

ELECTRIC PROPERTIES OF BIOPOLYMERS: PROTEINS

E. O. Forster
Corporate Research Laboratories
Esso Research and Engineering Company
Linden, New Jersey

A. P. Minton*
Polymer Department
Weizmann Institute of Science
Rehovoth, Israel

*PRESENT ADDRESS: National Institute of Arthritis and Metabolic Diseases, National Institutes of Health, Bethesda, Maryland.

I. INTRODUCTION

A knowledge of the electrical properties of proteins is of interest to researchers in the biological and physical sciences because such knowledge may lead to greater understanding of structure-function relationships as a fundamental component of all living systems. This area of study requires the combined skills of the biologist, chemist, physicist, and electrical engineer. Unfortunately it has only been in the last few years that cooperation among the various disciplines has begun to take place. It is therefore not surprising that publications on this subject appear in an unusually wide spectrum of journals and books. In undertaking the present review it was the authors' intent to compose from these diverse sources a picture of the current "state of the art" and thus help to guide future research efforts in fruitful directions. A similar attempt was made previously in Volume 1 of this series regarding synthetic polymers (1).

In this survey the response of proteins in solution and in the dry as well as hydrated states to the application of static and alternating electric fields will be discussed. The various interpretations of electrical properties in terms of the structural and hydration characteristics of the protein will be reviewed and the recent literature critically discussed. References to the older literature will be generally omitted but may be found in the review articles to which reference will be made. Because of an unexpected delay in publication of the volume of which this chapter is a part, the editor has made it possible for addenda covering very recent publications to be included. A note (Addendum i) in the text refers the reader to a corresponding entry numbered i in Sec. V.

II. DIELECTRIC PROPERTIES OF PROTEINS IN AQUEOUS SOLUTION

The dielectric properties of any sample system in an alternating electric field may be conveniently described in terms of the complex permittivity (relative to vacuum) ϵ^* which is defined as $\epsilon'(\omega) - i\epsilon''(\omega)$, where ω is the angular frequency which is equal to $2\pi f$, and f is the frequency of the applied field. The real part of the permittivity, $\epsilon'(\omega)$, called the dielectric constant, is equal to the ratio of capacitances of two identical capacitors with sample and vacuum, respectively, between the plates. The imaginary part of the permittivity, $\epsilon''(\omega)$, called the dielectric loss, is related to the ac conductivity of the sample through the relation $\epsilon''(\omega) = \sigma(\omega)/\omega$.

In the presence of a static field the sample system becomes polarized. On switching off the field the sample returns to a depolarized state through a variety of processes referred to as relaxation mechanisms. If we assume that the relaxations are first order with respect to concentration of the relaxing species, then the polarization decay process may be represented as a sum of exponentials,

$$P(t) = \sum_i C_i \exp(-t/\tau_i) = \sum_i C_i \exp(-t\omega_i)$$

where τ_i and ω_i are a time and frequency characteristic of the ith relaxation. Application of Fourier transforms to this relation leads to a representation of the polarization as a function of applied frequency. Conversion from polarization to permittivity results in a sum of terms of the well-known form introduced by Debye (2),

$$\frac{\epsilon^*(\omega) - \epsilon(\infty)}{\epsilon(0) - \epsilon(\infty)} = \sum_i \frac{C_i}{1 - i\omega\tau_i} \tag{1}$$

where $\epsilon(0)$ and $\epsilon(\infty)$ are the limiting permittivities at low and high frequencies respectively. Each relaxation is characterized by a maximum in the plot of ϵ'' versus ω and a lowering of the dielectric constant. In frequency regions between relaxation frequencies the dielectric constant levels off and the dielectric loss approaches zero.

Depending on whether the function of a protein is enzymic or structural, its physical conformation may be roughly described as either globular or extended. Whereas globular proteins are generally found singly or in weakly bound oligomeric complexes, extended proteins tend to form relatively stable linear or planar arrays of large numbers of molecules. In this section the properties of globular proteins will be of primary interest because of a more direct relationship between macroscopic and molecular properties, but some examples of the dielectric behavior of extended proteins in solution will be encountered and discussed in Sec. II. B. 4.

The dielectric properties of aqueous protein solutions have been studied in the frequency region extending from 0 to 3×10^{11} Hz. Solutions of globular proteins generally display three Debye-type dispersion regions corresponding to three basic kinds of relaxation mechanisms. These three regions, which are centered about frequencies of the order of 10^6, 10^8, and 10^{10} Hz, will be referred to as the low-, mid-, and high-frequency dispersions, respectively (Fig. 1).

At low protein concentrations the deviation between the properties of the solution and the properties of the solvent (pure water or dilute

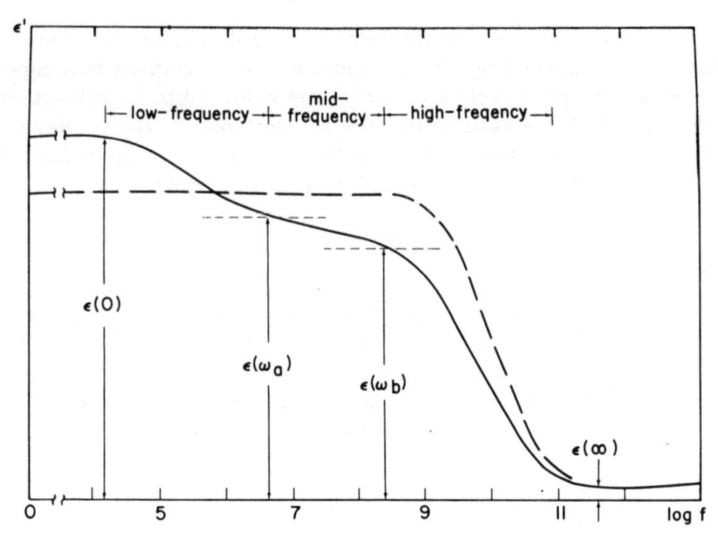

Fig. 1. The three dispersion regions of protein solutions. ϵ' is the real part of the permittivity (dielectric constant), $\epsilon(0)$, $\epsilon(\omega a)$, $\epsilon(\omega_b)$, and $\epsilon(\infty)$ are the limiting values of the dielectric constant between dispersions, and f is the frequency of the applied field. The dielectric constant of pure water (dashed line) is included for comparison.

electrolyte) becomes linear in protein concentration. Thus we define a quantity called the dielectric increment at infinite dilution as

$$\delta^*(\omega) = \lim_{c \to 0} \frac{\partial \epsilon^*(\omega)}{\partial c}$$

where ϵ^* and therefore δ^* are treated as complex quantities in the general case, but between dispersion regions their imaginary parts approach zero. Consequently the several quantities ϵ (i) appearing in Fig. 1 are essentially real quantities ($\epsilon(i) = \epsilon'(i) \approx |\epsilon^*(i)|$). In the idealized scheme to be discussed henceforth we shall be treating them as real quantities and the primed notation will be dropped for convenience. Because the molar concentration c in typical protein solutions is small, many authors prefer to express c in g/100 cm^3 of solution. The dielectric increment calculated by these authors must be multiplied by a factor of molecular weight/10 to obtain the dielectric increments that will be dealt with in this review.

The study of proteins in solution is directed toward an understanding of the magnitude of the dielectric increment and its dependence on experimental variables in terms of the molecular structure and function of the protein.

A. Experimental Methods

A thorough review of the basic methods and problems encountered in impedance (capacitance-conductance) measurements of moderately conducting samples in the low-frequency dispersion region has been presented by Schwan (3), and Mandel (4) has discussed the problem of electrode polarization in detail.

Several commercially available impedance bridges are suitable for measurements of moderate precision in the low-frequency region, subject to a limitation on the conductance of the sample cell, placing an upper limit on the ionic strength of the solutions to be studied. Many sample cells have been described in the literature, incorporating various features for the reduction of (or correction for) electrode polarization effects that arise at the low-frequency end of the range of measurement ($< 10^5$ Hz). These features include variable electrode separation (5), roughening of the electrode surface (6), coating of the electrodes with platinum black (7), and large electrode spacing (8). Residual errors may be corrected by calibration with an electrolyte solution of low-frequency conductivity equal to that of the protein solution under investigation (3).

Impedance bridges are generally utilized at frequencies up to 10^6 Hz. At higher frequencies significant errors arise from lead wire inductance, both within the instrument and between the instrument and sample cell. With short leads the Hewlett-Packard (formerly Boonton) RX-Meter 250-A, consisting of a self-contained oscillator, a Schering bridge, and a sensitive null detector, has a claimed utility up to a frequency of 2.5×10^8 Hz. Coaxial transmission line methods are employed in the region between 10^8 and 5×10^9 Hz and waveguide methods are employed at higher frequencies.

Hasted (9) has provided a general guide to microwave techniques for the study of lossy liquids, which includes methods applicable to the study of protein solutions, and Schwan (3) has discussed transmission line techniques. Recent studies have employed resonance methods, which depend on the change in resonant frequency brought about by a change in either the length of a resonant transmission line or on introduction of the test sample into a cavity (10, 11). Balance methods that measure the attenuation and phase shift of a signal transmitted through a test sample column of known length have also been found useful (12).

B. Low-Frequency Dispersion

The low-frequency dispersion will be discussed separately because the experimental methods employed and interpretations of data obtained from them are of a qualitatively different type than those associated with studies above this frequency range.

1. Classical Interpretation: Relaxation of Orientation Polarization

Until the early 1950's, the dielectric properties of protein solutions were interpreted on the basis of a model largely developed by Oncley (13) and co-workers. This model assumed that the protein molecules could be represented by dipolar hydrodynamically equivalent ellipsoids of rotation undergoing orientational relaxation in a continuous medium of well-defined viscosity. The basic features of this model are summarized below. A more detailed discussion may be found in Chaps. 6, 11, 21, and 22 of Cohn and Edsall (14).

The theory of Kirkwood (15) provides a representation of the static dielectric constant of a polar liquid $\epsilon(0)$, in terms of μ, the dipole moment of an isolated molecule, and $\bar{\mu}$, the dipole moment of the molecule in the field of its neighbors,

$$\frac{\epsilon(0) - 1}{3} = \frac{3\epsilon(0)}{2\epsilon(0) + 1} \cdot \frac{4\pi N}{3\tilde{V}} \left[\alpha + \frac{\mu \cdot \bar{\mu}}{3\,kT} \right] \tag{2}$$

where N is Avogadro's number, $\tilde{V}$ is the molar volume, α is the optical polarizability, k is Boltzmann's constant, and T is the absolute temperature. For a liquid of high dielectric constant (order of 10^2), $3\epsilon(0)/(2\epsilon(0) + 1) \approx 3/2$, and ignoring solvent-solute interactions, the formula for the dielectric constant of a polar solute 1 in a polar solvent 2 may be written

$$\epsilon(0) - 1 = \frac{6\pi N}{1000} \left\{ C_1\left[\alpha_1 + \frac{\mu_1 \cdot \bar{\mu}_1}{3kT}\right] + C_2\left[\alpha_2 + \frac{\mu_2 \cdot \mu_2}{3kT}\right]\right\} \tag{3}$$

where C_i is the molar concentration of component i.

Since it is known that proteins are hydrated in solution, a slightly more realistic model presupposes that each protein molecule is associated with n water molecules that are irrotationally bound; that is, they cannot rotate in an alternating electric field independently of the protein. One further assumes that n is large and that these molecules are oriented in no particular direction with respect to each other. Therefore the net dipole moment of these irrotationally bound water molecules is zero. Rewriting Eq. (3) to accommodate these assumptions, we obtain

$$\epsilon(0) - 1 = \frac{6\pi N}{1000}\left\{C_1\left[\alpha_1 + \frac{\mu_1\bar{\mu}_1}{3kT}\right] + n\,C_1\left[\alpha_2\right] + (C_2 - n\,C_1)\left[\alpha_2 + \frac{\mu_2\cdot\bar{\mu}_2}{3kT}\right]\right\} \tag{4}$$

An expression for $\delta(0)$, the static increment, may be obtained by differentiating Eq. (4) with respect to c_1 and noting that

$$\frac{\partial c_2}{\partial c_1} = -\frac{\tilde{V}_1}{\tilde{V}_2}$$

where $\tilde{V}_1$ and $\tilde{V}_2$ are the partial molar volumes of protein and water, respectively,

$$\delta(0) = \frac{6\pi N}{1000}\left\{\left[\alpha_1 + \frac{\mu_1\bar{\mu}_1}{3kT}\right] + n\left[\alpha_2\right] - \left(\frac{\tilde{V}_1}{\tilde{V}_2} + n\right)\left[\alpha_2 + \frac{\mu_2\cdot\bar{\mu}_2}{3kT}\right]\right\} \tag{5}$$

Equation (5) may be considered as the sum of two terms; a term due to the sum of the mid- and high-frequency dispersions and a term due to the low-frequency dispersion. Since the latter term vanishes at frequencies well above the low-frequency dispersion region, the former term shall be denoted by

$$\delta(\omega_a) = \frac{\partial\epsilon(\omega_a)}{\partial c_1}$$

the increment above the low-frequency dispersion; its magnitude will be discussed in Sec. I.C.1.

$$\delta(0) = \frac{6\pi N}{1000} \frac{\mu_1 \cdot \bar{\mu}_1}{3kT} + \delta(\omega_a) \qquad (6a)$$

$$\delta(\omega_a) = \frac{6\pi N}{1000} \left\{ \alpha_1 + n\alpha_2 - \left(\frac{\tilde{V}_1}{\tilde{V}_2} + n \right) \left[\alpha_2 + \frac{\mu_2 \cdot \bar{\mu}_2}{3kT} \right] \right\} \qquad (6b)$$

The difference between the increments above and below the low-frequency dispersion (Fig. 2) may then be expressed as

$$\Delta\delta_L = \delta(0) - \delta(\omega_a) = \frac{6\pi N}{1000} \frac{\mu_1 \cdot \bar{\mu}_1}{3kT} \qquad (7)$$

Making the approximation that μ and $\bar{\mu}$ are parallel vectors of similar magnitude, Kirkwood (16) defined an "effective dipole moment" $\tilde{\mu}$ such that $\tilde{\mu}^2 = \mu \cdot \bar{\mu}$. Then

$$\tilde{\mu} = \sqrt{\frac{500\ kT}{\pi N} \Delta\delta_L} \qquad (8)$$

Values of $\tilde{\mu}$ obtained from this expression for certain amino acids are in good agreement with values of μ obtained from the crystallographic structure, hence validating, at least in these cases, the above approximation.

When the static electric field is turned off, the molecules that were oriented in the field return to a state of random orientation through rotational diffusion. This process is referred to as relaxation of orientation polarization or rotational relaxation. Perrin (17) showed that an isolated dipolar ellipsoid undergoing rotational relaxation in an alternating applied field would display a frequency dependent dipole moment given by

$$\mu(\omega)^2 = \left(\frac{\mu_1^2}{1 + i\omega\tau_1} + \frac{\mu_2^2}{1 + i\omega\tau_2} + \frac{\mu_3^2}{1 + i\omega\tau_3} \right) \qquad (9)$$

where the μ_i are the components of the dipole vector along the three principal axes of the ellipsoid, and the τ_i are three relaxation times which may be related to the shape and size of the ellipsoid, as will be discussed below.

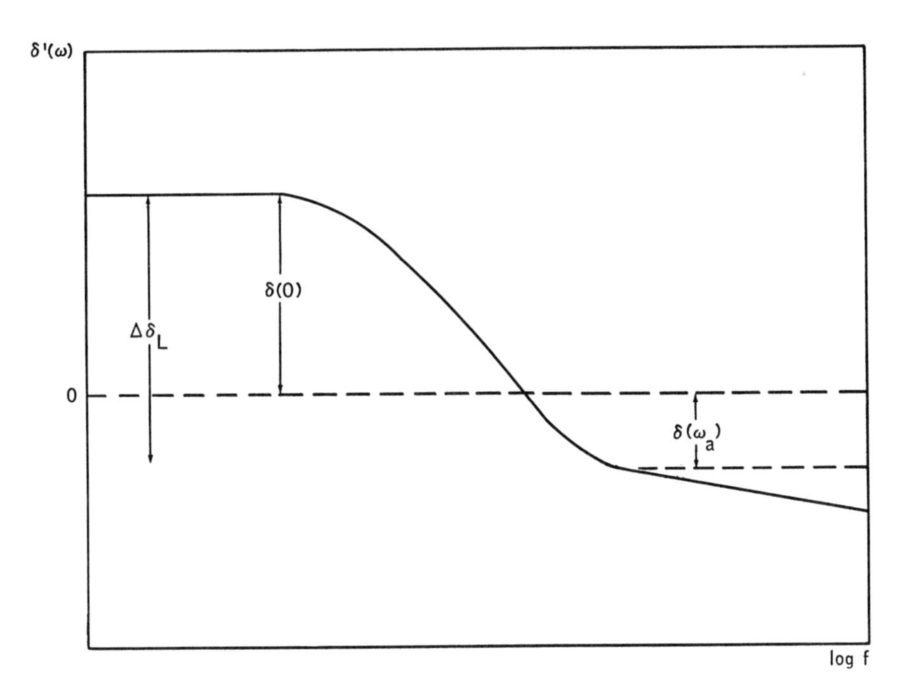

Fig. 2. Schematic diagram of a typical low-frequency dispersion. $\delta(\omega_a)$ is the upper asymptotic limit to the Debye-type contribution.

If we rewrite eq. (8) by introducing the complex square dipole moment obtained by Perrin (17) in place of the product of the real static dipole moments μ_1 and $\bar{\mu}_1$, we obtain an expression for the complex frequency dependent dielectric increment which may be separated into its real and imaginary parts called the dielectric and loss increments respectively.

$$\Delta\delta_L' = \frac{\pi N}{500kT}\left(\frac{\mu_1^2}{1+\omega^2\tau_1^2} + \frac{\mu_2^2}{1+\omega^2\tau_2^2} + \frac{\mu_3^2}{1+\omega^2\tau_3^2}\right) \tag{10a}$$

$$\Delta\delta_L'' = \frac{\pi N\omega}{500kT}\left(\frac{\mu_1^2\tau_1}{1+\omega^2\tau_1^2} + \frac{\mu_2^2\tau_2}{1+\omega^2\tau_2^2} + \frac{\mu_3^2\tau_3}{1+\omega^2\tau_3^2}\right) \tag{10b}$$

Perrin (17) derived the following relation between the relaxation times τ_i, τ_j, τ_k and the coefficients of rotational diffusion C_i, C_j, C_k:

$$\tau_i = \frac{1}{kT} \frac{C_j C_k}{C_j + C_k} \tag{11}$$

He further obtained expressions for the coefficients of rotational diffusion as a function of the lengths a, b, c of the three half axes of the ellipsoid and η, the viscosity of the suspending media. However, these functions incorporate complex elliptic integrals and consequently, although the coefficients of diffusion can be calculated from the three half-axial lengths, the half-axial lengths cannot be uniquely defined by the coefficients of diffusion.

For an ellipsoid of rotation (half axes a, b = c) Perrin's equations simplify considerably. If we denote the ratio of axial lengths b/a by ρ, then for a prolate ellipsoid ($\rho<1$),

$$\tau_i = \frac{8\pi}{3} \frac{\eta a^3}{kT} \left[\frac{1-\rho^4}{\left(\frac{2-\rho^2}{\sqrt{1-\rho^2}}\right) \ln\left(\frac{1+\sqrt{1-\rho^2}}{\rho}\right) - 1} \right] \tag{12a}$$

$$\tau_2 = \frac{16\pi}{3} \cdot \frac{\eta a^3}{kT} \left[\frac{1-\rho^4}{\left(\frac{2\rho^2-1}{\sqrt{1-\rho^2}}\right) \ln\left(\frac{1+\sqrt{1-\rho^2}}{\rho}\right) + \frac{1}{\rho 2}} \right] \tag{12b}$$

and for an oblate ellipsoid ($\rho>1$),

$$\tau_1 = \frac{8\pi}{3} \frac{\eta a^3}{kT} \left[\frac{1-\rho^4}{\left(\frac{2-\rho^2}{\sqrt{\rho^2-1}}\right) \tan^{-1}\left(\sqrt{\rho^2-1}\right) - 1} \right] \tag{12c}$$

$$\tau_2 = \frac{16\pi}{3} \frac{\eta a^3}{kT} \left[\frac{1-\rho^4}{\left(\frac{2\rho^2-1}{\sqrt{\rho^2-1}}\right) \tan^{-1}\left(\sqrt{\rho^2-1}\right) + \frac{1}{\rho 2}} \right] \tag{12d}$$

The ratio of τ_2 to τ_1 is a function of ρ only, independent (in the framework of this model) of the absolute dimensions of the ellipsoid, the viscosity of the suspending media, or the temperature.

The problem of determining whether the observed ratio of relaxation times is τ_1/τ_2 or τ_2/τ_1 is academic, since for all oblate ellipsoids ($\rho > 1$) the two relaxation times vary by less than 10%, a factor that cannot be distinguished experimentally at present. If two relaxation times can be resolved experimentally (for a recent example, see Moser et al. (6)), they differ by at least a factor of three or four and hence must be associated with a prolate ellipsoid. The larger of the two relaxation times is then τ_1. Knowledge of either τ_1 or τ_2, ρ, η, and T, in combination with Eq. (12a) or (12b), permits the calculation of a and b (=c).

Since Eqs. (10a) and (10b) show that the dielectric and loss increments yield identical information about molecular parameters, reference shall be made henceforth only to the dielectric (real) increment and we shall drop the primed notation for convenience. Equation (10a) may be rewritten in a form appropriate for a prolate ellipsoid of rotation,

$$\Delta\delta_L = \frac{\pi N}{500\,kT}\left(\frac{\mu_1^2}{1+\omega^2\tau_1^2} + \frac{\mu_2^2}{1+\omega^2\tau_2^2}\right) \tag{13}$$

where μ_1 is the component of the dipole moment along the long axis of the molecule and μ_2 is the component perpendicular to the long axis.

If the plot of δ versus frequency may be resolved into two Debye relaxations in this region (Fig. 3), then

$$\Delta\delta_L = \Delta\delta_1 + \Delta\delta_2 \tag{14a}$$

where

$$\Delta\delta_1 = \delta(\omega < \omega_1) - \delta(\omega_1 < \omega < \omega_2) \tag{14b}$$

and

$$\Delta\delta_2 = \delta(\omega_1 < \omega < \omega_2) - \delta(\omega > \omega_2) \tag{14c}$$

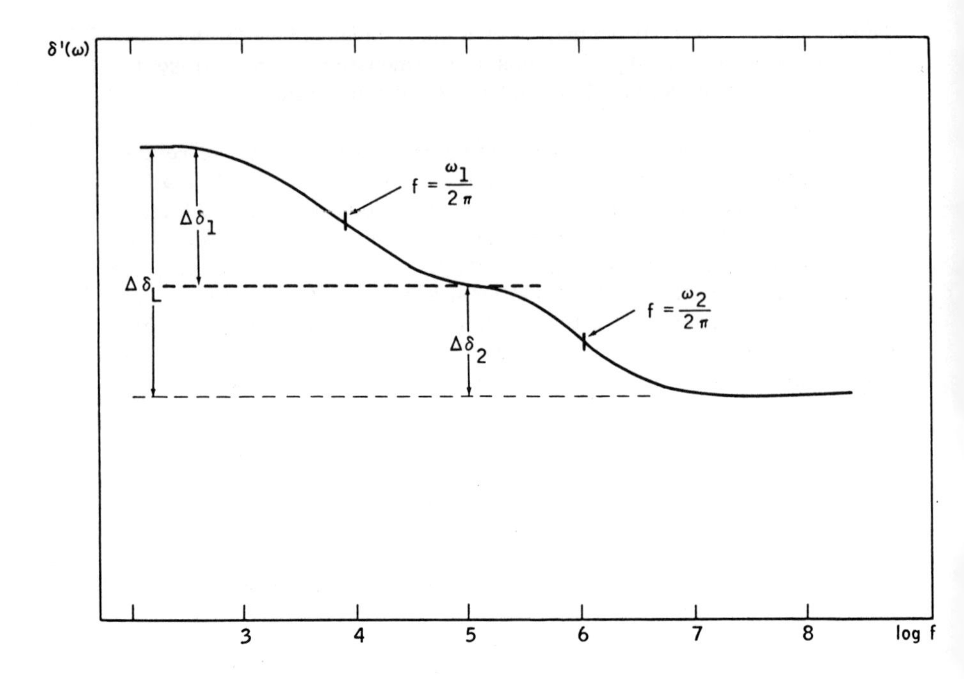

Fig. 3. A low-frequency dispersion displaying two widely-separated relaxation times. ω_i is the inverse of τ_i, the ith relaxation time.

If the two dispersions merge, these definitions do not suffice; but methods exist for extracting the individual Debye-type contributions (and therefore $\Delta\delta_1$ and $\Delta\delta_2$) from the total curve which is the sum of these contributions. From Eq. (13) it can be seen that

$$\Delta\delta_2/\Delta\delta_1 = \cos\theta \tag{15}$$

where θ is the angle between the dipole vector and the long axis of the prolate ellipsoid.

In summary, if the classical rotational relaxation model applies to proteins in aqueous solution, the low-frequency dispersion can yield information under certain conditions as to the shape and dimensions of a hydrodynamically equivalent spheroid of revolution, the magnitude of the dipole moment, and in the case of the prolate ellipsoid, the direction of the dipole moment with respect to the long axis.

2. Alternate Interpretations

Since the early 1950's there have appeared several theories of polarization in macromolecular solutions that do not involve orientation of the macromolecule in the field. It has been proposed that these polarization mechanisms may be partially or completely responsible for the observed dielectric behavior of macromolecular (and hence protein) solutions in the low-frequency dispersion region.

The treatments may be classified into two groups: Maxwell-Wagner and fluctuation theories. Maxwell-Wagner polarization arises at a boundary between two dielectric media of differing dielectric constant and conductivity. Since the current density induced by the applied electric field in the two media differs, a time-dependent charge appears at the interface. It is the relaxation of this boundary polarization that gives rise to the observed dispersion. Fluctuation phenomena, on the other hand, are due to the dynamic nature of the equilibrium between charged side groups of the macromolecule and counterions (including protons) in the solvent, leading to fluctuations in the electric moment of the molecule.

a. Maxwell-Wagner theories. The Maxwell-Wagner effect has been treated in a variety of heterogeneous systems. An extensive review by van Beek on the general topic of Maxwell-Wagner effects has recently appeared (18), and may be consulted for a more detailed approach. From equations given by van Beek, which are valid at low concentration, one would expect a suspension of ellipsoids of axial ratio a/b, with their a axes oriented along the field axis, to display a Debye-type dispersion in the dielectric increment with magnitude and relaxation time given by

$$\Delta\delta_L = \frac{\epsilon_1(\sigma_2 - \sigma_1)}{\sigma_1 + A_a(\sigma_2 - \sigma_1)} + \frac{\sigma_1\left\{\left[\sigma_1 + A_a(\sigma_2 - \sigma_1)\right](\epsilon_2 - \epsilon_1) - \left[\epsilon_1 + A_a(\epsilon_2 - \epsilon_1)\right](\sigma_2 - \sigma_1)\right\}}{\left[\sigma_1 + A_a(\sigma_2 - \sigma_1)\right]^2} - \frac{\epsilon_1(\epsilon_2 - \epsilon_1)}{\epsilon_1 + A_a(\epsilon_2 - \epsilon_1)} \tag{16a}$$

$$\tau = \epsilon_o \frac{\epsilon_1 + A_a(\epsilon_2 - \epsilon_1)}{\sigma_1 + A_a(\sigma_2 - \sigma_1)} \tag{16b}$$

Here ϵ_1 and σ_1 are the dielectric constant and conductivity of the suspending media, ϵ_2 and σ_2 are the same for the ellipsoids, ϵ_o is the absolute permittivity of free space, and A_a is a quantity called the depolarizing factor along the a axis. A_a is a function of the axial ratio and has a value ranging from zero to one which may be obtained from published formulas, graphs, and tables [see, for example, O'Konski (19)]. Fricke (20) has shown that a suspension of randomly oriented ellipsoids will behave as an even mixture of three Debye species, providing that the ellipsoids are internally isotropic ($\epsilon_{2a} = \epsilon_{2b} = \epsilon_{2c}$ and $\sigma_{2a} = \sigma_{2b} = \sigma_{2c}$). For ellipsoids of rotation, two of the contributions will be identical (b and c axes in the field direction) and will be functions of A_b, the depolarizing factor along the b axis, rather than A_a. For spheres, all three contributions will be identical and $A_a = A_b = 1/3$.

O'Konski (19) has developed a model of counterion polarization by introducing a thin (with respect to the dimensions of the ellipsoid) conducting layer surrounding the ellipsoid. This layer is characterized by a conductivity λ which is a function of the number, charge, and mobility of charge carriers in the counterion atmosphere (conducting layer). O'Konski found that the behavior of a suspension of ellipsoids with conducting layer oriented with their a axes along the field direction is described by Eqs. (16a) and (16b) except that σ_2 is replaced by σ_{2a}, where

$$\sigma_{2a} = \sigma_2 + \frac{2\lambda}{b} \tag{17}$$

In general the surface contribution to the total ellipsoidal conductivity is a function of the dimensions of the ellipsoid and its orientation with respect to the field axis. O'Konski also obtains the expressions for σ_{2b} to be used in calculating the properties of randomly oriented ellipsoidal suspension in a matter entirely analogous to that used for simple Maxwell-Wagner dispersion.

Schwarz (21) has increased the generality of this model by pointing out that the charge carriers in the counterion layer have a mobility that is limited by the activation energy of transfer from one equivalent binding site to the next and rates of diffusion between binding sites. As the applied field frequency increases, a phase lag will begin to appear at some point between the field and counterion response. Consequently he replaced the real surface conductivity of O'Konski by a frequency-dependent surface conductivity. The resulting analysis indicates that a suspension of spheres with a counterion layer would exhibit dielectric behavior according to the Maxwell-Wagner Eqs. (16a) and (16b) except that ϵ_2 is replaced by $\bar{\epsilon}_2$ and σ_2 is replaced by $\bar{\sigma}_2$, where

$$\bar{\epsilon}_2 = \epsilon_2 + \frac{1}{1 + \omega^2 \tau_2^2} \frac{e^2 R \rho_0}{\epsilon_0 kT} \tag{18a}$$

and

$$\bar{\sigma}_2 = \sigma_2 + \frac{\omega^2 \tau_2^2}{1 + \omega^2 \tau_2^2} \frac{e^2 R \rho_0}{kT} \tag{18b}$$

Here τ_2 is the relaxation time associated with the counterion diffusion, e is the electronic charge, R is the radius of the sphere, and ρ_0 is the mean counterion surface density. Schwarz has observed that his equations are equivalent to those of O'Konski in the region above the dispersion associated with the surface layer.

It is important to note that in the derivations of the Maxwell-Wagner formulas (Eqs. 16a and 16b) it was assumed that the dielectric properties of the individual components 1 and 2 were independent of frequency in the region where Maxwell-Wagner dispersion is observed in the suspension. If either of the components (or, additionally in the Schwarz treatment, the ion atmosphere) displays frequency-dependent dielectric behavior in this region, then Eqs. (16a) and (16b) do not necessarily hold and a more complex situation may result depending on the relative magnitudes of the dielectric parameters involved. However, Schwarz (21) has shown that experimental conditions may be adjusted to simplify the theory considerably.

b. Fluctuation theories. In 1952 Kirkwood and Schumaker (22) presented a statistical treatment of the effect of fluctuations in the dipole moment due to the dynamic nature of the equilibrium between basic side chains (for example, $-NH_3$, $-COO^-$) and their conjugate acidic forms. Except in highly acidic solutions, the number of basic sites would be expected to exceed the number of available protons, and consequently many different proton configurations of approximately equal energy exist in the absence of an applied field. However, in an applied field some of these configurations are of lower energy than others and hence preferred.

Kirkwood and Schumaker replaced $\mu \cdot \bar{\mu}$ in Eq. (7) by $\langle \mu^2 \rangle$, the average squared dipole moment, which is composed of two terms,

$$\langle \mu^2 \rangle = \langle \mu \rangle^2 + \Delta \mu^2$$

where $\langle \mu \rangle$ is the mean permanent moment and $\Delta \mu^2$ is the fluctuation contribution, given by

$$\Delta \mu^2 = \langle (\mu - \langle \mu \rangle)^2 \rangle$$

By taking into account the polarizing effect of the applied field and averaging over all proton configurations (neglecting proton-proton repulsion), they obtained the approximate relation,

$$\Delta \mu^2 = e^2 \sum_{\alpha} \frac{\nu_\alpha R_\alpha^2}{2 + K_\alpha / [H^+] + [H^+] / K_\alpha} \tag{21}$$

where e is the electronic charge, ν_α is the number of basic groups of species α, R_α^2 is the mean squared distance of these groups from the hydrodynamic center of the molecule, K_α is the intrinsic dissociation constant of the acid conjugate to the basic group, and $[H^+]$ is the proton concentration. Equation (21) requires correction for proton-proton repulsion as the proton concentration increases.

Using data on the pKs and numbers of different dissociating groups per molecule obtained from titration data, they evaluated this expression for several protein molecules, for spherical and ellipsoidal models on the assumption that charged groups of each type were spread uniformly over the surface of the molecule. Their results indicated that the fluctuation moment could account for most or all of the experimentally observed increment.

Working from the premise that the fluctuations in a very weak applied field are of the order of thermodynamic fluctuations, Takashima (7) has obtained an estimate of the fluctuation moment by considering the thermodynamic fluctuation in the occupation number at a single site (defined to be 1 when there is a proton associated with the site and 0 otherwise) and then summing over all sites,

$$\Delta\mu^2 = ekT \sum_\alpha \phi_\alpha \left[\frac{\nu_\alpha R_\alpha}{[H^+] + K_\alpha} \right]^2 \tag{22}$$

where ϕ_α is the number of sites of type α per unit surface area.

Recently Scheider (23) has examined the frequency-dependent behavior of a system with fluctuating dipole moment. By considering in detail the character of the relaxation of the fluctuating moment he has obtained the general expression,

$$\langle \mu(\omega)^2 \rangle = \sum_{i=1}^{3} \left\{ \bar{\mu}_i^2 \left(\frac{1}{1 + i\omega\tau_i} \right) + \frac{\overline{\delta\mu_i^2}}{1 + \tau_i/\tau_\delta} \left(\frac{1}{1 + i\omega\tau'_i} \right) \right\} \tag{23}$$

where $\bar{\mu}_i$ is the mean permanent dipole moment in direction i of the molecular internal coordinates, $\overline{\delta\mu_i^2}$ is the mean square amplitude of the fluctuation along i, τ_i is the rotary diffusion time of the molecule, τ_δ is a fluctuation time constant which will be defined below, and $\tau_i' = \tau_i\tau_\delta / (\tau_i + \tau_\delta)$. Scheider obtained the following expression relating τ_δ to the rate constants of proton association k_{21} and dissociation k_{12}:

$$\tau_\delta = \frac{1}{k_{21} + [H^+]\, k_{12}} \tag{24}$$

By considering the major contributing dissociation reactions in bovine serum albumin (BSA) at the isoionic point (the pH at which the molecule bears no net charge), Scheider concluded that in the case of BSA Eq. (23) reduces to

$$\langle \mu(\omega)^2 \rangle = \sum_{i=1}^{3} \left[\left(\bar{\mu}_i^2 + \overline{\delta\mu_i^2} \right) \left(\frac{1}{1 + i\omega\tau_i} \right) \right] \tag{25}$$

In this case the rotational relaxation times for the molecule are apparently unaffected by the fluctuation process, and in any case can be shifted at most by a factor of two toward smaller relaxation times.

c. Electrophoretic effect. The preceding polarization mechanisms can be applied to protein molecules bearing no net charge. However, a study of the dielectric properties of proteins away from the isoionic point must also include consideration of the possibility of effects arising from translational as well as rotational movement of the molecule in the alternating field.

As the applied frequency increases, at some point there will appear a phase lag between the field and the translational response of the molecules due to frictional forces acting on them, and consequently a dispersion will be observed if its magnitude is sufficiently great. This effect was treated by Schwan et al. (24) in connection with an experimental study of microscopic charge-bearing particles. They found a negative contribution to the dielectric increment at frequencies below the electrophoretic dispersion. An order of magnitude calculation for spheres of 2×10^{-7} cm radius reveals that even at moderately high charges the electrophoretic contribution to the dielectric increment is several orders of magnitude smaller than observed increment values throughout the low-frequency dispersion region.

3. Attempts to Distinguish between the Various Proposed Polarization Mechanisms

Following the appearance in the literature of alternative relaxation mechanisms (and hence possible alternate interpretations of the data) considerable confusion arose as to the actual relaxation process or processes responsible for a given set of data. Consequently doubt arose as to the validity of the structural parameters obtained from the classical Debye-rotor model. In recent years there have been several attempts to determine whether a particular polarization process is operative in a particular protein solution or to establish experimental conditions in which relaxation mechanisms other than orientation relaxation are inoperative.

The first such study was performed by Takashima (25) who studied the low-frequency dispersions of hemoglobin (Hb), ovalbumin, and human serum albumin (HSA) as a function of temperature below the freezing point. The behavior of all three proteins was qualitatively the same. The data were combined with those from earlier studies on these proteins above the freezing point. No sudden discontinuities were observed at the freezing point, but the dielectric increment of Hb gradually decreased by a factor of five or six and that of the albumins by two or three as the temperature was lowered from +10°C to -20°C. The relaxation time of Hb decreased by about a factor of four and those of the albumins by a factor of two in this same temperature range. Whereas the decrease of increment would seem to correlate with a progressive immobilization of rotors as the protein molecules are "locked" into an ice matrix, the decrease of relaxation time definitely does not. Furthermore, qualitatively the same behavior was observed in moist crystals of oxyhemoglobin, in which a rotational relaxation would appear to be out of the question. Takashima found that he could qualitatively account for the observed increments using the Kirkwood-Schumaker model. It should be mentioned that Takashima's data as reported in the literature are somewhat less complete than might be desired. He failed to indicate either the salt concentration (or ionic strength) of the solution or the static conductivity and its dependence on temperature. He did not report whether the dielectric properties were concentration dependent in the concentration range studied and no data are given in the important temperature region between +4°C and -4°C.

Takashima (26) studied the low-frequency dispersions of several proteins in solution as a function of solvent viscosity but using as solvent glycerin-water mixtures of different composition to which was added sufficient electrolyte to maintain the conductivity at the level of distilled water. As the viscosity increased, the dielectric increment dispersions of the globular proteins metmyoblobin, ovalbumin, and hemoglobin were found to shift to lower frequencies and broaden. The last two of these proteins exhibited two subisidary dispersions at higher viscosities, indicating that the broadening is due to the gradual separation of a larger viscosity-dependent dispersion and a smaller dispersion that is approximately independent of viscosity. The viscosity-dependent relaxation times of the three proteins were found to vary with viscosity in the manner predicted by Stokes' law. Takashima interpreted this as evidence that at least the greater part of the observed increment and relaxation is due to Debye-rotor behavior. He also found that the dielectric behavior of a larger protein, catalase, did not vary significantly with viscosity, and interpreted this as evidence that some other relaxation mechanism was dominant in this case.

In order to assess the contribution due to Maxwell-Wagner effect, Takashima (27) studied the dispersion of ovalbumin as a function of added ions. He observed that those ions which are known from other studies not to bind with protein (NH_4^+, $N(CH_3)_4^+$) had a negligible effect on the dispersion up to concentrations of 2 and 5 x 10^{-3} M, respectively. Some ions (Na^+, K^+, Hg^{2+}) lowered the increment slightly but did not affect the relaxation time, while others (Ag^+, Cu^{2+}, H^+, Ca^{2+}, Mg^{2+}) affected both the dielectric increment and the relaxation time in varying degrees but always by lowering the increment and decreasing the relaxation time. The Ca^{2+} ion was found to exhibit the greatest effect. By use of purified salts the change in pH with addition of salt was kept to a minimum (ΔpH < 0.3).

Takashima concluded that Maxwell-Wagner behavior was not the basis of the observed dispersion since those ions that do not bind to the protein did not affect the dispersion but did affect the conductivity of the suspending phase (Eqs. (16a) and (16b)). However, it should be pointed out that the magnitude of the dependence on solvent conductivity is related to the relative magnitudes of the dielectric parameters ϵ_1, ϵ_2, σ_1, and σ_2. (For example, if $\sigma_1 \ll \sigma_2$, then Eq. (16a) reduces to a form independent of σ_1.) Takashima's analysis of this dependence is incomplete as presented, but a simple argument may suffice here: For the case of a vanishingly small concentration of spheres, the Maxwell-Wagner formula for the relaxation time, Eq. (16b), reduces to

$$\tau = \epsilon_o \frac{2\epsilon_1 + \epsilon_2}{2\sigma_1 + \sigma_2} \tag{26}$$

It will be seen that unless $\sigma_2 \gg 2\sigma_1$, there will be a substantial dependence of the relaxation time on σ_1. In the absence of ions that bind to the surface of the protein it is difficult to see how σ_2 could greatly exceed σ_1 or even approach it, since the volume conductivity of protein is extremely low. Thus an increase in σ_1 by a factor of four should result in an observable decrease in the relaxation time that is not observed. Hence Takashima's conclusion stands with regard to simple Maxwell-Wagner effects.

However, no account was taken of either the generalizations of O'Konski (19) or Schwarz (21). For example, in the model of O'Konski Eq. (26) becomes

$$\tau_i = \epsilon_o \frac{2\epsilon_1 + \epsilon_2}{2\sigma_1 + \sigma_2 + 2\lambda/R} \tag{27}$$

where λ is the magnitude of the surface conductance and R is the radius of the equivalent sphere. Thus as R becomes smaller, $2\lambda/R$ may become large with respect to σ_1 and consequently there would not be a dependence of the relaxation time on σ_1 unless λ were a function of σ_1. In order to properly assess the meaning of Takashima's data with respect to the two more recent models, it will be necessary to numerically evaluate the expressions for $\delta(\omega_a) - \delta(\omega_b)$ and τ given by the O'Konski and Schwarz treatments and to evaluate, if possible, the derivatives of these quantities with respect to the concentration of electrolyte as well. This procedure will require a knowledge of the dependence of the number and mobility of ions in the counterion shell on the concentration of ions in bulk solution.

Lumry and Yue (28) have discussed the problem of avoiding effects due to polarization of the counterion atmosphere. Ion-atmosphere effects will diminish at lower salt concentrations, but it is known that the absence of counterions has a large effect on the structure and presumably the function of protein as well as on its hydrodynamic properties. High salt concentrations would remove these difficulties but technical considerations do not allow accurate measurement of dielectric dispersion in highly conducting liquids. Lumry and Yue attempted to circumvent these difficulties by measuring the dispersions of several proteins in solutions of 0 to 2% glycine and β alanine. These amino acids, which are zwitterionic (bearing both positive and negative charges) in the mid-pH range, provide an ionic environment for the protein similar to that of singly charged ion, but are not transported in the electric field and consequently do not introduce ion-atmosphere polarization or increase the conductivity of the solution. (They do contribute an orientation polarization but the dispersion region is orders of magnitude higher in frequency than that of the protein, Aaron and Grant (29).)

The authors found that the dispersion parameters $\Delta\delta_L$ and τ of the proteins metmyoglobin and carbonmonoxymyoglobin were not significantly altered by addition of 2% zwitterion, except that the dispersion data could be better fitted to a single relaxation time. The addition of zwitterion to chymotrpysinogen A lowered the relaxation time by about 10% and significantly reduced the dispersion of relaxation times. Reduced and oxidized cytochrome C were also examined but not at the isoionic point. Considerable concentration effect was observed in both the relaxation time and increment.

Lumry and Yue (28) present a considerable selection of data on these various preparations. However, there is simultaneous variation in protein concentration, salt concentration, and pH in addition to the presence or absence of zwitterion. Thus it is difficult to say in many instances whether an observed change in the increment or relaxation time is due to the change in protein concentration, salt concentration, or pH. The authors concluded that the comparison of behavior with and without added zwitterion provides information as to whether ion-atmosphere polarization or structural alterations in the absence of salt affect the dielectric dispersion. The zwitterion method does not eliminate the contribution from fluctuation moment, but Lumry and Yue estimated, using a kinetic argument, that proton transfer is generally too slow or too improbable (due to concentration factors) to provide a fluctuation moment comparable to the magnitude of the permanent moment.

In an attempt to check experimentally the predictions of the Kirkwood-Schumaker theory (22), Takashima (7) studied the pH dependence of the dielectric increment $\Delta\delta_L$ of ovalbumin and BSA. The theory predicts a characteristic dependence of fluctuation moment on pH which can be approximately calculated knowing the number and kind of titratable groups on the protein and making a rough estimate as to its shape (axial ratio). Taking special precautions to reduce electrode polarization arising from increased conductivity at pH values away from seven, Takashima succeeded in obtaining data in the region between pH 3 and pH 10. The data did not follow the predictions of the Kirkwood-Schumaker theory (22). However, Takashima's modification of the theory [Eq. (22)] predicts a pH dependence more in harmony with observed results. Thus he concluded that the fluctuation moment may indeed be a large contribution to the observed moment in contrast to the conclusions of Lumry and Yue (28).

Goebel and Vogel (30) obtained precise data on the low-frequency dispersion of deionized met-hemoglobin solutions in H_2O and D_2O as a function of temperature between 5^oC and 35^oC and hydrogen ion concentration between pH 6.3 and pH 7.4. The authors examined possible polarization mechanisms, including orientation polarization of permanent dipoles, fluctuation dipole moments, polarization of H-bonds in the protein, electronic polarization of atoms, and molecular polarizability due to displacement of surface charges (ion-atmosphere effects). The incremental contribution arising from each mechanism was considered and it was found that under the experimental conditions prevailing in this study, the total contribution of all mechanisms except the first two mentioned, using the maximum reasonable value of each, would provide less than one thirtieth of the observed increment. The viscosity dependence of the relaxation time (viscosity here changing with temperature) agreed with that predicted by Stokes' law for a rotating ellipsoid. The distribution of relaxation times observed was attributed to a distribution of microscopic viscosity values. Goebel and Vogel (30) concluded that the observed increment is, in fact, due to the orientation of a protein dipole moment, which is composed of permanent and fluctuation moments. An assessment of the relative contributions of these two components was postponed to a later publication, which has not appeared in the literature as of the date of writing (Sec. V, Addendum 1).

4. Applications of the Dielectric Method

No attempt will be made to summarize here the large number of studies on the low-frequency dielectric dispersion of proteins in solution which appeared in the literature before 1955. Instead the interested reader is referred to the extensive review by Oncley (13) and for later contributions to the review by Smyth (31). The data in all of the studies mentioned in these reviews have been interpreted in the classical way to obtain values for shape and size parameters of the various proteins.

A recent structural study is that of Moser et al. (6) who used the dielectric method in conjunction with birefringence relaxation techniques in order to study the properties of BSA at the isoionic point in a deionized solution. Fractionation techniques were employed to prepare highly homogeneous solutions of monomeric, dimeric, and oligomeric species. Measurements were made as a function of concentration and temperature, and all results were extrapolated to infinite dilution. The dielectric constant or increment dispersions were analyzed in terms of a two-term Debye relaxation according to the equation

$$\frac{\delta(\omega) - \delta(\omega_a)}{\delta(0) - \delta(\omega_a)} = \sum_i \frac{A_i}{1 + \omega^2 \tau_i^2} \tag{28}$$

where A_i is the fractional contribution to the increment associated with relaxation i. From the ratio and magnitude of the relaxation times it is possible to calculate the dimensions of an equivalent ellipsoid of rotation, subject to the conditions and limitations reviewed in Sec. II.B.1, but the authors found that slight variation in the results obtained from the dielectric method led to substantial variation in the parameters so obtained. In order to reduce the uncertainty, the longitudinal relaxation time was fixed independently by measurement of the birefringence relaxation. Thus the dielectric curves could be interpreted satisfactorily, yielding a well-defined axial ratio and half-axial dimensions. With these results the value of $\delta(\omega_a)$ was deduced to obtain an estimate of bound water by a method to be discussed in Sec. II. C. 1. The authors considered contributions from counterion migration in the field and calculated that this effect did not contribute to the observed dispersion. They concluded that the observed moment was due to the sum of permanent and fluctuation moments and argued from qualitative considerations that the latter was only a minor effect. On the basis of general agreement with size and shape parameters obtained from other methods, the authors concluded that the classical interpretation of relaxation times was justified in this case (Sec. V, Addendum 2).

Several studies have appeared in which the objective was not to obtain absolute values of size and shape parameters but rather changes in them as a result of variations in the protein environment.

Takashima and Lumry (5) measured the low-frequency dispersion of horse and bovine hemoglobin solutions as a function of the partial pressure of oxygen with which the hemoglobin was in equilibrium. They found a marked dependence of the increment, relaxation time, and Cole-Cole distribution parameter α (a measure of the spread of relaxation times) on the oxygen content of the system. All three of these quantities appeared to reach peak values at partial pressures corresponding to approximately 25% and 75% oxygen saturation, the ratio of oxygen bound to binding capacity. The authors were unable to interpret this phenomenon but suggested that the results indicated marked alterations in the molecular structure as oxygen was bound successively to each of the four binding sites in the molecule.

Recently, Hanss and Banerjee (8), in an attempt to resolve some of the contradictions inherent in Takashima and Lumry's earlier results, restudied the effect of oxygenation on the low-frequency dispersion of solutions of human and horse hemoglobin. The authors employed experimental techniques that allowed them to work at lower protein concentrations and higher ionic strengths than those of the earlier study and were able to measure the entire range of oxygenation in the same solution. Contrary to the results of the earlier study, Hanss and Banerjee found no significant dependence of the dispersion on the degree of oxygenation, although the precision of their method was greater. The results of Hanss and Banerjee have quite recently been confirmed by the careful study of Schlecht et al. (32) who measured the effect of oxygenation upon the low-frequency dispersion of horse hemoglobin prepared in a variety of ways and, in addition, performed a duplicate of the Takashima-Lumry experiment using the original instrumentation. They reported a slight monotonic increase in increment with increasing degree of oxygenation, but in no case was the effect reported by Takashima and Lumry observed.

Takashima (33) measured the low-frequency dispersion of deionized solutions of BSA and ovalbumin as a function of added urea, which is known to denature these proteins. He found that the dielectric increment of BSA increased by a factor of eight and the mean relaxation time increased by a factor of two in 8 M urea. The increment of ovalbumin increased by a factor of about 3.5 and the mean relaxation time by a factor of 10. The concentration dependence of these alterations was complex and differed considerably from BSA to ovalbumin. Hendricks et al. (34) restudied the urea dependence of the low-frequency dispersion of deionized BSA solution. Their results, which are in semiquantitative agreement with those of Takashima, indicate that the dielectric increment increases by about a factor of ten and the mean relaxation time increases by about a factor of three in the presence of concentrated urea. From a classical interpretation, Takashima (33) and Hendricks et al. (34) concluded that the urea denaturation of BSA is accompanied by an increase of both molar volume and charge assymmetry in the molecule. The changes in the properties of ovalbumin with added urea occur, unlike those of BSA, in a relatively small interval between 6 M and 8 M added urea. Takashima suggested that this might reflect aggregation rather than unfolding on denaturation of ovalbumin.

The low-frequency dielectric dispersion of acid-soluble tropocollagen has recently been studied by Hanss et al. (35) as a function of temperature. They found that at protein concentrations below 0.2 g/1, two distinct types of dielectric behavior could be distinguished. The dielectric increment of native collagen varied linearly with $f^{-1/2}$, where f, the frequency of the applied field, varied between 2×10^4 and 10^7 Hz. However, if the collagen solution was heated to above 60^oC for one hour and then recooled, a simple Debye-type dispersion with a critical frequency of the order of 10^6 Hz was observed. Native collagen is known to exist in a rodlike form composed of three polypeptide chains wound into a rigid triple helix. Other rodlike polyelectrolytes, such as DNA, display a similar dielectric behavior in this region, which is attributed to diffusion-controlled relaxation of counterion polarization along the length of the molecule. Although it is tempting to attribute the collagen results to the same mechanism, the authors pointed out that, unlike DNA, the dielectric increment of collagen was insensitive to variations in the amount of simple electrolyte (KCl) present in solution. It was suggested that the observed frequency dependence could be interpreted as the high-frequency tail of a lower frequency (that is, 10^3 Hz) dispersion about a specified distribution of orientation relaxation times.

Since dielectric dispersion is marked by both changes in capacitance and conductance, mention should be made here of studies of the conductance dispersion of protein solutions by Lapaeva (36, 79). It was reported that the conductance of several protein solutions exhibited dispersion (toward higher values) at characteristic frequencies ranging from 10^2 to 3×10^6 Hz. However, on examination of Lapaeva's plots of conductivity versus frequency it is immediately apparent that these dispersions are far too narrow to be described by a Debye equation, and hence cannot be attributed to exponential decay of polarization of any type. It would appear that what is being observed is an artifact, or if not, an entirely different (and as yet untreated) electric phenomenon. On the chance that these dispersions may be real, perhaps this work deserves to be repeated.

C. Mid- and High-Frequency Dispersions

Although the number of proteins that have been studied above the low-frequency dispersion is small, recent studies have definitely established a mid-frequency dispersion characterized by an increment $\Delta\delta_M = \delta(\omega_a) - \delta(\omega_b)$ considerably smaller than those characterizing the low- and high-frequency dispersions. This dispersion cannot be described by a Debye equation with one or a small number of relaxation times but rather requires a broad distribution of relaxation times. The dispersion has been attributed to orientation relaxation of either rotatable side chains of the protein and/or water which is associated with the protein in such a manner as to allow rotation, but with a higher activation energy resulting in a longer relaxation time. [For a brief discussion of orientation relaxation as a rate process, see Glasstone, Laidler, and Eyring (80).]

Although this water has been called "bound water" in the literature, it seems more appropriate to refer to it as associated water, to make clear the distinction between this species and irrotationally bound water, which is so tightly bound to the protein that its relaxation time is effectively infinite (that is, the molecules are not free to rotate in the field, independent of the protein, even at the lowest frequencies). There already exist in the literature examples of confusion of the two types of bound water. (For example, Grant (12) has compared his estimate of the amount of associated water per gram protein with Rosen's (37) estimate of irrotationally bound water in VSA. Rosen, in turn, has compared his estimate with that of Buchanan, et al. (38). However, at the ultrahigh frequencies of measurement in the last study, associated water becomes effectively irrotationally bound as well (Sec. II.C.2). Thus Buchanan et al. have measured the total contribution of both, that is, water which is truly irrotationally bound and water which is defined here as associated water. It is important to note that only the mid-frequency dispersion distinguishes irrotationally bound from associated water.)

1. The Determination of Irrotationally Bound Water

From Eq. (6b) it may be seen that the dielectric increment above the low-frequency dispersion $\delta(\omega a)$ is a function of the number of irrotationally bound water molecules per protein molecule as well as the parameters α_1, α_2, and $\mu_2 \cdot \bar{\mu}_2/3KT$. These last three may be independently evaluated experimentally as follows:

1. The expression $(\alpha_2 + \mu_2 \cdot \bar{\mu}_2/3kT)$ can be evaluated at a particular temperature from the static dielectric constant of pure water and the Kirkwood equation, Eq. (2).

2. The value of α_2 may be obtained from the dielectric constant of water at a frequency well above its orientation polarization dispersion (for example, 10^{12} Hz) and the Clausius-Mossotti equation,

$$\frac{\epsilon - 1}{\epsilon + 2} = \frac{4\pi N}{3\tilde{V}} \alpha_2 \tag{29}$$

where N is Avogadro's number and $\tilde{V}$ is the molar volume of water.

3. Knowing the value of α_2, the value of α_1 may be obtained by measurements of the dependence of the dielectric constant of a protein solution on the protein concentration at very high frequencies (for example, 10^{12} Hz) and the following Clausius-Mossotti equation for mixtures:

$$\left[\frac{\partial\left(\frac{\epsilon - 1}{\epsilon + 2}\right)}{\partial c_1}\right] = \frac{4\pi N}{3000}\left[\alpha_1 - \frac{\tilde{V}_1}{\tilde{V}_2}\alpha_2\right] \tag{30}$$

As an alternative to the foregoing procedures, Eq. (6b) may be considerably simplified by assuming that both α_1 and $n\alpha_2$ are very small with respect to

$$\left(\frac{\tilde{V}_1}{\tilde{V}_2} + n \right) \left[\alpha_2 + (\mu_2 \bar{\mu}_2/3kT) \right]$$

With this assumption we can obtain directly from Eq. (6b) and the Kirkwood equation for pure water,

$$\delta(\omega_a) = - \left(\frac{\tilde{V}_1}{\tilde{V}_2} + n \right) \frac{\tilde{V}_w}{1000} \left(\epsilon_w(0) - 1 \right) \tag{31}$$

where $\epsilon_w(0)$ is the static dielectric constant of water at the same temperature and $\tilde{V}_w$ is the molar volume of water.

Using essentially the same reasoning but different parameters, Oncley (13) obtained an equation equivalent to Eq. (31) which he used in conjunction with measured values of $\delta(\omega_a)$ to obtain estimates of the amount of irrotationally bound water in various proteins (Oncley (39)), which ranged from 0.1 to 0.3 g H_2O/g protein.

2. The Determination of Associated Water

If we assume that the mid-frequency dispersion is entirely due to rotational relaxation of associated water, we can estimate the amount of associated water and its mean relaxation time. However, if there is a nonnegligible contribution from rotational relaxation of rotatable side chains, the problem becomes much more complex. By and large, investigators of the mid-frequency dispersion have preferred to side step this difficulty by adopting the initial assumption on the grounds that "reasonable" results may be so obtained.

If the assumption is valid, then the amount of associated water may be calculated in two ways. The first method is based on Kirkwood's analysis. Equation (6b) may be rewritten to include the contribution from associated water,

$$\delta(\omega_a) = \frac{6\pi N}{1000} \left\{ \alpha_1 + n_1 \alpha_2 + n_2 \left[\alpha_2 + \frac{\mu_2 \cdot \bar{\mu}_2}{3kT} \right] - \left(\frac{\tilde{V}_1}{\tilde{V}_2} + n_1 + n_2 \right) \left[\alpha_2 + \frac{\mu_2 \cdot \bar{\mu}_2}{3kT} \right] \right\} \tag{32a}$$

where n_1 is the number of irrotationally bound water molecules, n_2 is the number of associated water molecules per molecule of protein, and ω_a is a frequency above the low-frequency dispersion of the protein and below the dispersion due to associated water (Fig. 1). This equation assumes that below its dispersion, associated water contributes an orientation polarization per unit volume close to or equal to that of free (bulk) water, a conclusion that has been arrived at through different lines of reasoning by Schwan (40) and Grant (41). If it is further assumed that the mid-frequency dispersion corresponds to the progressive decline with increasing frequency of the orientation polarization contribution of associated water, then

$$\delta(\omega_b) = \frac{6\pi N}{1000}\left\{ \alpha_1 + (n_1 + n_2)\,\alpha_2 - \left(\frac{\tilde{V}_1}{\tilde{V}_2} + n_1 + n_2 \right) \left[\alpha_2 + \frac{\mu_2 \bar{\mu}_2}{3kT}\right]\right\} \tag{32b}$$

where ω_b is a frequency above the dispersion due to associated water and below that due to bulk water (Fig. 1). The difference between the quantities represented by Eqs. (32a) and (32b) is then

$$\Delta\delta_M = \delta(\omega_a) - \delta(\omega_b) = \frac{6\pi N}{1000}\, n_2\left[\frac{\mu_2 \cdot \bar{\mu}_2}{3kT}\right] \tag{33}$$

Since

$$\left[\frac{\mu_2 \bar{\mu}_2}{3kT}\right] \approx \left[\alpha_2 + \frac{\mu_2 \cdot \bar{\mu}_2}{3kT}\right]$$

Eq. (33) may be combined with the Kirkwood Eq. (2) for pure water to obtain

$$\Delta\delta_M = \frac{\tilde{V}_w}{1000}\, n_2\, (\epsilon_w(0) - 1) \tag{34}$$

A second method employs a formula for the high-frequency permittivity of heterogeneous mixtures derived by Fricke (20) as given by van Beek (18). For a suspension of randomly oriented ellipsoids of rotation of dielectric constant ϵ_2 in a suspending media of dielectric constant ϵ_1,

$$\epsilon - \epsilon_1 = \frac{\epsilon_1 V}{3}\left[\frac{\epsilon_2 - \epsilon_1}{\epsilon_1 + A_a(\epsilon_2 - \epsilon_1)} + \frac{2(\epsilon_2 - \epsilon_1)}{\epsilon_1 + A_b(\epsilon_2 - \epsilon_1)}\right] \tag{35}$$

where A_a and A_b are the depolarizing factors along the axes of the ellipsoid (Sec. II.B.2a) and V is the volume fraction of ellipsoids. The validity of this equation is limited to the region $V < 0.1$. At the frequency ω_a, ϵ_1 is the static dielectric constant of water and ϵ_2 is the dielectric constant of anhydrous protein above its orientation dispersion region. ϵ_2 may be calculated from the Clausius-Mossotti Eq. (29) and the value of α_2 obtained as described in Sec. II.C.1. An estimate of α_2 may also be obtained from a measurement of the dielectric constant of "dry" protein crystals.

At the frequency ω_b, ϵ_1 is still the same but ϵ_2 is now the dielectric constant of a protein ellipsoid with a shell of associated water that can no longer orient in the field, which is a function of the dielectric constants of both protein and water above their respective orientation dispersion regions. The exact form of this function is unimportant because since the two dielectric constants are similar, $\epsilon_2(\omega_b)$ is only slightly different from $\epsilon_2(\omega_a)$. This change can only vary the value of $(\epsilon_2 - \epsilon_1)$ by 1% to 2% at the most since ϵ_1 is so much greater. Substituting typical values of $\epsilon_w = \epsilon_1 \approx 78$ and $\epsilon_2 \approx 5$ into Eq. (35), we obtain

$$\epsilon_w - \epsilon = \epsilon_w V\left\{\frac{1}{3}\left[\frac{1}{1.07 - A_a} + \frac{2}{1.07 - A_b}\right]\right\} \tag{36}$$

It is seen that as the dielectric constant decreases with increasing frequency the apparent volume fraction of the suspended ellipsoids increases. This is due to the addition of a shell of immobilized associated water molecules which will be assumed to be of uniform thickness. Representing the quantity in curly brackets by B, the difference equation may then be written

$$\epsilon(\omega_a) - \epsilon(\omega_b) = \epsilon_w V_{Shell} \cdot B \tag{37}$$

Assuming that the molar volume of associated water is close to that of bulk water, it then follows that

$$\Delta\delta_M = \delta(\omega_a) - \delta(\omega_b) = \frac{\epsilon_w n_2 \tilde{V}_w}{1000} B \tag{38}$$

Using the table of depolarizing factors provided by O'Konski (19) the quantity B may be calculated as a function of axial ratio (Fig. 4).

It is interesting to note that although the two equations, Eq. (34) and (Eq. (38), were obtained in quite different ways — the first from Kirkwood's statistical treatment of fluctuations in the microscopic environment of a molecule, and the second from entirely macroscopic considerations of the Maxwell-Wagner effect — the two equations are of the same form; and for approximately spherical molecules, the values of n_2 calculated from them differ by less than 40%. This variance stems from the different approximations and assumptions made in deriving the two expressions and represents a theoretical uncertainty. As shall be shown shortly the experimental uncertainty in the measured value of $\Delta\delta_M$ is considerably greater.

3. Studies of the Mid-Frequency Dispersion

Grant (12, 41) has studied the mid-frequency dispersion in ovalbumin and Bovine Serum Albumin (BSA) solutions at the isoionic point. Because of the smallness of the dispersion increment, it was necessary

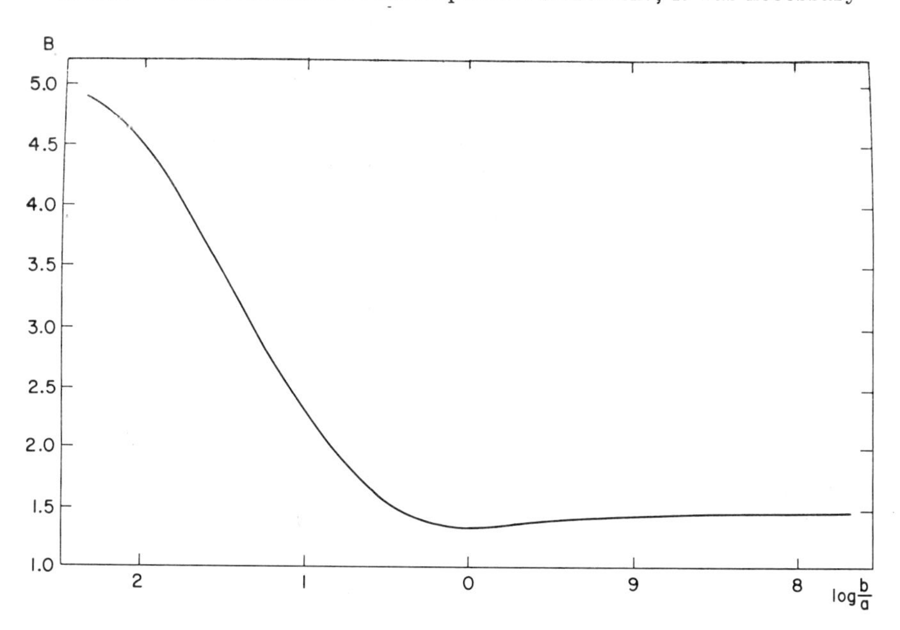

Fig. 4. Dependence of the parameter B occuring in equation (37) upon the axial ratio of the hydrodynamically equivalent ellipsoid of revolution.

to carry out these studies at relatively high protein concentrations (> 6 gprotein/100 ml solution). Since this is approximately 10 times the concentration of solutions used in the study of the low-frequency dispersion, we cannot assume that the increment is independent of the concentration. However, Grant found that for two solutions of BSA of 6.4% and 18% protein, the factor $\frac{(\epsilon - \epsilon_w)}{C}$ differed by less than 6%.

Since the dispersion was too broad to display clearly defined values of $\epsilon(\omega_a)$ and $\epsilon(\omega_b)$, Grant obtained them by extrapolation of data given by other workers for BSA solutions of the same concentration and pH in the low- and high-frequency regions. In the 6.18% BSA solution the values used were $\epsilon(\omega_a) = 74.9$ and $\epsilon(\omega_b) = 72.8$. There is a difficulty in this procedure that should be pointed out. Grant extrapolated his low-frequency data to obtain $\epsilon(\omega_a)$ using a dispersion relation based on a single relaxation time. The careful study of Moser et al. (6) indicated that such a dispersion relation was inadequate. Thus an uncertainty of at least 1% (the nominal accuracy of experimental measurement) and likely greater should be attributed to Grant's estimate of $\epsilon(\omega_a)$. Similarly, the high-frequency data have been estimated by the original investigators to be accurate to $\pm$ 2% to 3% and any extrapolation of this data is unlikely to be more precise. Thus $\Delta\epsilon_M = \epsilon(\omega_a) - \epsilon(\omega_b)$ may vary from 0 to 5. Since the amount of associated water is approximately proportional to $\Delta\delta_M$ (see Eqs. (34 and 38)), which is in turn proportional to $\Delta\epsilon_M$, only qualitative significance may be attributed to Grant's estimate of associated water (Sec. V, Addendum 3).

Schwan and his students Li (42) and Pennock (43) have studied the mid-frequency dispersion of hemoglobin solutions. It was observed that the increment increased (in a negative sense) almost linearly with frequency. The rate of negative increase decreased with increasing protein concentration above 10%. Such a frequency dependence is compatible with a broad range of relaxation times described by the distribution function $f(\ln\tau) = L$, where L is a constant, and the distribution is terminated at relaxation times corresponding ($\tau = 1/2\pi f$) to frequencies of 4×10^7 and 3×10^9 Hz, respectively.

Schwan has analyzed the data in terms of the orientational relaxation of both rotatable side chains (10) and associated water (40), and finds that the observed phenomena may reasonably be described by either mechanism alone or a combination of both. On the assumption that the mid-frequency dispersion is due entirely to associated water,

Schwan has calculated that the associated water has a low-frequency increment and relaxation time intermediate between those of ice and pure water, and estimates the associated water to be 0.3 ± 0.1 g/g Hb. Pennock's more recent treatment (43) yields a value of 0.2 ± 0.05 g/g Hb.

4. Studies of the High-Frequency Dispersion

The high-frequency dispersion observed in protein solutions is attributed to the rotational relaxation of bulk water in the solution water which is not affected by the presence of protein. Consequently it may be described by a Debye-type expression with a single relaxation time very close to that of pure water. The dielectric increment at the lower end of this region, $\delta(\omega_b)$, will be due to the percentage of volume occupied by the protein, together with its irrotationally bound and associated water (which is effectively irrotationally bound above the mid-frequency dispersion). The dielectric increment at the upper end, $\delta(\infty)$, will be due to electronic and atomic polarization. Hydration values calculated with the aid of Eq. (37) or equivalent expressions are dependent on B and hence the axial ratio. Figure 4 reveals that B is only sensitive to the axial ratio if the hydrodynamically equivalent ellipsoid of rotation is oblate rather than spherical or prolate. Hydration values calculated for an axial ratio of one represent maximum values.

The high-frequency dispersion of several proteins in solution was studied by Haggis et al. (44) and Buchanan et al. (38), who measured the dielectric constant and loss at three frequencies in this region. Their estimates of the total bound water at these frequencies, which corresponds to the sum of the associated and the irrotationally bound water, range from 0.05 ± 0.05 g H_2O/g protein for β lactoglobin to 0.18 ± 0.05 g H_2O/g protein for lysozyme.

If we compare the estimate of total bound water on hemoglobin by Buchanan et al (38) of 0.1 to 0.2 g H_2O/g Hb to the sum of the estimates of Oncley (39) and Pennock (43) for irrotationally bound (0.1 to 0.3 g H_2O/g Hb) and associated (0.15 to 0.25 g H_2O/g Hb) water, respectively, it is seen that the three estimates are inconsistent. Either one of the separate estimates is too high or the estimate of the sum is too low. Consequently a re-examination of the assumptions made in calculation and/or measurement seems to be called for. Perhaps the inconsistency could be resolved by attributing the mid-frequency dispersion largely to rotatable side chains instead of associated water, thus greatly reducing the estimate of the latter species.

Recently von Casimir et al. (45) have measured the difference in dielectric constant and loss of equimolar solutions of oxy- and deoxy-hemoglobin at 9.5 GHz in order to obtain information about the relative water binding properties of the two species. There are two ways in which the high frequency (microwave) dielectric properties of the solution may be varied by a change in the state of oxygenation of the hemoglobin: The configuration of the protein may change in such a way as to alter the volume fraction of hemoglobin and/or the number of irrotationally bound or associated water molecules per protein molecule may change due to alterations in the surface characteristics of the protein. In addition, the authors assert that changes in hydration can also shift the critical frequency of the bulk water (high-frequency) dispersion, thus altering the time constant appearing in the single-term Debye equation which they used to describe the dispersion. This assertion is, however, inconsistent with the authors' definition of water of hydration as water that is excluded from the polarization mechanism. If the water were entirely excluded, it could not contribute to the dispersion at all and its presence would manifest itself as a lowering of the magnitude of the bulk water dispersion. If, on the other hand, the water of hydration were not totally excluded but still capable of orientation relaxation with a somewhat longer time constant, the water dispersion could not be satisfactorily described by a one-term Debye equation. Thus the equation used by von Casimir et al. (45) to interpret shifts in critical frequency and dielectric increment in terms of molecular parameters appears inapplicable.

Von Casimir et al. (45) found that the dielectric constant of a 10.7% solution of deoxyhemoglobin decreased by 0.022 ± 0.005 units on oxygenation. Accepting the definition of water of hydration given by them, we find that the apparent volume fraction of hemoglobin (protein and total bound water) increases by 0.2% of its original value on oxygenation. The measurements do not permit a separation into changes in intrinsic protein volume and the amount of total bound water as a result of oxygenation.

III. DIELECTRIC AND DC CONDUCTION PROPERTIES OF DRY AND HYDRATED SOLID PROTEINS

A. Experimental Methods

The fundamentals of electrical measurements on solid synthetic macromolecules have been recently discussed by Forster (1). Because the electrical properties of solid proteins are in many ways similar to those of synthetic macromolecules, similar experimental techniques and instrumentation may frequently be employed.

It is not presently feasible to grow single protein crystals to a size permitting the direct attachment of electrodes to them. Early experiments were conducted on samples of dry gel but in general more uniform results were obtained with samples of compressed polycrystalline powder. Cardew and Eley (46) found that a compression of 80 kg/cm^2 was sufficient to provide ohmic contact of the electrodes with the sample and to reduce to a minimum the effects of intergranular resistance within the sample. A volume correction factor is required in order to calculate specific conductance.

B. Dielectric Properties

Bayley (47) measured the dielectric constant and loss of powdered insulin, salmine sulfate, bovine serum albumin (BSA), and chymotrypsinogen in the frequency range between 3×10^2 and 10^4 Hz. He found that after the proteins were dried in vacuum for several hours, their dielectric constants and losses decreased greatly, the dielectric constants reaching final values of about two. He attributed these decreases to the removal of atmospheric water that had been adsorbed on the powder particles.

Rosen (37) investigated the dielectric constant and loss of packed powders of BSA, silk fibroin, β casein, and whale myoglobin as a function of the amount of adsorbed water in the frequency range between 5×10^4 and 2×10^6 Hz. The dry samples were found to display frequency-independent dielectric constants between 1.25 and 2 and losses close to zero. Rosen found that both the dielectric constant and loss of a sample at a given frequency were approximately linearly dependent on the weight fraction of adsorbed water. Up to a "critical" hydration value this dependence was characterized by a very small frequency-independent slope. Beyond this value a new dependence, linear over the restricted range of his measurements, was established which displayed a large slope that decreased with increasing frequency. The critical hydration varied only slightly with frequency and this variation was within the limits of standard error. Furthermore this quantity was, within experimental error, very nearly the same for all four proteins measured, with a mean value of 0.24 ± 0.02 g H_2O/g protein. Rosen also found that in BSA this behavior was unchanged by denaturation of the protein, indicating that the phenomenon is not dependent on the secondary structure of the protein.

Rosen (37) proposed that the critical hydration is determined by the number of water molecules that may be bound sufficiently tightly to a protein molecule to prevent their contributing an orientation polarization. By assuming that one water molecule can be so bound to each peptide bond and polar side chain, with the aid of data on the amino acid content of the four proteins studied, Rosen was able to calculate values for the critical hydration in fair agreement with the experimental values. As Rosen pointed out, however, it is extremely unlikely that all such sites are accessible to water molecules. Rosen's observation that the critical hydration is nearly the same for all four proteins and is insensitive to denaturation is more consistent with the simpler hypothesis that the density of binding sites and mean size of the powder particles for all four proteins studied are approximately the same.

Takashima (48) and Takashima and Schwan (49) have measured the dependence of the dielectric properties on added water in the frequency range between 2×10^1 and 5×10^6 Hz in crystalline powders of ovalbumin and BSA. Their measurements cover a broader range of hydration values and dielectric constants and have less absolute precision than those of Rosen (37). However, the results for BSA appear to agree with those of Rosen in the region covered in both studies. As a result of the broader range of measurement a new feature was observed. As the added water content approached 30% by weight, the dielectric constant began to level off at around 175, which is considerably greater than the dielectric constant of pure water. Qualitatively similar behavior was observed in ovalbumin. The dry powder exhibited a very small dielectric constant of about two to three throughout the frequency range. Addition of water did not substantially affect the dielectric constant until the content of adsorbed water reached 10% to 11% by weight; then the dielectric constant began to increase relatively rapidly with further increases in adsorbed water content until at around 20% it began to level off at values greater than 200.

Brausse et al. (50) have recently measured the dielectric properties of powdered lyophilized horse hemoglobin as a function of hydration in the frequency range between 10^2 and 3×10^6 Hz. The results of this study are qualitatively similar to those mentioned in the preceding paragraph. In hemoglobin the static dielectric constant begins to increase rapidly at a water content of about 15% by weight and begins to level off at about 30% to values of 300 to 600 depending on temperature. These authors also measured the water vapor adsorption isotherms of the lyophilized hemoglobin and calculated that the critical hydration (that hydration value corresponding to the onset of strong dependence of dielectric constant on hydration) is equal to two to three times the hydration corresponding to a BET (Brunauer, Emmet, Teller) monolayer. (It is assumed here that water is diffused throughout the entire powder particle and not limited to the surface. If, however, the preponderance of the water is restricted to the surface, then a 10% adsorption would correspond to a layer of water thousands of Angstroms thick surrounding the particle.)

The dependence of dielectric constant on added water in dried protein is of a qualitatively different type than that observed by Takashima and Schwan (49) in crystalline powders of amino acids and dipeptides, indicating that the behavior of the protein is not merely a generalized surface effect. The size of the dielectric constant at the highest water content measured would seem to indicate that a counterion (proton) polarization and/or fluctuation mechanism may be operative. Direct current conductivity results which may lend support to this suggestion will be discussed in Sec. III.C.2.

Algie (51) has recently investigated the dielectric behavior of dry solid keratin from rhinoceros horn and wool over a wide range of frequencies (10^{-6} to 10^{6} Hz) and temperatures (-14°C to 125°C). He has identified six dispersion regions with relaxation times ranging from 3×10^{-8} sec to $> 10^{9}$ sec at 50°C, and attributes these to orientation relaxation of successively larger aggregates of polymer subunits, polymer chains, and crystallites. These relaxations are characteristic of the solid rather than of the individual protein molcule, and similar relaxations have been observed in dielectric and mechanical relaxation studies on solid synthetic polyamides such as nylon 6-6 (1).

C. Conduction Properties

The early suggestion of Szent-Gyoergi (52) that the dc conduction properties of protein systems might have biological significance has stimulated a number of studies. It has been found that dry proteins, like other organic polymers, are semiconductors, and that the variation of the conductivity with temperature may be described by an Arrhenius-type equation,

$$\sigma(T) = \sigma_0 \exp(-E/2kT) \tag{39}$$

where $\sigma(T)$ is the specific conductance at temperature T°K, σ_0 and E are two empirical parameters referred to, respectively, as the pre-exponential factor and the activation energy for conduction, and k is the Boltzmann constant. A wide variety of proteins have been shown to vary only slightly in their conduction properties. For dry proteins E is found to be approximately 2.5 eV and σ_0 to lie in the region between 10^2 and 10^4 ohm^{-1} $-cm^{-1}$. On adsorption of water the activation energy decreases until at water contents of approximately 15% by weight the specific conductance may be as much as 10^{10} times greater than that of the dry state.

Studies of conductivity are directed toward an understanding of the conduction process or processes in sufficient detail to permit the quantitative calculation of the activation energy E and σ_0, and to account for the dependence of E on the amount of adsorbed water. For reviews of the literature before 1962 the reader is referred to the publications of Eley (53) or Rosenberg (54).

1. Dry Protein

Eley and Spivey (55) interpreted the conductance data for a wide variety of dry proteins in terms of a model for intrinsic conduction that combines features of the potential well theory for π -electron conductivity in hydrocarbons and the band theory for crystalline semiconductors. According to this model the activation energy is a measure of the energy difference between a localized ground state (or valence band) and a delocalized excited state (or conduction band). Once an electron is in the conduction band, it may then be transported throughout the protein system via the interconnecting network of hydrogen bridges that is to be found in both the pleated sheet and α -helical protein structures. This model received support from an early electronic structure calculation by Evans and Gergely (56) who predicted an energy gap of about 3.1 eV in such a hydrogen-bonded network in fair agreement with observed activation energies (~2.5 eV). However, Taylor (57) observed that the failure of many proteins to absorb light in the region near 4000 Å ruled out the general identification of the activation energy of conduction with an electronic energy gap of this magnitude. A more recent and complete calculation by Suard et al. (58) has raised the probable value of the energy gap to about 5 eV, which is in accordance with spectroscopic data. Finally, recent measurements of the microwave Hall effect in proteins by Trukhan (59) have yielded values for the mobility of charge carriers that are 10^3 to 10^4 times smaller than those calculated from the intrinsic model by Cardew and Eley (46) and Eley and Spivey(55). These considerations show the intrinsic conduction model to be unrealistic.

In view of the fact that experiments have been carried out on powders or gels rather than single crystals, Kasha (60) suggested that the observed conduction properties might be more appropriately attributed to a defect or impurity mechanism than to an intrinsic mechanism. Eley and Leslie (61) and Niven (62) have observed that the deuteration of proteins reduces the conductivity by a factor of 1.6 to 1.8, and the latter has taken this as evidence of protonic rather than electronic conduction.

Eley (63) has recently postulated that the observed behavior might be accounted for if the limiting electron transfer process occurs between the protein and the electrode rather than within the protein itself. If this is the case, then the activation energy for this step ought to be dependent on the work function of the particular metal used as in the electrode. This would be contrary to the results of recent measurements using several electrode materials by Eley and Thomas (64).

In order to reconcile this discrepancy Eley (63) invoked a phenomenon proposed earlier by Green (65). According to this model the tendency of electrons in the electrode and in the surface of the polymer toward attainment of equilibrium (equality of chemical potential) leads to the formation of a double layer at the electrode-protein interface. This double layer in turn reduces the dependence of electrode characteristics on the work function of the electrode material. There are presumed to exist at the surface of the semiconductor states characterized by energy levels that are somewhat different from those of the interior. These states compensate for the high potential gradient at the interface by having a lower conduction (and valence) band energy if the charge carriers are negative or a higher valence (and conduction) band energy if the charge carriers are positive. Thus the presence of such surface states reduces the activation energy of conduction to a value below that of the intrinsic energy gap of the semiconductor.

Eley (63) stated that ,"Thermal energy will be more than adequate to move the carriers into the semiconductor interior from the surface," and thus obviates the necessity of accounting for the energy differences between surface and interior states. But this assumption is equivalent to the statement that band energies in surface states differ from those in the interior by one or two times kT (< 0.1 eV) at most. Thus Eley's surface state model is incapable of reconciling the observed activation energies of conduction ($\sim$2.5 eV) with a 5-eV energy gap for intrinsic semiconduction.

The very sign of the charge carrier itself is an as yet unresolved problem. Rosenberg (66) has concluded from an experiment on slightly hydrated hemoglobin (described in Sec. III.C.2) that the charge carriers in dry protein are electrons. However, Trukhan's microwave Hall coefficient determinations (59) indicate that the charge carriers in dried hemoglobin are positive.

Although the mechanism of conduction in dry proteins is still unknown, the similarity in conduction properties between many different types of proteins as well as such lipoprotein structures as chloroplasts and the dried rod segments of sheep's eyes (54) indicate clearly that the mechanism is not closely linked to the fine structure of the individual protein.

2. Hydrated Protein

The conductivity of hydrated proteins has been the subject of more investigations than that of dry proteins because of the greater resemblance of hydrated protein to biologically active material. It is hoped that these studies can provide insight into the interaction of water and protein at their mutual interface.

Early studies on wool and keratin by Baxter (67) and King and Medley (68), respectively, established the proportionality,

$$\ln \sigma = \alpha m + \text{constant} \tag{40}$$

where m is the amount of absorbed water. This relationship holds at low values of m over five or six orders of magnitude in σ and then levels off to a "saturation value" of σ. King and Medley proposed an ionic conduction mechanism and applied Bjerrum's statistical theory of ion pairing to qualitatively explain their observed data. They proposed that the leveling off was due to complete ionization of the salts that had been absorbed in the protein.

Rosenberg (54) measured the dependence of conductivity of deionized hemoglobin powder on adsorbed water content and found that the following empirical relation was obeyed up to water contents of 7.5% by weight:

$$\sigma = \sigma_0 \exp\left[-\frac{(E_d - \gamma_m)}{2kT} \right] \tag{41}$$

where E_d is the dry-state activation energy. Eley and Leslie (61) have confirmed this relation at low hydrations, with slightly different values of α and σ_0 than those obtained by Rosenberg (54), and extended the measurements to higher hydration values where the conductivity begins to level off. Equation (41) is entirely equivalent to Eq. (40) providing $\sigma = 2kT\,\alpha$. Equation (41) seems to imply that the effect of water is to reduce the activation energy of charge carrier production. This is not an unequivocal interpretation since conductance is proportional not only to the number but also the mobility of charge carriers. Equation (41) might have just as easily been written as

$$\sigma(T,\mu) = A\mu(m)\exp(-E_d/2KT)$$

where $\mu(m)$ is the mobility of charge carriers such that $\mu(m) = \mu_0 e^{\alpha m}$.

Since the ionic conduction proposal of King and Medley (68) was not applicable in the case of deionized Hb, Rosenberg (54) proposed a more general model to explain the dependency of conductivity on water concentration at low values of hydration. In this model the addition of water or other material with a high dielectric constant reduces the attractive potential (by reducing coulomb interaction) that binds the charge carrier to the protein and consequently lowers the activation energy for production of a charge carrier. Rosenberg obtained from this model the following equation:

$$\sigma(T,\epsilon_1) = \sigma_0 \exp\left(\frac{-E_d}{2kT}\right)\exp\left[\frac{e^2}{2kTR}\left(\frac{1}{\epsilon_0} - \frac{1}{\epsilon_1}\right)\right] \tag{42}$$

where e is the electronic charge, R is a constant called the "radius of polarization," ϵ_0 is the static dielectric constant of the dry protein, and ϵ_1 is the effective static dielectric constant in the immediate vicinity of the protein which is a function of the water content m. By equating Eqs. (41) and (42) Rosenberg (54) obtained the following expression for ϵ_1:

$$\epsilon_1 = \frac{\epsilon_0}{\left(1 - \dfrac{2KTR\,\alpha\,\epsilon_0 m}{e^2}\right)} \tag{43}$$

Rosenberg claimed that since ϵ_1 increases with increasing m, at some point $1/\epsilon_0$ will become large with respect to $1/\epsilon_1$ and the value of the exponential in brackets will approach a constant value independent of water content. He further cited agreement with observed leveling off of conductivity at high water content as support for the correctness of his mechanism. However, Rosenberg's Eq. (42) cannot predict leveling off since it is merely an alternative expression of Eq. (41) which is valid only up to 7.5 wt% of absorbed water. Furthermore Eq. (43) predicts a singularity in ϵ_1 when $m = e^2/2kTR \alpha \epsilon_0$ and a negative effective constant when m exceeds that value. Using Rosenberg's numerical values for the various parameters, the singularity in ϵ_1 can be shown to correspond to a hydration value of approximately 8 wt% of adsorbed water.

It has been shown furthermore that formic acid (68) and methanol (69) are more effective per mole adsorbed in increasing the conductivity than is water, which has a significantly higher dielectric constant than either of the two liquids (78 versus 58 and 32, respectively). Thus one may question the basic hypothesis underlying Rosenberg's model.

Eley and Leslie (61) attribute the enhancement of conductivity by adsorbed water to the donation of electrons by the adsorbed water itself. In order to explain the observed fact that the conductivity increases exponentially with adsorbed water rather than linearly, they proposed that the activation energy (in this case the energy of ionizing an electron from a bound-water molecule) varies with the concentration of bound water due to interaction between the adsorbed water molecules. The difficulty with this model is the extremely high ionization potential of an isolated water molecule (12.56 eV). The difference between this value and observed activation energies is about 10 eV, which is two to three times larger than the energy of a typical electrostatic chemical bond. Consequently it does not seem likely that adsorption can lower the ionization potential by anywhere near the required amount.

A series of experiments have been performed to determine the nature of the charge carriers in hydrated proteins based on the principle that ionic (or protonic) conduction, unlike electronic conduction, should lead to electrolysis at the electrodes of a conductivity cell. King and Medley (68) found that the passage of current through keratin with at least 15 wt% adsorbed water is accompanied by the evolution of nearly an electrochemical equivalent of H_2 gas. However, the expected amount of O_2 gas was not observed, somewhat qualifying their conclusion of ionic conductivity. Rosenberg (66) calculated that current passed through hemoglobin with 7.5 wt% adsorbed water should lead to hydrolysis if the conduction was protonic. If this were to take place in a sealed cell, the water content and consequently the conductivity should decrease as the reaction proceeded. On performing the experiment Rosenberg observed no decrease in current with time and interpreted this as evidence of electronic conduction in hemoglobin with 7.5 wt% adsorbed water, and consequently electronic conduction in dry hemoglobin. Marićić et al. (70) passed current through Hb with 9.2 wt% adsorbed water and observed no evolution of gas. At 18 and 40 wt% adsorbed water hydrogen was evolved. However, the amount varied from experiment to experiment. The investigators had difficulty with drift due to polarization but only at the lower hydration value. Combining these experiments it would seem that a protonic mechanism does not become effective until values of the hydration exceed 10 wt%.

Eley and Leslie (69) have measured the adsorption isotherms for bovine plasma albumin and calculated that a concentration of adsorbed water of around 7 wt% corresponds to a BET monolayer. Thus the proton conduction mechanism appears to take effect only as the water concentration approaches two to three BET monolayers. Eley and Leslie (69) suggested that this may be due to a requirement that the water molecules be close enough to each other to form hydrogen-bonded chains, and in this way transfer protons in a concerted fashion which serves to greatly lower the activation energy of proton transfer. For a discussion of this process see Klotz (71).

The above suggestion of Eley and Leslie correlates well with the results obtained by Takashima (48), Rosen (37), Takashima and Schwan (49), and Brausse et al. (50) on the dielectric constants of hydrated proteins (Sec. III.C.2). The sudden onset of proton mobility as a result of the formation of hydrogen-bonded chains of water molecules could manifest itself both as a corresponding increase in dc conductivity and in the appearance of a proton fluctuation moment of the type considered by Kirkwood and Schumaker (22) and Takashima (7).

IV. SUMMARY

A. Proteins in Solution

Since the 1950's the classical Debye-rotor interpretation of low-frequency dielectric dispersion in protein solutions has been challenged by alternate relaxation theories based on relaxation of the counterion shell and fluctuation of the proton distribution. Counterion effects are, however, unimportant in solutions of low ionic strength; if electrolyte is necessary, a zwitterion (for example, amino acid) may be used in its place. Analysis of fluctuation phenomena reveals that the orientation times characteristic of the Debye-rotor interpretation are only slightly shifted or not at all. However, the fluctuation moment may provide a substantial contribution to the total observed electric moment and consequently the permanent moment defies direct measurement.

The relaxation time or times may be analyzed in terms of a hydrodynamically equivalent ellipsoid of revolution. The difficulty of independently establishing two relaxation times in the case of a multiple dispersion has been noted. If there is only a very small component of the dipole along one of the axes, as may well be the case in a protein symmetrically constructed of subunits (for example, hemoglobin), only one relaxation time will be observed and no information may be obtained about the shape of the molecule. This would also be the case for a protein hydrodynamically equivalent to an oblate ellipsoid of revolution, the two relaxation times of which differ by less than 10% at any axial ratio.

In any case knowing the dimensions of an equivalent ellipsoid of revolution is of questionable value, as the actual shape of the molecule need not resemble that of the equivalent ellipsoid. It has been suggested (28) that perhaps the greater value of the dielectric method may be as a monitor of shape changes as related to biological function. The results of Hanss and Banerjee (8) and Schlecht et al. (32) on oxy- and deoxyhemoglobin do not appear to support this view. It is known from other studies that a significant change in the three-dimensional molecular conformation (and hence shape) takes place in Hb on uptake of oxygen, but in spite of these changes precise measurements revealed only a small alteration in the low-frequency dielectric dispersion.

Measurements of the mid- and high-frequency dispersions may be capable of yielding information on the amount of water that is irrotationally bound to the protein in solution as well as the amount of water that may be weakly bound. However, it is difficult at present to assess the relative contributions from the relaxations of associated water and rotatable polar side chains. Other difficulties in interpretation arise because of the small size of the mid-frequency increment and the extremely broad character of the dispersion.

B. Dry and Hydrated Solid Protein

Dry powdered protein is characterized by a semiconductivity with an activation energy in the region of 2.5 ± 0.3 eV and a room temperature conductance of $< 10^{-16}$ (ohm cm)$^{-1}$. The dielectric constant is about two and does not appear to display frequency dependence. Neither of these properties seems to be a sensitive function of protein structure. The conduction process in dry protein is unknown but an intrinsic mechanism appears to be unlikely. The dielectric constant may be attributed to a combination of atomic and electronic polarization.

Hydrated proteins also display semiconducting behavior. At a given temperature the conductance increases logarithmically with adsorbed water content up to a limit of about 10% adsorbed water and begins to level off beyond that value. Electrolysis experiments appear to indicate that the conduction is primarily electronic below 10% but predominantly protonic above 15% to 20% adsorbed water. A satisfactory explanation for the dependence of electronic conduction on water content has not yet been arrived at. The onset of proton conduction in the 10% to 20% adsorbed water region appears to correlate with the large increase of dielectric constant (to values of the order of several hundreds) in this hydration range, leading to the hypothesis that this constant is due to a proton fluctuation moment.

The biological significance of protein conduction has been questioned [see discussion following Rosenberg (72)] because the activation energy in moist protein, which is of the order of 1 eV, is quite high with respect to activation energies normally encountered in biological processes.

C. Outlook

In view of the equivocal nature of the information that can be obtained from dielectric studies of protein structure, and in view of the growing indication that dc conduction properties appear to be governed by electrode or surface effects rather than intrinsic protein properties, one cannot help but wonder whether the effort required to obtain precise and accurate experimental data is justified. Likewise it may be asked whether information on protein hydration can be obtained in a more direct and precise way through the use of other techniques (for example, nuclear magnetic resonance or vapor adsorption kinetics).

Although the dielectric properties of synthetic polymers and smaller polar molecules are more or less well understood it must be recalled that this understanding was achieved through the use of simple but realistic models. It remains to be seen whether a protein structure inherently more complex than that of the synthetic polymers can be satisfactorily represented by a model that is both simple and realistic. It should be noted that dc conduction properties of synthetic polymers and even simpler organic substances are still far from being understood.

It is felt that Hasted's conclusion (9) regarding the state of research on the dielectric properties of water in 1961 may be applied with only minor modifications to the field of electrical properties of proteins in 1970. This review is therefore concluded with the following paraphrase of Hasted's remarks.

It is clear that serious attempts have been made over the last two decades to relate the electrical properties of protein in its various states to the physical constants of the protein molecule. The object in view has been to throw light on the physical properties rather than the electrical properties; and now that powerful new tools have become available, there is a tendency to leave the electrical studies behind. There remains, however, a large volume of semitechnical work directed towards specific electrical problems. This is not an entirely satisfactory state of affairs. Further progress will require the development of more realistic models for use in the dielectric and dc conduction theories. Only then can the electrical studies of protein, or indeed of any substance, be considered as satisfactorily grounded theoretically.

V. ADDENDA

1. Schlecht et al. (73) have studied the low-frequency dielectric dispersions of horse met-, oxy-, and deoxyhemoglobin; human met-, oxy-, and deoxyhemoglobin A; human oxy- and deoxyhemoglobin F; bovine methemoglobin; and sperm whale metmyoglobin. Using independently obtained data on the molecular volume and shape of hemoglobin and the viscosity of water, these workers were able to closely estimate the dielectric relaxation time by means of Perrin relations analogous to Eqs. (12a-d). They further showed that the relaxation times of methemoglobin and metmyoglobin differ by a factor of approximately four, in agreement with the Perrin relations, assuming that

their molecular volumes also differ by this factor. Although their methods of protein preparation were such as to insure a high degree of homogeneity, they still observed a considerable distribution of relaxation times for which they were unable to account satisfactorily. They concluded, as did Goebel and Vogel (30), that the low-frequency dispersion is due to orientation relaxation of a molecular dipole moment that is made up of permanent and fluctuation contributions.

In view of the apparent inability of the dielectric method to distinguish between permanent and fluctuation contributions to the total dipole moment, an independent estimate of their magnitudes would be useful in assessing their relative importance and determining the dependence of their ratio on the detailed protein structure. Orttung (74) has attempted to calculate these for horse oxyhemoglobin and modifications thereof using theoretical considerations (75) and a structural model of the molecule based on X-ray crystallographic data. A detailed examination of Orttung's theory is not within the scope of this review, but it should be mentioned that numerous approximations are employed, not all of which are entirely justified. Foremost among these is the assumption of the general validity of the Onsager local field. Orttung has performed calculations of the total dipole moment, composed of permanent and proton fluctuation moments, using alternate approximate schemes of calculation of the average proton occupation number at a given binding site. The best agreement with experiment is obtained by neglecting proton-proton interaction. As Orttung notes, there are several other unexplained discrepancies between his calculated results and experimental observations. Assuming Orttung's calculations to be at least semiquantitatively correct, they show a substantial fluctuation contribution at all pH values between 4. 5 and 9, decreasing somewhat with pH. They further show a considerable difference between the fluctuation contributions to the total dipole moments of horse and human oxyhemoglobin. This implies that generalizations to proteins of different structure altogether are unjustifiable.

2. Hendrickx et al. (76) have measured the low-frequency dispersions of monomeric BSA and transferrin. The temperature of measurement is not reported nor are the parameters of the two-term dispersion relations to which they fit their data. Although their article contains several references to the earlier study of Moser et al. (6) there is no mention of the obvious discrepancy between the results of the two studies on BSA. A value for $\Delta\delta_L$ extrapolated to zero concentration may be calculated from the graphs provided by Hendrickx et al.

which is only 40% of the value reported at 25 degrees and 30% of the value reported at 1 degree by Moser et al. (6). Since the data of Moser et al. are in qualitative agreement with the results of Grant et al. (77) on less well-characterized solutions of BSA, this discrepancy would seem to be beyond normal variability in results due to differences in preparatory methods and casts doubt on the reliability of the results of Hendricks et al. (76). These authors have studied the variation in relaxation time of solutions of BSA and transferrin in solvents of varying viscosity (water-glycerol mixtures) and concluded that the results are compatible with the classical rigid-rotor model of orientation relaxation.

3. Grant et al. (77) have restudied the low- and mid-frequency dispersions of concentrated (>7%) solutions of Bovine Serum Albumin (BSA) in an attempt to obtain a more precise estimate of the quantity that they call bound water and which is referred to here as associated water. Unfortunately this study suffers from much the same shortcomings as Grant's earlier study (12). Instead of using a single-term dispersion relation to extrapolate the low-frequency data, they used a dispersion based on a distribution of relaxation times about a single mean value. Since a 2% solution of BSA obtained from the same source has been shown to contain as much as 10% protein as dimers and higher oligomers (78) it is likely that the 3- to 15-fold more highly concentrated solutions used by Grant et al. contain an even higher percentage of aggregated protein. Thus the dispersion relation used by Grant et al. appears no more appropriate than that used by Grant (12). It therefore comes as no surprise that their value of $\epsilon(\omega_a)$, 75.9, differs by more than 2% from that obtained from the data of Moser et al. (6), who used a more appropriate two-term dispersion relation of the form of Eq. (28).

Grant et al. (77), like Grant (12), have re-extrapolated the microwave data of Buchanan et al. (38) to obtain a value of $\epsilon(\omega_b)$. But whereas the original investigators estimated $\epsilon(\omega_b)$ to be 71.6 for a 6% solution and Grant (12) estimated it to be 72.8 from a second fitting to the same data, Grant et al. (77) estimated it to be 71.0 from a third fit to the same data. In view of the estimated accuracy of the original data (±2% to 3%) one must admit to considerable uncertainty (± >2%) in any estimate of $\epsilon(\omega_b)$ from the data of Buchanan et al. (38). Thus a more cautious assessment of the data of Grant et al. reveals that $\Delta\epsilon_M \equiv \epsilon(\omega_a) - \epsilon(\omega_b)$ could vary anywhere from approximately one to eight. Further taking into account the ambiguity inherent in present interpretations of $\Delta\epsilon_M$, it is clear that the estimate by Grant et al. of the amount of associated water (0.18 to 0.64 g H_2O/g protein) is of no more than qualitative significance.

REFERENCES

(1) E. O. Forster, in Physical Methods in Macromolecular Chemistry I, (B. Carroll, ed.), Dekker, 1969, p. 165.

(2) P. J. F. Debye, Polar Molecules, reprinted by Dover Publications, New York, 1960.

(3) H. P. Schwan, Physical Techniques in Biological Research VI, Academic Press, New York, 1963, p. 377ff.

(4) M. Mandel, Protides of the Biological Fluids, 13, 415 (1965).

(5) S. Takashima and R. Lumry, J. Am. Chem. Soc. 80, 4238 (1958).

(6) P. Moser, P. G. Squire, and C. T. O'Konski, J. Phys. Chem., 70, 744 (1966).

(7) S. Takashima, Biochem. Biophys. Acta, 79, 531 (1964).

(8) M. Hanss, and R. Banerjee, Biopolymers, 5, 879 (1967).

(9) J. B. Hasted, Progress in Dielectrics, 3, 101 (1961).

(10) H. P. Schwan, Adv. in Biol. and Med. Phys., 5, 188ff (1957).

(11) P. O. Vogelhut, Nature, 203, 1169 (1964).

(12) E. H. Grant, J. Mol. Biol., 19, 133 (1966).

(13) J. L. Oncley, in Proteins, Amino Acids, and Peptides, (E. J. Cohn and J. T. Edsall, eds.), Reinhold, New York, 1943, p. 543ff.

(14) E. J. Cohn and J. T. Edsall, Proteins, Amino Acids, and Peptides, Reinhold, New York, 1943.

(15) J. G. Kirkwood, J. Chem. Phys. 7, 911 (1939).

(16) J. G. Kirkwood, in Proteins, Amino Acids, and Peptids, (E. J. Cohn and J. T. Edsall, eds.), Reinhold, New York, 1943, p. 294ff.

(17) F. Perrin, J. Physique et Radium, VII-5, 497 (1934).

(18) L. K. H. van Beek, Progress in Dielectrics, 7, 69 (1967).

(19) C. T. O'Konski, J. Phys. Chem., 64, 605 (1960).

(20) H. Fricke, J. Phys. Chem., 57, 934 (1953).

(21) G. Schwarz, J. Phys. Chem., 66, 2636 (1962).

(22) J. G. Kirkwood and J. G. Schumaker, Proc. Natl. Acad. Sci., 38, 855 (1952).

(23) W. Scheider, Biophysical J., 5, 617 (1965).

(24) H. P. Schwan, G. Schwarz, J. Maczuk, and H. Pauly, J. Phys. Chem., 66, 2626 (1962).

(25) S. Takashima, J. Am. Chem. Soc., 80, 4474, 4478 (1958).

(26) S. Takashima, J. Polymer Sci., 56, 257 (1962).

(27) S. Takashima, J. Polymer Sci., A1, 2791 (1963).

(28) R. Lumry and R. H. Yue, J. Phys. Chem., 69, 1162 (1965).

(29) M. W. Aaron and E. H. Grant, Trans. Faraday Soc., 59, 85 (1963).

(30) W. Goebel and H. Vogel, Z. Naturforsch., 19b, 292 (1964).

(31) C. P. Smyth, Dielectric Behavior and Structure, McGraw-Hill, New York, 1955, Chap. XIII.

(32) P. Schlecht, H. Vogel, and A. Mayer, Biopolymers, 6, 1717 (1968).

(33) S. Takashima, Biochim. Biophys. Acta, 79, 531 (1964).

(34) H. Hendrickx, V. Blaton, R. Verbruggen and H. Peeters, Protides of the Biological Fluids, 14, 483 (1963).

(35) M. Hanss, D. Herbage, and P. Comte, J. Chim. Phys., 65, 176 (1968).

(36) L. Lapaeva, Biokhimya, 30, 358 (1965).

(37) D. Rosen, Trans. Faraday Soc., 59, 2178 (1963)

(38) T. J. Buchanan, G. H. Haggis, J. B. Hasted, and E. G. Robinson, Proc. Roy. Soc., A213, 379 (1952).

(39) J. L. Oncley, Ann. N.Y. Acad. Sci., 41, 121 (1941).

(40) H. P. Schwan, Ann. N.Y. Acad. Sci., 125, 344 (1965).

(41) E. H. Grant, Ann. N.Y. Acad. Sci., 125, 418 (1965).

(42) K. Li, Ph. D. Dissertation, Univ. of Pennsylvania, Philadelphia, 1955.

(43) B. E. Pennock, Ph. D. Dissertation, Univ. of Pennsylvania, Philadelphia, 1967.

(44) G. H. Haggis, T. J. Buchanan, and H. B. Hasted, Nature, 167, 607 (1951).

(45) W. von Casimir, N. Kaiser, F. Keilmann, A. Mayer, and H. Vogel, Biopolymers, 6, 1717 (1968).

(46) M. H. Cardew and D. D. Eley, Disc. Faraday Soc., 27, 115 (1959).

(47) S. T. Bayley, Trans. Faraday Soc., 509 (1951).

(48) S. Takashima, J. Polymer Sci., 62, 233 (1962).

(49) S. Takashima, and H. P. Schwan, J. Phys. Chem., 69, 4176 (1965).

(50) G. Brausse, A. Mayer, T. Nedetska, P. Schlecht, and H. Vogel, J. Phys. Chem., 72, 3098 (1968).

(51) J. E. Algie, Kolloid Z., 233, 13 (1968).

(52) A. Szent-Gyoergi, Science, 93, 609 (1941).

(53) D. D. Eley, Horizons in Biochemistry, (M. Kasha and B. B. Pullman, eds.), Academic Press, New York, 1963 p. 361ff.

(54) B. Rosenberg, J. Chem. Phys., 36, 816 (1962).

(55) D. D. Eley and D. I. Spivey, Trans. Faraday Soc., 56, 1432 (1960).

(56) M. G. Evans and J. Gergely, Biochim. Biophys. Acta, 3, 188 (1949).

(57) P. Taylor, Disc. Faraday Soc., 27, 239 (1959).

(58) M. Suard, G. Berthier, and B. Pullman, Biochim. Biophys. Acta, 52, 254 (1961).

(59) E. M. Trukhan, Biophysics (Russ), 11, 468 (1966).

(60) M. Kasha, Rev. Mod. Phys., 31, 162 (1959).

(61) D. D. Eley and R. B. Leslie, Adv. in Chem. Phys., 7, 251 (1964).

(62) C. D. Niven, Can. J. Phys., 39, 657 (1961).

(63) D. D. Eley, J. Polymer Sci., 17, 73 (1967).

(64) D. D. Eley and D. W. Thomas, Trans. Faraday Soc., 64, 2459 (1968).

(65) M. E. Green, J. Phys. Chem., 69, 3510 (1965).

(66) B. Rosenberg, Nature, 193, 364 (1962).

(67) S. Baxter, Trans. Faraday Soc., 39, 207 (1943).

(68) G. King and J. A. Medley, J. Colloid Sci., 4, 1, 9 (1949).

(69) D. D. Eley and R. B. Leslie, Electronic Aspects of Biochemistry, Proc. Intl. Sym., Ravello, Italy (1963), p. 105.

(70) S. Maričič, G. Pifat and V. Pravdic, Biochem. Biophys. Acta, 79, 293 (1964).

(71) I. M. Klotz, in Horizons in Biochemistry, (M. Kasha and B. Pullman, eds.), Academic Press, New York, 1963, pp. 527-8.

(72) B. Rosenberg, Oxidases and Related Redox Systems, Proc. Symposium, Amherst, Mass. 1964, 1, 72-9.

(73) P. Schlecht, A. Mayer, G. Hettner, and H. Vogel, Biopolymers, 7, 963 (1969)

(74) W. H. Orttung, J. Phys. Chem., 73, 418 (1969)

(75) W. H. Orttung, J. Phys. Chem., 72, 4058, 4066 (1968)

(76) H. Hendrickx, R. Verbruggen, M. Y. Rosseneu-Motreff, V. Blaton, and H. Peeters, Biochem. J., 110, 419 (1968)

(77) E. H. Grant, S. E. Keefe, and S. Takashima, J. Phys. Chem, 72, 4373 (1968)

(78) P. G. Squire, P. Moser, and C. T. O'Konski, Biochem., 7, 4261 (1968)

(79) L. A. Lapaeva, Biophysics (Russ.), 13, 1136 (1968)

(80) Glasstone, Laidler, and Eyring, Theory of Rate Processes, McGraw-Hill, New York, 1941, pp. 547-551.

Chapter 4

THERMAL METHODS

Emanuel P. Manche

Division of Natural Sciences and Mathematics
York College of the City University of
New York, Jamaica, N.Y.

and

Benjamin Carroll

Department of Chemistry
Rutgers University
Newark, New Jersey

I. INTRODUCTION

Thermal methods of analysis may be defined as a set of techniques to describe the physical or chemical change of a substance as a function of the temperature. The latter is programmed usually at a linear rate. Of the many physical properties selected to observe the change, the heat capacity of a substance and its weight are the most frequently used. A list of the various thermal methods that have thus far been reported is given in Table 1.

Thermal analysis data may reveal a variety of information. It yields a "fingerprint" which may uniquely characterize the polymer and assess its thermal stability. Thermal analysis data may also permit the evaluation of the kinetic parameters for the chemical changes that may have taken place during the heating process. The effects of radiation upon a polymer may be observed (1). First- and second-order transition temperatures and latent heats of transformation are additional kinds of information that have been obtained.

The simultaneous use of two or more thermal techniques provides valuable additional information; publications have appeared describing the apparatus and the results of various combinations of techniques that complement one another. In this section the principal area of discussion will center around differential thermal and thermogravimetric analysis, although other thermal techniques will also be considered.

The application of polymers to ever-increasing temperatures has brought added interest to this field. Commercial instruments for the various thermal techniques are readily available. Specialized journals [for example Thermochimica Acta (Elsevier Publishing Co.) and the Journal of Thermal Analysis (Publishing House of Hungarian Academy of Sciences)] and societies [for example, North American Thermal Analysis Society (NATAS) and the International Confederation for Thermal Analysis] have appeared recently to take care of the outpuring of papers in this field.

TABLE 1

Thermal Technique	Physical Property Measured as a Function of Temperature
Thermogravimetric analysis (TGA)	Change in mass
Differential thermogravimetric analysis (DTGA)	First derivative of change in mass
Fractional thermogravimetric analysis	Change in mass of uncondensed fraction
Differential thermal analysis (DTA)	Temperature difference between sample and reference
Derivative differential thermal analysis (DDTA)	First derivative of temperature difference
Differential scanning calorimetry (DSC)	Energy required to keep sample and reference at same temperature
Electrothermal analysis (ETA)	Change in electrical conductivity
Thermodilatometric analysis (TDA)	Change in volume
Thermomicroscopy (Optical thermal analysis)	Intensity of polarized light
Thermometric analysis	Temperature as a function of time
Thermometric titrimetry	Temperature as a function of time or titrant volume
Torsional braid analysis (TBA)	Change in dynamic mechanical behavior
Pyrolysis	Analysis of degradation products
Effluent gas analysis (EGA)	Qualitative and/or quantitative analysis of gases

Thermal Technique	Physical Property Measured as a Function of Temperature
Mass spectrometric thermal analysis (MTA)	Ion current of fragments
Thermoparticulate analysis (TPA)	Detection of condensation nuclei
Thermovolatilization analysis (TVA)	Change in vacuum pressure
High temperature X-ray diffraction analysis	Change in diffraction intensities and pattern
Thermoluminescence (TL)	Light emission
Dynamic reflectance spectroscopy (DRS)	Reflectance of sample

II. EXPERIMENTAL METHODS

A. Differential Thermal Analysis

Thermal analysis is usually traced back to the now-classic work on the thermal behavior of clays by Le Chatelier in 1887 (2). In 1899 Roberts-Austen (3) suggested that a two-thermocouple system be employed, from which the technique derived its name. The first application to polymers is attributed to Brasseur and Champetier (4) for their study in (1947) of the mechanism of polycondensation reactions. However, the general usefulness of differential thermal analysis for polymers has been recognized mainly during the past decade.

In differential thermal analysis (DTA) the difference in temperature, ΔT, between a sample and a reference is observed while both are being heated or cooled continuously through a preselected temperature program. The use of ΔT rather than the temperature permits greater sensitivity to small temperature changes. To obtain meaningful data from the enthalpic effects attending a chemical or physical transformation requires that the thermal characteristics of the sample and reference cells be alike. A frequent practice is to dilute the sample with a large quantity of reference material.

The DTA method employs two series-connected thermocouples as shown in Fig. 1. One junction is immersed in the sample cells, S, while the other is similarly placed in the reference cell, R. Both cells are then subjected to a heating or cooling program. The difference in temperature between sample and reference cell, ΔT, is observed as a function of temperature, preferably the sample temperature although the furnace (block) temperature T_b is frequently used. A plot of ΔT versus T on an X-Y recorder is convenient for routine analysis. The use of a two-pen strip-chart recorder is preferred when precise information on the rate of heating is required.

1. Differential Thermal Analysis Instrumentation

In general a well-designed DTA instrument should detect any phenomenon that is accompanied by an enthalpic change. Until a few years ago, one usually designed and constructed his own instrument. As a result, a great variety of designs led, in many cases, to a variety of published results on the same systems. This problem has been considered by MacKenzie and Farquharson (5) and others. The introduction of complete commercial DTA packages helped eradicate conflicting results to a considerable degree. In 1965 a Committee on Standarization was established at the First

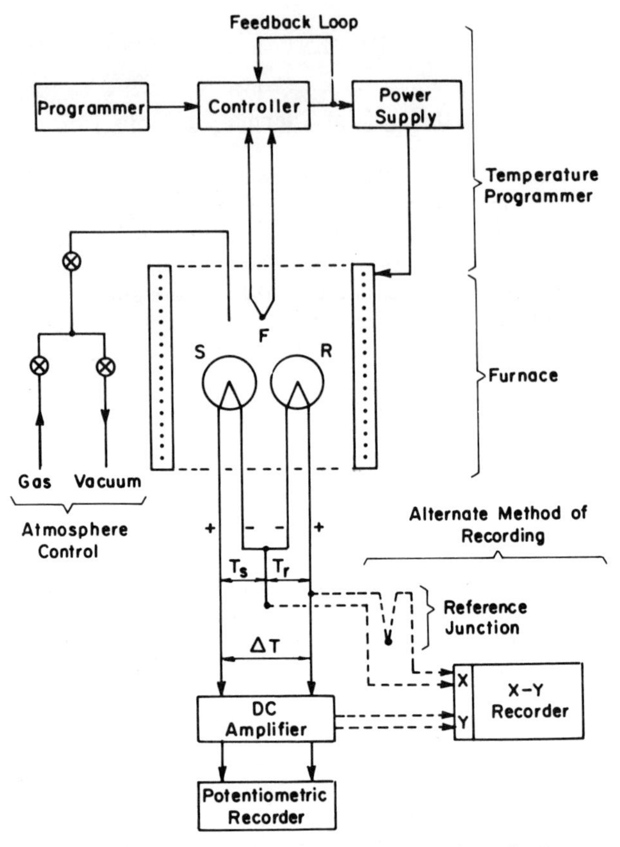

FIG. 1. Block diagram of differential thermal analysis apparatus. The X-Y recorder serves as alternate recording method.

F = furnace (block) temperature
R = reference cell
S = sample cell

International Conference on Thermal Analysis (6). A set of recommendations were put forward by this committee for reporting DTA as well as thermogravimetric details (see Appendix I). The trend in the recent literature is to describe in detail the furnace and sample holder only because of the ease of availability of the other basic components such as potentiometric recorders, furnace temperature programmers, and the availability of newly developed high-gain, low noise-level, dc amplifiers. With this in mind those who wish to construct their own apparatus should consult the excellent books by Garn (7) and Wendlandt (8). Instrumentation will be considered here with emphasis on the more recent developments.

a. Sample and reference holder. Differential thermal analysis sample holders have been built from a variety of materials, the choice of material being dependent on the temperature range of interest as well as the nature of the sample. The materials of construction have been many and varied ranging from glasses, to metals, to ceramic materials, and even to the sample itself (9). Of the dozens of reported examples, we may cite glass (10), fused quartz (11), aluminum (12), brass (13), nickel (14), platinum (15), copper (15a), and ceramics (16, 17). In addition a multitude of geometries and arrangements have also been suggested. This is perhaps the most important variable of the instrument and no doubt the one responsible for the wide variations in some reported data. In a properly designed furnace the heat distribution to the sample and reference side must be uniform and matched, otherwise thermal lags will occur that may affect the symmetry of the recorded peaks (18). Moreover, if a third thermocouple is used to monitor the temperature of the system, nonuniformity in heat distribution within the cell will give rise to errors in the measured temperature. Some typical sample holders are illustrated in Fig. 2. Broadly classified, types (a) to (d) have one leg of the differential thermocouple in intimate contact with the sample, whereas the one represented in (e) to (h) do not. In the former group the undesirable characteristic is thermocouple contamination; a situation that is avoided with the second group. In the latter group there is the advantage of ease with which the sample container can be cleaned or completely discarded for a new onc.

Type (a) and its variations have been the most common design, with the block constructed of metal or ceramic materials. Type (b) is made up usually of a fine thermocouple inserted in a capillary sample tube (19). Harold and Planje (20) fashioned the sample and reference holders from bimetallic cups in such a manner that the sample and reference materials were, in effect, placed into thermocouples which were then embedded in a refractory block. A particularly elegant setup is that described by Mazieres (21) whereby three identical and matched thermocouples have their junctions fashioned into small cups to take microgram quantities of sample. Such a design does not employ the commonly used homogenizer block, and since the small sample is contained within the thermocouple itself, intimate and continuous contact results between sample and thermocouple. This assures rapid heat transfer with minimal sample thermocouple thermal gradients.

A holder that permits the passage of a gas through the sample material is depicted in (d). Here a porous ceramic plate, located at the bottom of the container, allows a desired gas to stream into and through the powdered sample. In this manner it is possible to sweep away gas that may be generated within the sample, thereby maintaining a known composition of gas

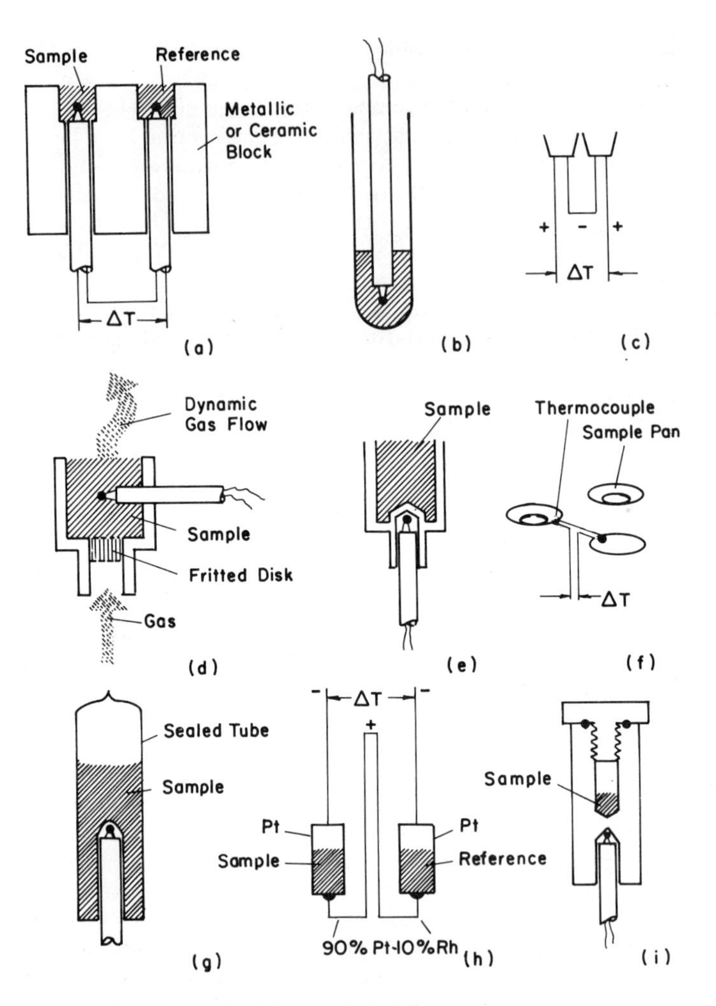

FIG. 2. Various types of sample holders.

around the particles of the sample at all times (22, 23). It is suitable only for powders or granular substances that do not melt during the experiment. In (e) we have a sample cup that is easily removable for sample loading or cleaning. It fits on top of the thermocouple. Although this type is much less sensitive than the one described above, it is well suited for liquids or materials that will melt in the temperature range of interest (23). An interesting and versatile type is shown in (f). These are very lightweight disposable pans that rest on circularly shaped differential couples and are suitable for all forms of test substances. The pans can be readily molded from a variety of metal foils and may be discarded after each experiment. Direct contact between sample and thermocouple is avoided. This is of particular importance in the case of polymers that carbonize and contaminate the thermocouple. Small sample weights are used; these range from 0.1 to 10 mg. The sensitivity and resolution is good. Excellent base line stability from one run to the next, after initial adjustment, has been reported (15). The latter three types of sample holders are commercially available (24).

In (g) we have a sealed sample holder that can be made of silica (25) or even Pyrex tubing for relatively low-temperature thermal analysis. Garn and Kessler (26) used sealed tube sample holders constructed from commercially available noble metal tubes. These holders are desirable in those instances in which the samples generate high vapor pressure or the generated gas is not easily handled. A diagrammatic representation of sealed holders is given in (h). In this particular arrangement (27) the sample is sealed in a small platinum capsule, a platinum wire is welded to one end of the tube and a Pt/Pt-10% Rh wire fused to the other end. This now forms one leg of the differential thermocouple; the other is similarly made with a reference material. This type can reach a temperature of 1750°C. Illustration (i) is a schematic of a sample holder used in the study of explosives (28). An isochoric holder is made by tightening a stainless steel socket-head cap against a dead-soft copper washer to seal in the sample and products. An isobaric sample holder is made from a similar design but modified to have a porous metal nozzle which serves the dual purpose of a gaseous heat exchanger and a retaining filter for the reacting propellant.

Multiple sample holders have also been described. These arrangements make it possible to study two or more samples at one time (29, 30). Exact comparison of thermograms for two or more samples under identical heating rates are thus possible. Commercial instruments are available which have provisions for multiple cell holders (24, 31). One of these (31) is available with a nine-cavity holder. This permits the simultaneous study of as many as four samples under pessure, vacuum, or a dynamic gas-flow method, and to temperatures as high as 1600°C. Each sample reference material is provided with a differential thermocouple with its own high-gain, chopper-stabilized dc amplifier, the output of which is fed into a unique four-channel strip-chart recorder.

Apparatus that utilized thin-film thermocouples supported on quartz and useful to 500°C has been described by King, Camilli, and Findeis (32). The low mass of the thermocouples, the small sample size, together with other design features, greatly minimize heat transfer problems. As a result flat base lines and high-precision measurements are the distinguishing features of this system. Calorimetric data may be obtained by direct area measurement followed by a correction for thermoelectric power. This approach compares favorably to differential scanning calorimetry (DSC) to be described in a later section and is probably superior to any other DTA technique reported to date.

b. Heating-cooling systems. Systems covering a range of -190°C to 3600°C have been reported. High-temperature DTA systems are available that are capable of detection of phase changes of incandescent materials (33-35); clearly, their usefulness is limited to the inorganic field. The most common arrangement is one in which the homogenizer block is heated by an external source. Other designs employ internally heated blocks. The prevailing consideration of a DTA furnace is symmetry, so that heat transfer occurs in a uniform manner. Also it is desirable to keep the temperature gradient at a minimum. The most widely used type of furnace heating employs resistance elements, the choice of element being dependent on the temperature range of interest. Other reported techniques include infrared lamps (36) and induction heaters (34, 37, 38).

Furnaces with subambient temperature capabilities have been achieved by supplying cooled nitrogen via a helically wound tube around the block (19), circulating heat-exchange liquid (39), and with liquid nitrogen (21, 39-44).

c. Temperature programmer. An essential component for DTA is the programmer for increasing or decreasing the system temperature at a constant, uniform rate. Indeed, it is impossible to do precise DTA work without a linearly varying temperature program (45). A small but sudden departure from linearity may give rise to an apparent second-order transition on the thermogram, whereas a nonuniformity or fluctuation in the temperature program may give rise to specious peaks, which would falsely indicate the occurrence of thermal phenomena (18). The most satisfactory means of achieving temperature control is through the use of a servo system. A number of systems have been described that utilize proportioning control. Here the controller compares the temperature signal from the furnace with the desired temperature signal (program) and adjusts the power imput according to the magnitude of the deviation. A common control system has been the use of a saturable-core reactor with a cam-driven programmer. The present trend is to use silicon-controlled rectifier circuits as controllers. These are superior in a number of ways, being relatively small and lightweight and can control a wide range of loads. Commercial units of varying degrees of sophistication are available today that can control any furnace without need of impedance matching considerations. Control systems are available that utilize a feedback loop that not only provides continuous and stepless power supply to the furnace, but additionally permit modification of the action according to the amount and rate of change of signal.

d. Temperature measuring systems. The most common means of measuring temperature and differential temperature is with the use of thermocouples. The choice of couple materials depends largely on the maximum temperature to be achieved, the chemical reactivity of the sample, as well as the desirability of using standard types that generate a high thermoelectric potential. Generally base metal thermocouples such as chromel-alumel and iron-constantan are preferred; on the other hand the use of noble metal-type couples such as platinum-platinum (10% rhodium) are required if chemical inertness and high temperatures are criteria. An interesting thermocouple material is Platinel II (46). The emf versus temperature characteristic of this couple closely resembles that of chromel-alumel having an emf output six times that of platinum-platinum (10% Rh) at room temperature and four times at 1000°C. Other detection devices have included thermopiles (30, 47, 48), thermistors (49-51). Apparatus utilizing thermopiles is mainly intended to increase the output signal from the differential thermocouple and to achieve a high signal-to-noise ratio. Similarly, matched thermistors have been used with limited success when incorporated in a dc bridge obviating the need of a dc amplifier. With possibly one exception (52), detection devices other than thermocouples have not enjoyed great popularity.

e. Atmosphere control. The usefulness of atmosphere control has been demonstrated by numerous studies (23, 54, 54a, 55-59). Stone (23) compared the results obtained for certain reactions by the use of dynamic as well as static atmospheres. Furnace designs have been reported to accomplish atmosphere control some of which have been recently reported in connection with polymer work (13, 42, 53, 54). Garn (54a) has described a sample holder-furnace assembly for use with self-generated atmospheres at sub- and supra-atmospheric pressure. The self-generated atmosphere technique eliminates the need for actively controlling the atmosphere within the sample, provided that the atmosphere desired is that of the decomposition products.

f. High-pressure differential thermal analysis. Recent interest in phase transitions at elevated temperatures and pressures has resulted in a number of variously designed high-pressure furnaces for DTA. Devices have been described in which the pressure is generated with a piston-cylinder type press. Very high pressures are attainable; the limits are set primarily by the crushing strength of the pistons. Description of pressure equipment as well as various types of assemblies suitable for high-pressure investigation are available (60, 66).

Cohen et al. (62) discuss the techniques and the problems of material, design, and operation as well as an assessment of the quality of data obtainable by DTA using piston-type cylinder apparatus. For the cobalt-bonded tungsten carbide piston, 45 to 50 kb (1 kb = 986.9 atm = 1019.7 kg cm^{-2}) is the upper limit for single-stage devices. Much higher pressure, however, can be achieved by the technique of double-staging (60). A description of a high-pressure DTA apparatus has been reported by Savill and Wall (63) who used the technique to follow the course of liquid-phase organic reactions at pressures up to 10 kb. In another apparatus of similar design the melting and pyrolysis of polytetrafluoroethylene was investigated up to 30 kb and 800°C (64). More recently an apparatus was described to provide optimum geometry and thermal characteristics for DTA in a cell that can be conveniently pressurized and manipulated (65); however, the pressures so attainable are, as a consequence, sacrificed. The pressure-transmitting medium is nitrogen gas which is compressed over oil by means of a high-pressure intensifier. Figure (3) is a schematic of this cell. A clear advantage of this design lies in the fact that there is no measurable change in pressure as a result of heating the cell during a run, since the volume of the reservoir is 125 times the volume of the cell.

This apparatus was used for experiments on melting and crystallization of folded-chain crystals of polyethylene and poly-(ethylene-butene-1) copolymer, and melting of extended chain polyethylene crystals. The reported precision in transition temperature measurements are ±1°C for investigations up to 3.7 kb and 200°C. By proper design this type of high-pressure DTA cell could be made to operate up to 6.5 kb and 500°C.

2. Procedural Details

a. Heating rate. Optimum heating rates are not generally ascertainable a priori, but a useful value for initial examination is 10°C/min. This

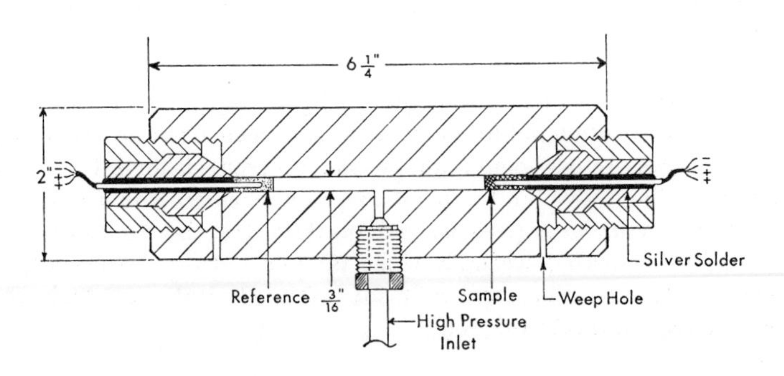

FIG. 3. High Pressure DTA cell. [Reprinted from Ref. (65), p. 378, by permission of Wiley.]

rate is the most commonly used and it has been recommended (5) that it be standardized for DTA with a maximum variation of 1°C/min. The recommendation has not been universally accepted and with the trend of incorporating variable heating rate controls in commercial instruments this goal appears remote.

The effect of heating rate on the peak temperatures and peak areas has been fairly well investigated (15, 18, 19, 66-68). It has been shown that the heating-rate effects are interdependent with other parameters such as thermocouple placement and sample size. In general, an increase in heating rate accentuates the ΔT peak, because more of the reaction takes place in the same interval of time. The peak maximum temperatures, T_m, are generally shifted to higher temperatures with higher heating rates accompanied by a slight increase in peak areas. With peak shifting, decreased resolution of two adjacent peaks may result in one of the peaks being completely masked. In the case in which the differential temperature ΔT is measured as a function of the sample temperature T_s rather than the block temperature T_b or reference material T_r, Vassallo and Harden (19) have observed that variation of heating rate does not shift T_m. The latter study was carried out at heating rates ranging from 5°C to 80°C/min. A similar finding was also reported by Barrall and Rogers (18).

b. System temperature measurement. The shape of the DTA curve and the assignment of transition temperatures depend on the system temperature used. There are three possibilities. The system temperature may be taken as T_b, T_r, or T_s. The effect of choice of thermocouple placement is well illustrated by Smyth (69) who based his calculations on heat-flow considerations. If the differential temperature ΔT is plotted against T_s, the peak temperature would correspond to the transition temperature. On the other hand if ΔT is plotted against sample surface temperature, which in effect corresponds to the temperature of the furnace block, then the point of initial departure from the base line would correspond to the transition temperature. However, the latter point would be difficult to determine accurately; the peak temperature would occur well above the transition temperature. Finally, if ΔT is plotted as a function of T_r, neither the point of initial departure from the base line nor the peak maximum temperature would correspond to the inversion temperature. The effects of choice of system temperature and heating rate are illustrated in Fig. 4. Apparently the most desirable manner of recording DTA curves is a plot of ΔT as a function of T_s, where the system temperature is that of the sample. The location of T_m then becomes independent of the heating rate.

c. Sample requirements. Sample factors that can lead to base line deviation and peak asymmetry may be listed as (1) mass, (2) thermal

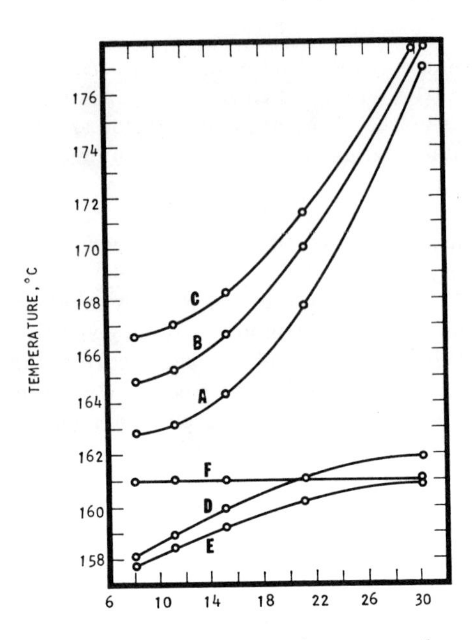

FIG. 4. Effect of location of system thermocouple and heating rate on T_m using successive runs on different sample weight of 8.3% salicilic acid on carborondum.

Case 1: (A) 0.0952 g; (B) 0.1125 g; (C) 0.1555 g.
Case 2: (D) 0.0952 g; (E) 0.0822 g.
Case 3: (F) 0.0952 g.

Case 1: Block temperature selected.
Case 2: Reference cell temperature selected.
Case 3: Sample temperature selected.

[Reprinted from Ref. (18) by permission of The American Chemical Society.]

conductivity, (3) heat capacity, (4) particle size, and (5) packing density. The effect of diluent in the sample is thus also significant. The temperature distribution in the sample and reference material has been the subject of several studies (68-73). An investigation by Ke (74) on high polymers clearly shows that as sample is increased, the peak temperature progressively shifts toward higher temperatures. To minimize the major problems of sample preparation, as small a specimen and holder assembly as possible consistent with other design factors, such as sufficient dc amplifier sensitivity, should be used.

Another consideration is the achievement of a zero base line drift.

In theory this can be achieved by matching the thermal diffusivity of the reference material with that of the specimen. This can be seen from the expression derived by Arens (71) and modified by Mackenzie and Farmer (75). Assuming metal specimen holders, a constant rate of temperature increase, and no reaction taking place, we may write

$$T = 1/4(dT/dt)R^2[(1/a) - (1/a')]$$

where dT/dt = rate of temperature increase
R = radius of cylindrical cell
a = thermal diffusivity of inert material
a' = thermal diffusivity of sample

It can be seen that if a and a' are not equal, a straight base line with a nonzero slope will result. A change of either diffusivity during a run will lead to a base line drift. It will be noted, according to the above equation, the greater the heating rate, the larger the drift. Although an effort is usually made to match the thermal properties of the reference to that of the specimen, the effective density and thermal conductivity depend on particle size and packing. Even when an initial match is made, it would be highly fortuitous for a given sample to retain its original value for thermal diffusivity after undergoing either a reaction or a crystalline rearrangement. Dilution of the specimen with inert material has been a technique often used to ensure that the thermal properties of the sample and reference cells are made as similar as possible. Dilution of sample to values of 10% down to 0.1% have been common. Some generally used diluents have been alumina, silica, silicon carbide (76,77), and silicone oils. "Sandwich" techniques have been used to equalize the thermal properties of specimen and reference compartment. A modified "sandwich" method found useful for polymer work has been described (78). Here the sample is admixed with one third of the inert material and then sandwiched between the other two thirds, with the thermocouple residing at the center of the admixed one third. Often it has been observed that the "inert" filler has entered into the chemical reaction at the elevated temperature (76).

B. Differential Scanning Calorimetry

The acquisition of calorimetric information directly rather than indirectly from thermometric data, using differential scanning techniques, is a more recent development (52, 79, 80, 81, 82). The fundamental operational basis common to these techniques is that the termperatures of the sample and reference are maintained at an equal or fixed differential level during the analysis, and the variation in power required to maintain this level during a transition is measured. Differential scanning calorimetry (DSC) is sometimes referred to as differential enthalpic analysis (DEA)

The system of Watson et al. (52) has two sample pans, one for sample and the other for reference material, mounted on separate heating coils. A resistance thermometer is permanently attached to each pan. A control unit programs the system's temperature at a preset uniform rate while keeping the sample and reference temperatures equal to each other. This is done by regulating the power supplied to each coil. The system is commercially available as a bench-type package (83). A block diagram is shown in Fig. 5. The system is divided into two separate "control loops," one for average temperature control, the second for differential temperature control. In the average temperature loop a programmer provides an electric signal that is proportional to the desired temperature of the sample and reference holders. The signal is compared with the average signal received from a pair of platinum resistance thermometers. A summing netword provides the average "error" signal. It is the differential temperature control loop that distinguishes conventional DTA and differential scanning calorimetry (DSC) as typified by the circuit of Fig. 5. Signals representing the sample and reference temperatures, measured by the same pair of resistance thermometers, are fed to the differential temperature amplifier via a comparator circuit. The output of the latter circuit determines the magnitude and direction of the power increment to the

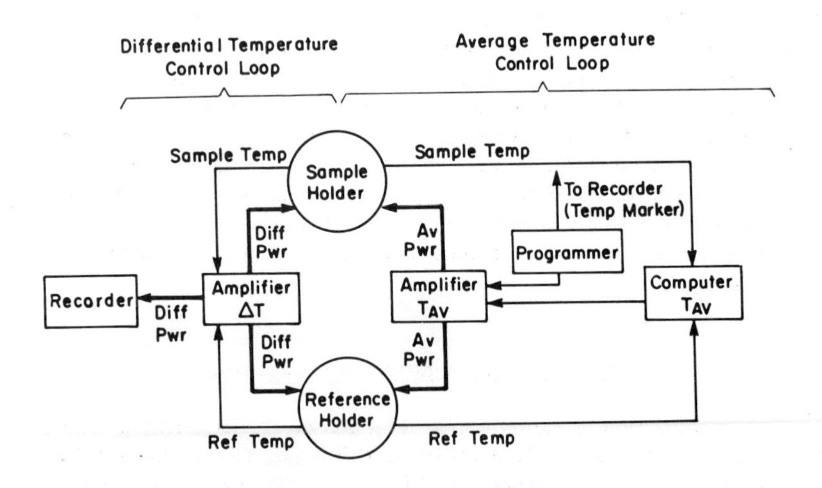

FIG. 5. Block diagram of differential scanning calorimeter. [Adopted from Ref. (52), p. 1234, by permission of The American Chemical Society.]

required heater to correct any temperature difference. The signal proportional to this differential power and its direction is fed to a recording potentiometer giving results resembling those obtained from conventional DTA. Both amplifiers operate on a time-sharing basis with each connected to the heaters for half of the time. Each loop is made to operate independently of the other.

A differential scanning calorimeter with radiant heating has been described in a recent publication by Hill and Slessor (84). The system uses focused radiation from projection lamps, and because of its remoteness from the sample holding assembly, the latter enjoys a measure of versatility. A special feature is the provision of electric calibration, which allows absolute measurements to be made up to 350°C.

C. Comparison of Differential Thermal Analysis and Differential Scanning Calorimetry Techniques

A generalized theory for the analysis of dynamic thermal measurement has been reported by Gray (85) and is considered here. The model used is shown in Fig. 6. The thermal analysis cell consists of the sample and its container at temperature T_s, a source of thermal energy at temperature T_b, and a path having a thermal resistance R, through which the thermal energy flows to or from the sample at a rate dq/dt. The total heat capacity of the sample plus container is designated C_s, which, together with R, are assumed constant for the range of interest. At any instant the

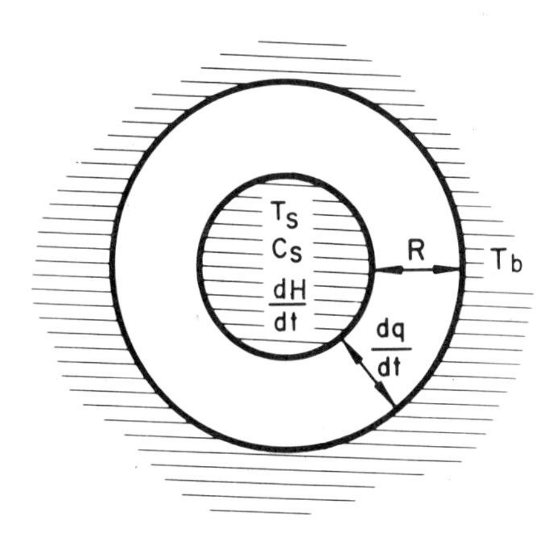

FIG. 6. Thermal analysis cell schematic. [Reprinted from Ref. (85), p. 210, by permission of Plenum Press.]

sample is generating energy at a rate dH/dt which can either increase the sample temperature or be lost to the surroundings. As energy must be conserved, the sum of the two effects must equal dH/dt, so that

$$dH/dt = C_s(dT_s/dt) - (dq/dt) \tag{1}$$

However,

$$dq/dt = (T_b - T_s)/R \tag{2}$$

so that Eq. (1) can be rewritten as

$$dH/dt = C_s(dT_s/dt) + (T_s - T_b)/R \tag{3}$$

By writing a similar equation to Eq. (1) for the reference cell where dH/dt = 0, subtracting it from Eq. (1), it may be shown that the instantaneous rate of heat generation by the sample in DTA is

$$\begin{aligned} R(dH/dt) &= \underset{\text{I}}{T_s - T_r} + \underset{\text{II}}{R(C_s - C_r)(dT_r/dt)} \\ &\quad + \underset{\text{III}}{RC_s\, d(T_s - T_r)/dt} \end{aligned} \tag{4}$$

Here R(dH/dt) can be considered as the sum of three terms in units of temperature.

In Fig. 7 these terms can be considered as representing heights on the thermogram. Distance I is the differential temperature continuously recorded by the instrument and distance II represents the base line displacement from the zero signal due to heat capacity mismatch between sample and reference. The third term is the slope of the curve at any point of interest multiplied by a constant, RC_s, so that this distance is equal to the height of the constructed triangle.

In the case of DSC, Eq. (1) and the first derivative of Eq. (2) yield an equation relating dH/dt to the measured quantities I, II, and III,

$$dH/dt = \underset{\text{I}}{-dq/dt} + \underset{\text{II}}{(C_s - C_r)(dT_b/dt)} - \underset{\text{III}}{RC_s(dq^2/dt^2)} \tag{5}$$

Similarly, Eq. (5) can be represented as distances as shown in Fig. 8.

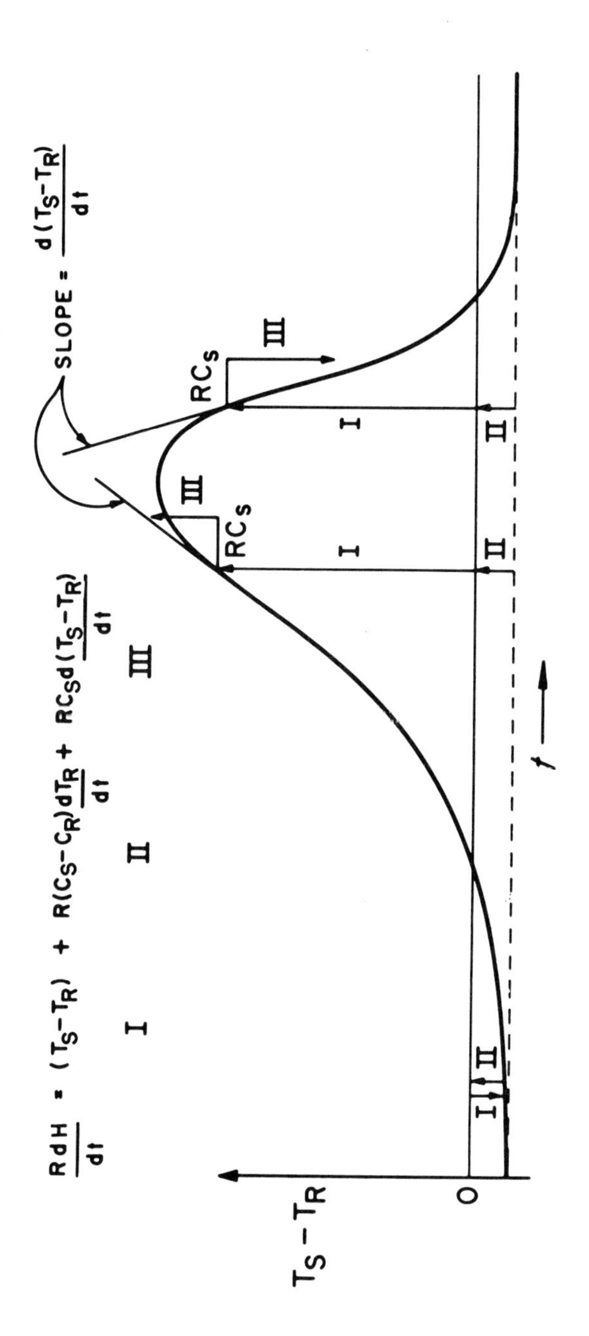

FIG. 7. Graphical determination of R(dH/dt) from experimental DTA curve. [Reprinted from Ref. 85, p. 210, by permission of Plenum Press.]

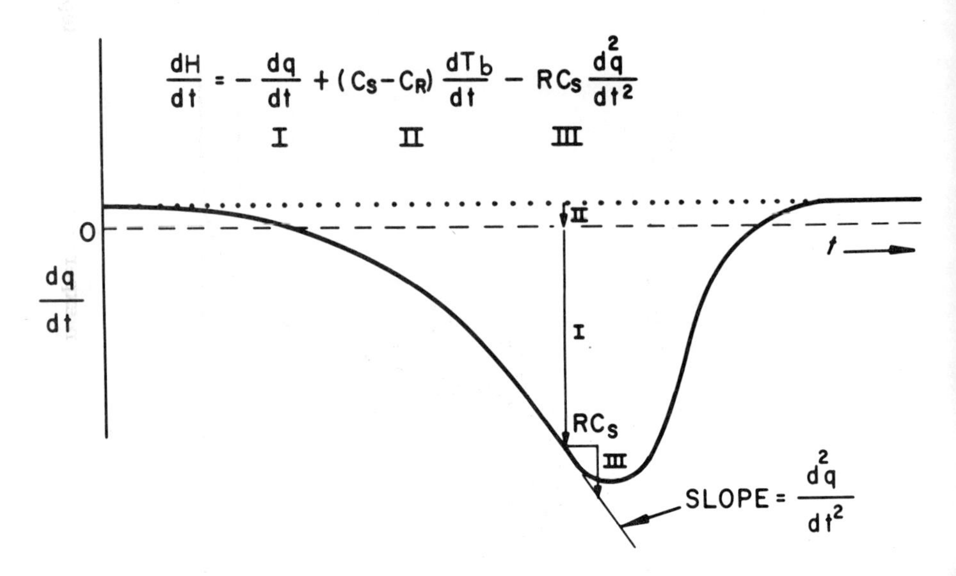

FIG. 8. Graphical determination of dH/dt from experimental DSC curve. [Reprinted from Ref. 85, p. 212, by permission of Plenum Press.]

The third term in these equations is a correction for instrumental thermal lag and in the ideal instrumental design should be made negligibly small compared with the first two terms.

Although a DSC record is analogous to that obtained in DTA, two major differences are evident from Eqs. (1) and (4). The first arises from the fact that the factor R occurs only in the third term of Eq. (5). It is therefore possible to reduce the thermal time lag constant RC_s by reducing R without loss of sensitivity for DSC. In DTA, on the other hand, the recorded signal (T_s - T_r) is proportional to R, so that the sensitivity decreases with a decrease of thermal resistance and becomes zero when R is equal to zero. The second major difference is that the area under a DSC trace is a direct measure of ΔH. Thus for calorimetric measurements a knowledge of R is not required. This means that a single point calibration of a DSC to convert areas to calories applies to the entire temperature range of interest. In DTA, the calibration is temperature dependent.

D. Temperature Calibration of Apparatus

Various methods are available for calibration purposes. The most common method has been the use of pure compounds with sharp melting or transition temperatures. A list of substances that have been found useful for calibration is given in Appendix II. Of these, the low-fusing metals are excellent calibration standards and are recommended because of their availability in highly pure form in addition to their relative ease of handling.

Another technique has been the use of an internal standard. For example, a small amount of natural quartz added to the reference material should yield a peak at 573°C (inverted) which is due to the α-β inversion. Care must be exercised in the use of the reference material. Keith and Tuttle (86) have investigated the inversion of quartz and found a variance in the inversion temperature, apparently due to solid solution of small amounts of foreign ions. Additional methods make use of calibrated platinum resistance thermometers as well as calibrated thermocouples.

E. Data Processing in Differential Thermal Analysis and Differential Scanning Calorimetry

1. First-order Transition Temperatures

The physical transformations of high polymers that have been studied by DTA include melting, crystallization from the melt, crystalline disorientation, crystal-crystal transition, and "cold" crystallization. Among the chemical reactions that have been investigated are polymerization, oxidation, vulcanization, crosslinking, curing, effects of ionizing radiation, and thermal or oxidative degradation.

Consider Fig. 9 which is a plot of ΔT as a function of increasing system temperature illustrating a transition temperature. The transition temperature is at point A if ΔT is measured as a function of block temperature T_b (which is essentially the surface temperature of the sample), and at point B if ΔT is measured against the temperature of the sample material T_s. The latter system, as we have already mentioned, is the more desirable one when working with polymers.

In the case of differential scanning calorimetry the temperature of transition is usually taken at point A, that is, the point of first observable deflection from the previously observable base line. The purity of a sample has been related to the melting-point depression and the melting range, which may be revealed by an analysis of the peak shape (87).

The heat of transition associated with first-order transitions may be readily determined by obtaining successive thermograms in the temperature region of interest on equal weights of reference standard and sample

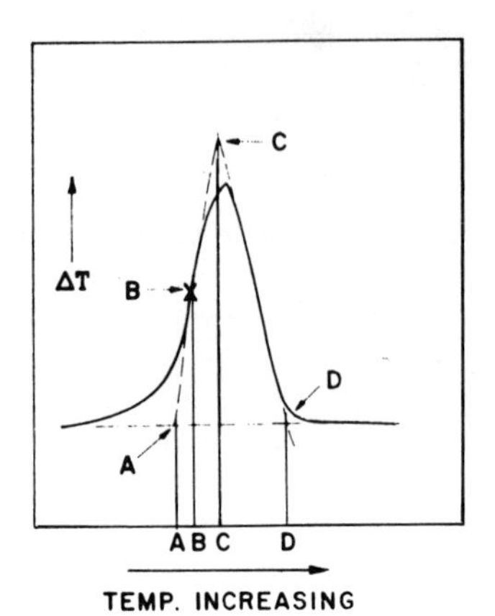

FIG. 9. Estimation of transition temperature in DTA. [Reprinted from Ref. (19), p. 134, by permission of The American Chemical Society.]

under identical experimental conditions. The enthalpy may be obtained from the relationship,

$$\Delta H_1 = \Delta H_2 (A_1/A_2) \tag{6}$$

where ΔH_1 = enthalpy change of the sample, cal/g
ΔH_2 = enthalpy change of the standard, cal/g
A_1 = area of the sample peak
A_2 = area of the standard peak

If instrumental conditions are not kept the same, Eq. (6) must be corrected accordingly. For example, in DSC Eq. (6) should be multiplied by

$$(W_2/W_1)(C_2/C_1)(R_1/R_2) \tag{7}$$

where W_1 = weight of sample
W_2 = weight of standard
C_1 = chart speed for sample
C_2 = chart speed for standard
R_1 = range setting of the sample
R_2 = range setting of the reference

(Here R is related to the instrument sensitivity; however, the larger the value of R, the greater signal attenuation.)

2. Second-Order Transition Temperatures: Glass Transitions

The glass transition has been attributed to a second-order transformation and is due to the onset of motion of chain segments in the amorphous region of a polymer. This is accompanied by abrupt changes in the specific heat, thermal-expansion coefficient, and free volume (88). The glass transition, in fact, is a property characteristic of any amorphous substance (89) and is indicated by a shift in the base line (70, 90, 91), representing a discontinuity in specific heat of the sample.

Although several methods are available for measuring glass transition temperatures, T_g, only the differential thermal and differential enthalpic methods of analysis will be considered here. A number of approaches have been put forward for the interpretation of DTA in terms of T_g. One of these is based on an analysis of the heat transfer problem. Strella (70), assuming a cylindrical model with a surface heated at a constant rate, obtained an expression that implies a sigmoidal shape for ΔT as a function of the surface temperature T_s. Accordingly the maximum value of ΔT at the glass-transition should be linearly dependent on heating rate, and that the inflection point indicating the glass transition should rise in temperature with the heating rate. Further, Strella deduced that the glass transition temperature could be determined by extrapolating the log T_{g_i} versus heating rate to zero heating rate, the T_{g_i} values being taken as the inflection point of a number of curves obtained at different heating rates. This procedure has been used on samples of poly(methylmethacrylate) and shown in Figs. 10 and 11. The value thus obtained was 110°C. This was also obtained by Rogers and Mandelkern (92) using a different technique.

Another approach is that given by Keavney and Eberlin (91). They reported the glass transition temperature of a number of polymers as the intersection of extrapolated lines.

The hole theory of liquids has been extended to describe the glass transition (93, 94, 95). Based on this approach Wunderlich and Bodily (13) describe a modified DTA apparatus to allow dynamic measurements involving complete heating and cooling cycles around the glass transition temperature. This technique, which the authors refer to as "dynamic differential thermal analysis (DDTA)," was used to study the T_g of polystyrene.

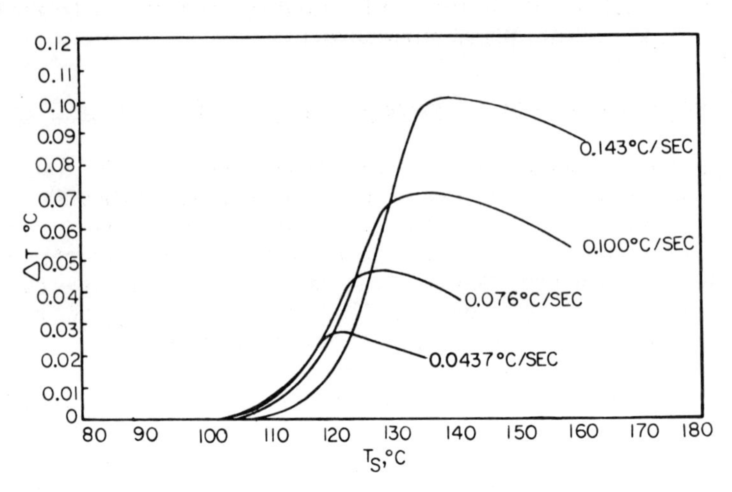

FIG. 10. Differential temperature versus surface temperature for a poly(methylmethacrylate) heated at several rates. [Reprinted from Ref. (70), p. 570, by permission of Wiley.]

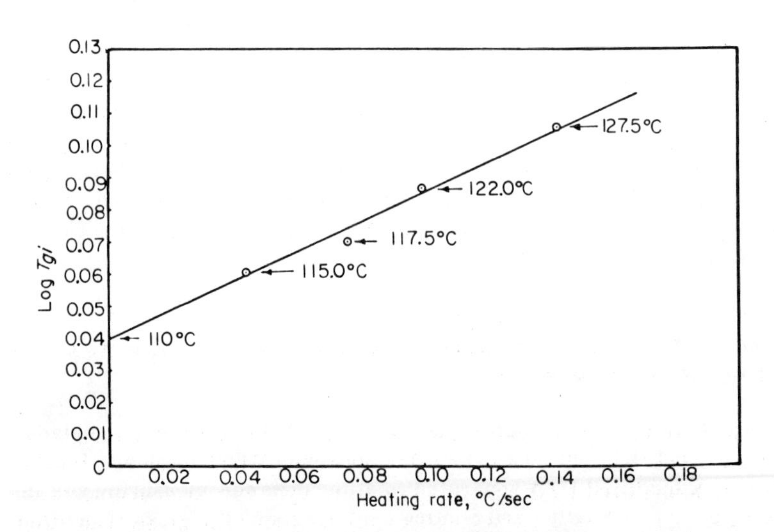

FIG. 11. Logarithm of the indicated glass transition temperature versus heating rate for poly(methylmethacrylate). [Reprinted from Ref. (70), p. 570, by permission of Wiley.]

In any case the magnitude of the shift associated with the glass transition is much smaller than that of thermogram peaks involving latent heat. For example, Keavney and Eberlin reported a base line shift of approximately 0.1°C for ΔT due to T_g in a sample of polyvinyl chloride, whereas a signal two orders of magnitude greater results in a first-order transition on an equivalent amount of the same sample. Hence for meaningful T_g data an essential requirement is equipment that possesses high sensitivity and high base-line stability.

A most convenient technique for obtaining the glass transition temperature is differential scanning calorimetry (DSC). A change in heat capacity will result in a change in the power necessary to keep the sample and reference at the same temperature. As a result the base-line shifts to a new position leading to a graphical display similar to that observed with DTA. However, the DSC trace lends itself with comparative ease to quantitative interpretations (52).

F. Applications

1. Physical Transitions

a. Melting. Melting in high polymers usually takes place over a range of temperatures since crystallizable polymers consist of regions of crystallites surrounded by amorphous regions with the crystallites having a range of sizes. The melting point is usually taken as the peak temperature T_m since this represents the point that most closely approximates the loss of crystallinity.

Polyolefins possess high degrees of crystallinity. Nonetheless, different physical and chemical properties can arise among these polymers because of variations in molecular weight, molecular-weight distribution, and branching. DTA traces of two high-density and one low-density polyethylene are shown in Fig. 12 (96). The two high-density polyethylenes show approximately a 15°C melting range, as measured from the base-line deviation to the peak. The low-density material, however, yields a lower initial melting point and a broader range. This has been attributed to the wide distribution of crystallite size.

b. Heats and entropies of fusion and degree of crystallinity. The term "crystallinity," as applied to polymers, has not been unambiguously resolved and is a reflection of the complexity of the subject (97). It has been suggested that the term be modified to indicate the means used to detect it. The heat of fusion, ΔH_f, can be obtained by measuring the area under a well-defined thermogram using an instrument that has been calibrated with

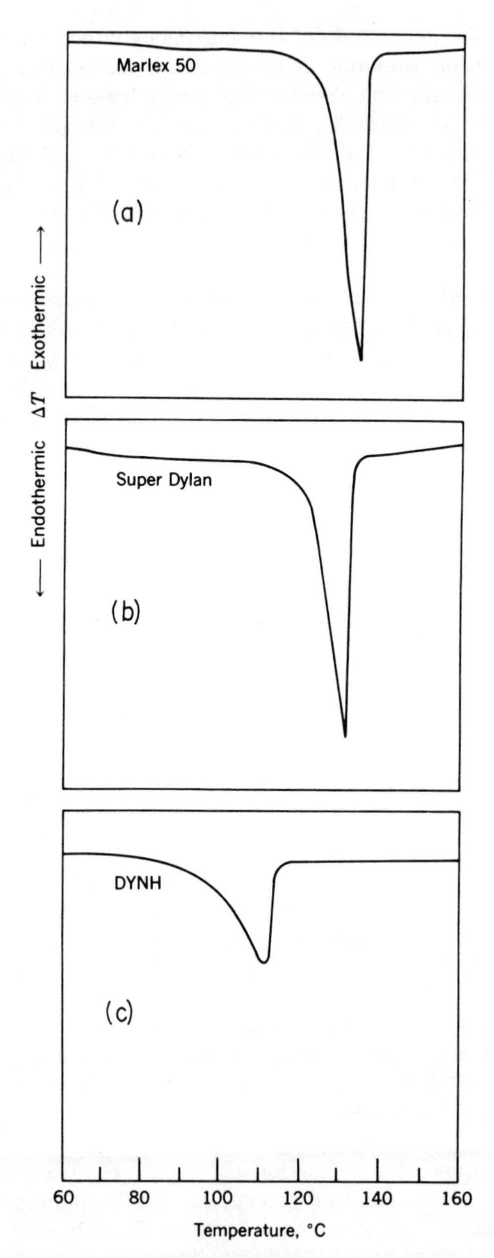

FIG. 12. Thermograms of polyethylenes. (a) (b) High-density form, (c) low-density form. [Reprinted from Ref. (96), p. 17, by permission of Wiley.]

a substance whose heat of fusion is well known. The entropy of fusion is calculated from the equation,

$$\Delta S_f = \Delta H_f / T_m \tag{8}$$

Ke (96) having calibrated his DTA instrument with benzoic acid (heat of fusion taken as 33.9 cal/g) obtained the heats and entropies of fusion for the three polyethylene polymers shown in Fig. 12. The values are given in Table 2.

For isotactic polypropylene, ΔH_f was found to be 15.4 cal/g with a value for T_m of 169°C. The smaller value of ΔH_f for polypropylene results in a low value for the entropy of fusion, this being 0.035 eu/g. The value is exactly that calculated by Gee (98) using other methods. The low entropy of fusion is attributed to the presence of methyl side groups in the polypropylene chains.

The fraction of crystallinity x can be obtained from either measurements of heats of fusion or specific heat measurements. This value may be taken as (99)

$$x = \Delta H_f^+ / \Delta H_f^\circ \tag{9}$$

where ΔH_f^+ is the heat of fusion of the sample and ΔH_f° is the heat of fusion of 100% crystalline phase.

The experimental values for crystallinity in Table 2 are based on the ΔH_f° for crystalline dotriacontane ($C_{32}H_{66}$), this value being the sum of the heats of rotational transition and normal melting. If the degree of crystallinity of a polymer is known by some other technique such as dilatometry, x-ray diffraction, or infrared analysis, then the heat of fusion of the polymer in a hypothetically perfect crystalline state may be calculated from the observed heat of fusion.

Once the degree of crystallinity for a given polymer has been determined, the relationship of crystallinity to temperature is obtainable from a plot of the measured areas under the thermogram between successive temperature intervals, $\int_{T_1}^{T_2} T\, dT$. The decrease of crystallinity with increasing temperature is shown in Fig. 13 for the three polyethylenes described in Fig. 12.

An alternative experimental technique for obtaining crystallinity without using heats of fusion is given by Eq. (10),

TABLE 2

Heats of Fusion, Entropies of Fusion, and Degree of Crystallinity of Polyethylenes (96) as determined by Differential Thermal Analysis

Polyethylene	M.P., °C	ΔH_f, cal/g	ΔS_f, eu/g	Crystallinity (%) EXPL.[a]	LIT.[b]
Marlex 50	135	58.6	0.144	91	93
Super Dylan	130	52.2	0.130	81	65-85
Bakelite DYNH	112	33.6	0.087	52	40-60

[a] Calculated by comparing the area of the respective endothermic peak with that of the double peak of dotriacontane.

[b] Obtained by other methods. Raff, R.A.V. and Allison, J.B., Polyethylene, Wiley-Interscience, New York, 1956, H. N. Friedlander et al. (eds.) in Advances in Petroleum Chemistry and Refining, Vol. 1, Wiley-Interscience, New York, 1958, p. 527.

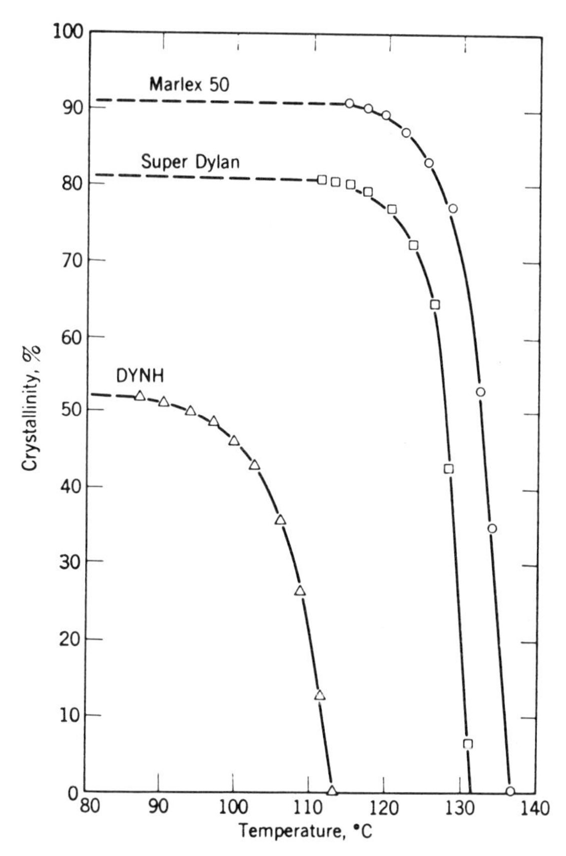

FIG. 13. Effect of temperature on the crystallinity of polyethylenes. [Adopted from Ref. (96), p. 21, by permission of Wiley.]

$$x = (C_{p_\ell} - C_p)/(C_{p_\ell} - C_{p_s}) \tag{10}$$

Here C_p is the measured specific heat of the semicrystalline polymer and the subscripts ℓ and s signify the amorphous and perfectly crystalline states, respectively. Equation (10) is not practical if the specific heats of the amorphous and crystalline phases are very close to each other, as they are in a number of cases below the glass transition temperature. Another requirement is that the specific heat measurements be made in a temperature range in which the crystallinity is constant, that is, $dx/dT = 0$. If this condition is not met, then Eq. (10) must be corrected by adding the term,

$$-\Delta H_f^\circ (dx/dT)/(C_{p_\ell} - C_{p_s}) \quad (11)$$

The experimental technique that best lends itself to the alternative approach is that described by O'Neill (100) for direct measurement of specific heats as a function of temperature. This can be readily done with DSC.

Knox (101) has used the DSC apparatus to measure rates of crystallization.

c. Effect of thermal history. The degree of crystallinity as well as the crystallite-size distribution is dependent on the thermal history of the sample. Figure 14 shows the effect of thermal history on the melting thermogram of high-density polyethylene samples (74). A similar behavior may be observed for branched polymethylenes (104). The effect of thermal history on the crystallinity and, consequently, on the thermogram is especially noticeable for polyesters. This was clearly shown by Ke (102) in an investigation on polyethylene terephthalate.

d. Effect of molecular weight. Ford et al. (103) have used the DSC to study the relationship between melting characteristics and molecular weight of two differently prepared linear polyethylenes. Their results are shown in Fig. 15.

e. Block copolymer structure. Thermal analysis is well suited to study random and block copolymers. The DTA trace of a crystalline ethylene-propylene block copolymer is shown in Fig. 16. The two separate melting peaks indicate the presence of two distinct types of crystallites, one attributable to ethylene and the other to propylene sequences. (It also shows the effect of the thermal history of the specimens.) For the fast annealed sample the crystalline melting points of 130°C and 165°C are very close to melting points of the corresponding homopolymers prepared by the same catalyst system. A more recent DTA study of block and random ethylene-propylene copolymers was made by Barrall et al. (12) showing the effect of the ethylene content, the block length, and degree of randomness.

f. Polymer mixtures. Differential thermal analysis can detect a physical mixture of polymers if their melting points are sufficiently wide apart. Figure 17 is a thermogram of a mixture of equal amounts of high-density experimental polyethylene and isotactic polypropylene. The shape and area of each of the peaks are characteristic of the respective component, the areas being proportional to the amount of each material present, although the peaks indicate slightly lower melting points than observed with the pure substances. The explanation for this shift is given by

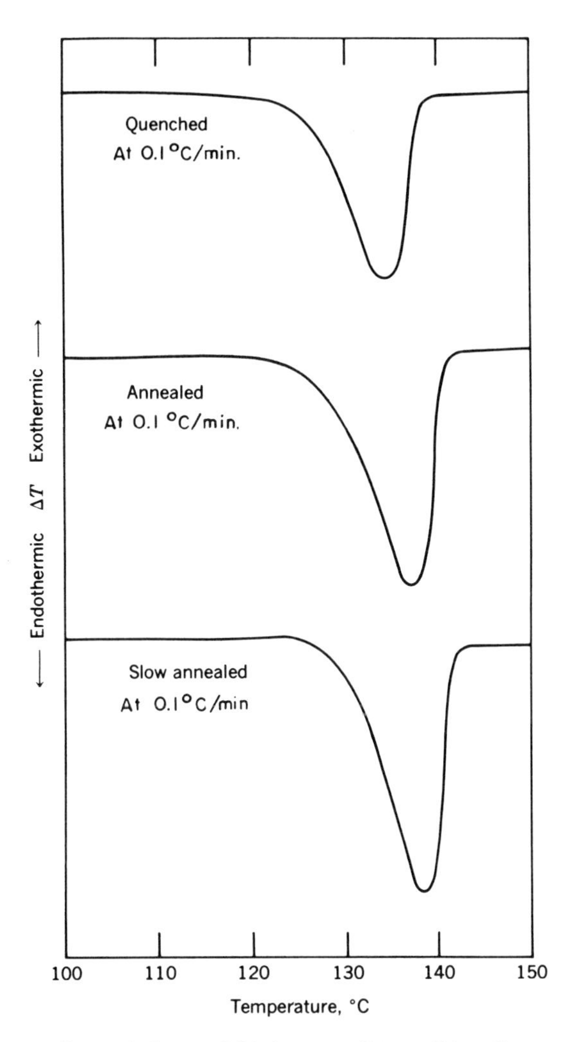

FIG. 14. Effect of thermal history on the melting thermogram of Marlex. [Reprinted from Ref. (74), p. 1456, by permission of Wiley.]

Flory's theory of melting-temperature depression of polymer-diluent mixtures (105).

g. Diluent effects. Studies of the effects of diluents on the melting of a polymer are of both theoretical and practical importance, permitting a

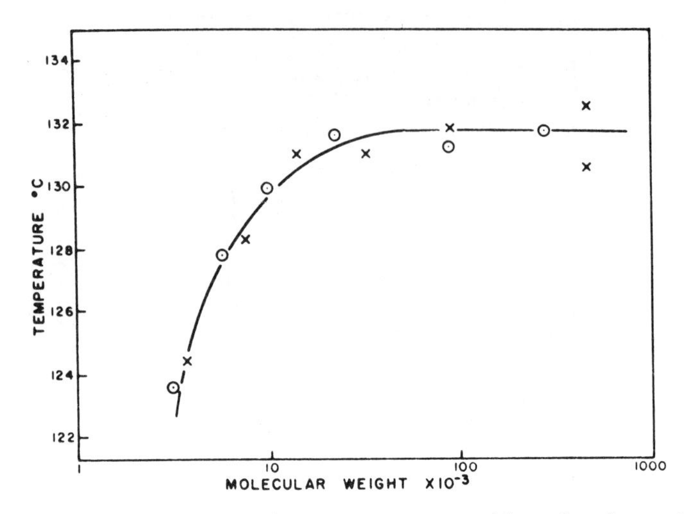

FIG. 15. Variation of melting temperature with molecular weight of two polyethylenes prepared from different catalyst systems.

⊙ Fractions from polyethylene No. 1
x Fractions from polyethylene No. 2

[Reprinted from Ref. (103), p. 43, by permission of Plenum Press.]

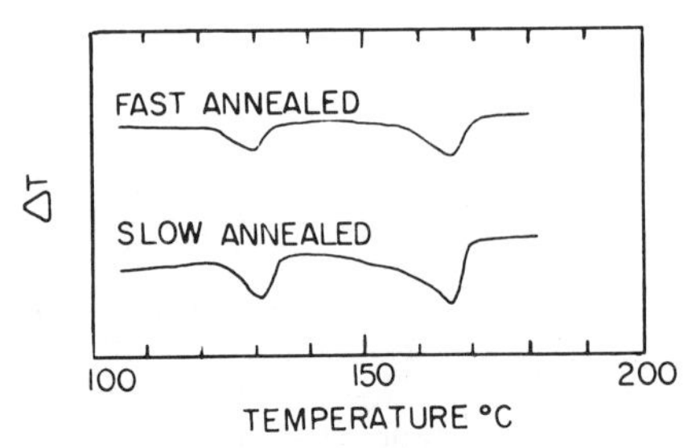

FIG. 16. Thermograms of a crystalline ethylene propylene block copolymer. [Reprinted from Ref. (104), p. 55, by permission of Wiley.]

better knowledge of the nature and magnitude of polymer-diluent interaction. Figure 18 (106) shows a number of mixtures of high-density polyethylene with dotriacontane as the diluent. Progressive depression of the crystalline melting point occurs with the increasing volume fraction of the diluent.

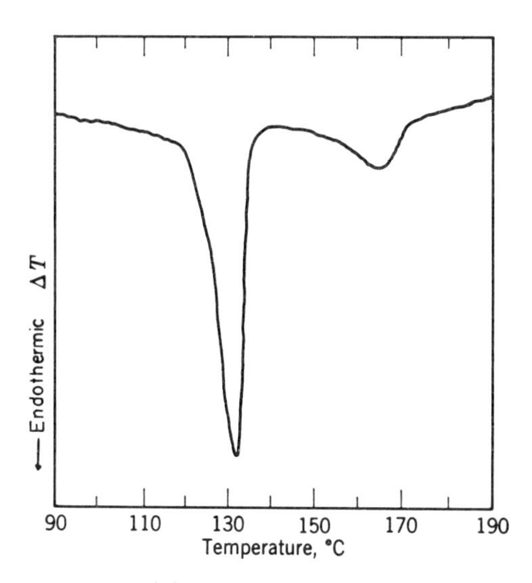

FIG. 17. Thermogram of a mixture of equal amounts, by weight, of a high-density polyethylene and isotactic polypropylene. [Adopted from Ref. (96), p. 18, by permission of Wiley.]

The Flory expression (105) for the lowering of the crystalling melting point of a polymer by the addition of a diluent may be written as

$$(1/T_m) - (1/T_m^\circ) = (R/\Delta H_\mu)(V_2/V_1)(\Phi_1 - X\Phi_1^2) \tag{12}$$

where T_m = melting temperature of pure polymer
T_m° = melting temperature of polymer-diluent mixture
ΔH_μ = heat of fusion per polymer repeating unit
R = gas constant
V_2/V_1 = ratio of the molar volumes of the polymer repeating unit and the diluent
Φ_1 = volume fraction of the diluent
X = thermodynamic interaction parameter

In calculating the melting-point depression via Eq. (12), X is replaced by BV_1/RT, where B is the interaction-energy density characteristic of the polymer-solvent pair. After rearrangement we obtain

$$[(1/T_m) - (1/T^\circ_m)]/\Phi_1 = (R/\Delta H_\mu)(V_2/V_1)[1 - (BV_1/R)(\Phi_1/T_m)] \tag{13}$$

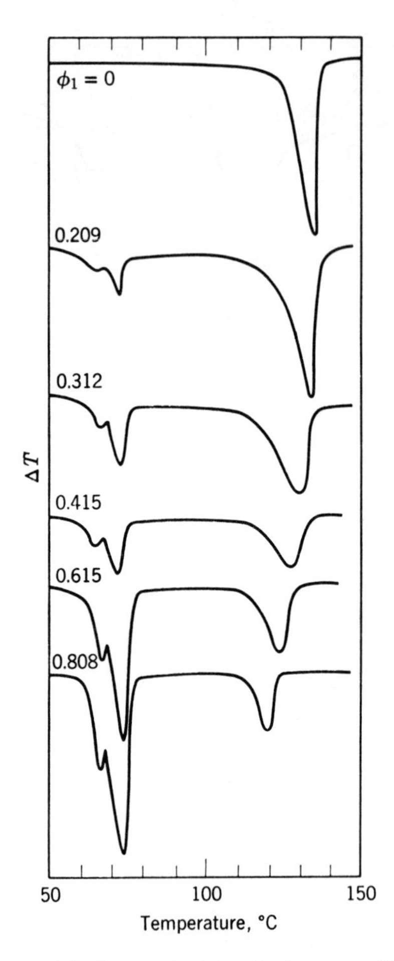

FIG. 18. Differential thermal analysis traces of high-density polyethylene alone and with increasing volume fractions, Φ, of dotriacontane. [Reprinted from Ref. (106), p. 81, by permission of Wiley.]

The intercept of the plot of

$$(1/T_m) - (1/T^\circ_m)/\Phi_1 \text{ vs } \Phi_1/T_m$$

leads to the heat of fusion. This is illustrated in Fig. 19 for high-density polyethylene (106). More recently differential scanning calorimetry has been used (107) to measure heats of fusion based on Eq. (13).

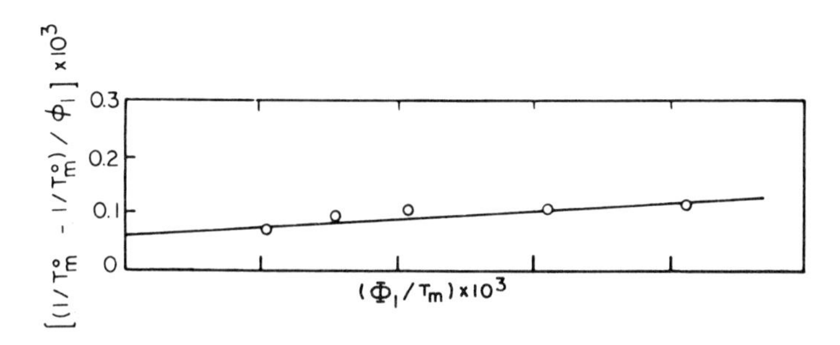

FIG. 19. Plot of melting-point depression of the mixtures against volume fraction of the diluent for high-density polyethylene and dotriacontane using DTA. [Adopted from Ref. (106), p. 83, by permission of Wiley.]

h. Crystallization. Polymer crystallization can be followed with DTA by measuring the latent heat. An example of this technique is provided by the work of Ke (102) in his study of saturated linear polyesters. This is illustrated in Fig. 20 for poly-(ethylene terephthalate) quenched from the melt as well as annealed. Four transitions appear to take place in the quenched samples, these being glass transition, "cold" crystallization, premelt crystallization and melting. These phenomena are observed in the quenched sample that is a completely amorphous polymer. For the quenched polymer the heats of transition are for "cold" crystallization, 7.9 cal/g, for premelt crystallization, 6.4 cal/g, and for fusion, 14.0 cal/g. A sum of the heats of the two crystallizations nearly equal the value of the heat of fusion indicating the completely amorphous character of the original material. On the other hand the annealed polymer displays a single peak with a larger value for the heat of fusion, viz., 16.7 cal/g, corresponding to a greater degree of crystallinity developed during annealing.

i. Glass transitions. The glass transition for a polymer has already been shown (Fig. 10). This value depends on molecular weight, on internal strain, and to some extent on the heating rate. The glass-transition temperature dependence on the molecular weight for atactic polypropylene is illustrated in Fig. 21 (108). It is evident that T_g increases rapidly in the region of low molecular weight and levels off at higher molecular weight. Keavney and Eberlin (91) in an earlier investigation determined T_g on a series of polyacrylonitrile samples ranging in molecular weight from 1500 to 300,000. They found that for a molecular weight above 60,000, T_g remains constant. These similar behaviors are attributed to the increased mobility of the chain ends with decreasing molecular weight of the polymers.

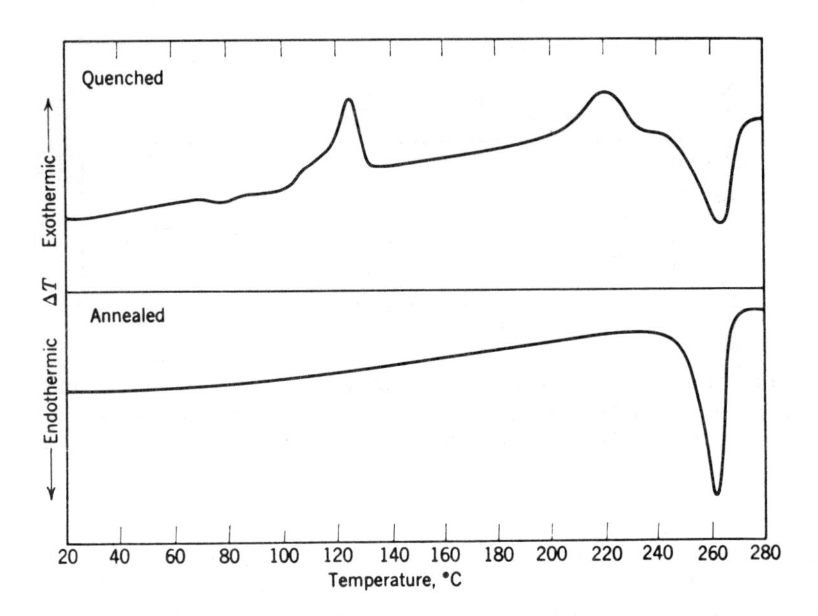

FIG. 20. Thermograms of poly(ethylene terephtalate). [Reprinted from Ref. (102), p. 625, by permission of Wiley.]

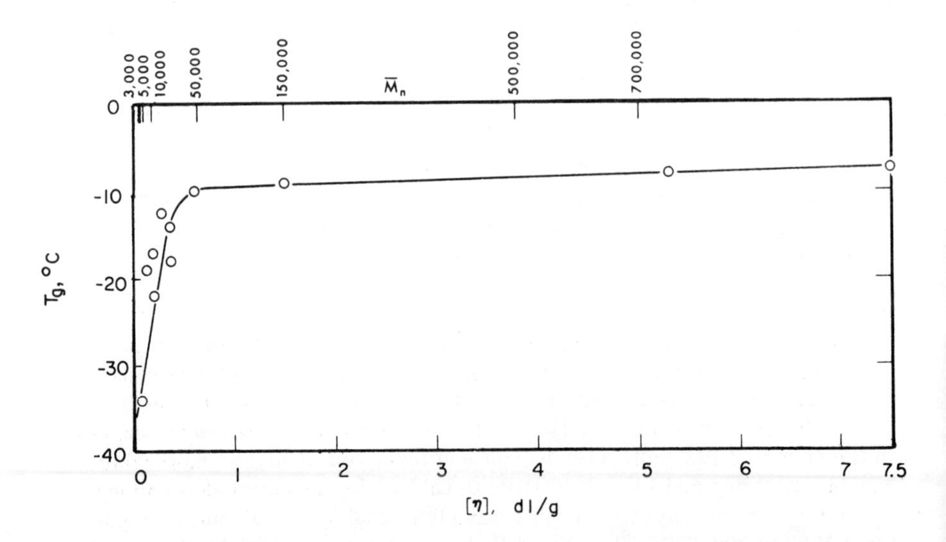

FIG. 21. Glass-transition temperature of atactic polypropylene as a function of molecular weight. [Reprinted from Ref. (108), p. 168, by permission of Wiley.]

2. Chemical Reactions

Polymerizations, degree of cure, oxidation, vulcanization, cross-linking, polymer-polymer reaction, transformation due to high-energy irradiation, thermal denaturation of biopolymers, or thermal degradation are among those chemical reactions that have been studied by DTA. For instance, in Fig. 22 the effect of differing cure cycles for Vibrin 135, a mixture of glycolmaleate polyester and triallyl cyanurate polymer, is discerned (16). Murphy and co-workers concluded that the strong exothermic reactions in the sample treated at room temperature, Curve (A), indicated incomplete cure. The two peaks, one at 150°C and the other at 320°C, also indicated that the glycol-maleate polyester and the triallyl cyanurate cure independently. Thermogram (B) showed that the polyester cure is more nearly complete, although the low-temperature peak is still appreciable in size. Thermogram (C), on the other hand, indicated a completed low temperature cure and a substantially reduced amount of uncured material in the higher curing phase. The same method (109) was used to evaluate the relative efficiency of two catalysts, tert-butyl perbenzoate and benzoyl peroxide. When the latter was used as a catalyst and then cured at 80°C for 24 hours, the 180°C exotherm completely disappeared. This indicated that the unsaturation in the polyester and two allyl double bonds in triallyl cyanurate had completely reacted at this low curing temperature.

a. Oxidation. DTA has been used to study the oxidation of polymers at elevated temperatures and the efficiency of antioxidants. As an illustration, Fig. 23 shows the DTA traces of an isotactic poly(4-methylpentne-1) with and without an antioxidant. With 1% Santonox, the polymer gave a well-defined peak at 238°C. The pure material yielded a smaller melting peak and, in addition, a large exotherm peaking at 186°C, indicating oxidation with the surrounding air. Because of the high reactivity of most polyolefins, a trace amount of antioxidant is usually added to eliminate undesirable peaks that may complicate the thermogram interpretation. The use of controlled atmospheres (58, 109-111) is a common practice in the DTA studies.

b. Thermal degradation. The application of DTA to the study of thermal degradation of polymers is based on the fact that the degradation is always accompanied by considerable amounts of heat changes. As an example, Paciorek et al. (112) used DTA to compare the thermal stability of polyethylene and various fluorinated polymers. Figure 24 shows the thermograms for polyethylene (PE), poly(vinyl fluoride) (PVF), poly(vinylidene fluoride) (PVF_2), polytetrafluoroethylene (PTFE), and a vinylidene fluoride perfluoropropene copolymer (VFPPC). The results indicate that the stability of these polymers is in the order:

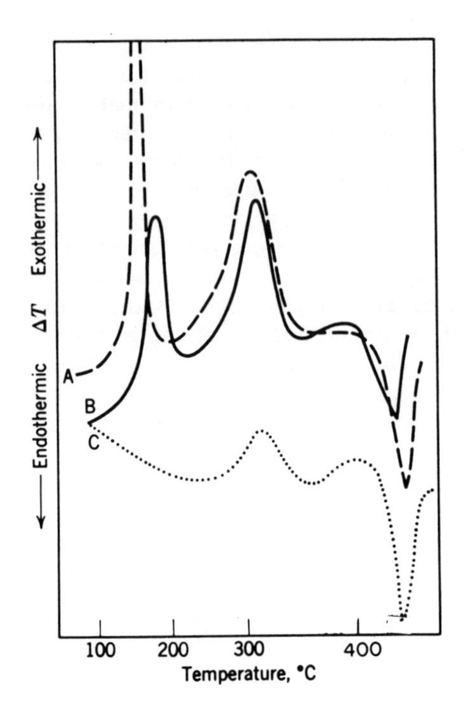

FIG. 22. Differential thermal analysis thermograms for Vibrin 135 polyester subjected to various curing conditions: (A) uncured sample; (B) sample cured at 80°C for 24 hours; (C) same resin as (B), post baked at 180°C for 24 hours. [Reprinted from Ref. (16), p. 448, by permission of Wiley.]

PTFE > PE > VFPPC > PVF_2 > PVF. The thermograms also reveal that there is considerable energy release accompanying the degradation of PVF and PVF_2, a moderate energy release for PTFE, whereas an energy intake is indicated for PE. The relative rate of decomposition is also indicated. From these observations it appears that a large and rapid energy release is associated with the presence of both hydrogen and fluorine in the polymer.

c. High-energy radiation. The first application of DTA to the study of effects of radiation on polymers goes back to 1960, when Murphy and Hill (113) investigated the effects of radiation on poly(vinyl chloride). More recently, Fock (114) studied the effects of ionizing radiation on polytetrafluorethylene and carbon black-filled polytetrafluoroethylene. The usefulness of the technique is illustrated in Figs. 25 and 26, where the

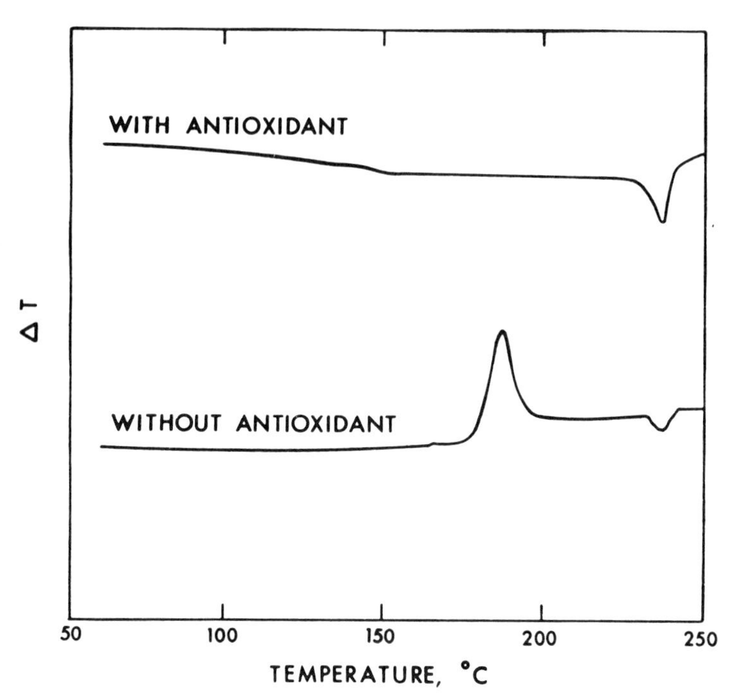

FIG. 23. Thermograms of poly(4-methylpentene-1) with and without antioxidant. [Reprinted from Ref. (74), p. 1458, by permission of Wiley.]

comparative melting behavior of unirradiated and irradiated polyethylene single crystals, grown at 77°C and 95°C, was investigated (115) with a differential scanning calorimeter. Additional examples may be found in the chapter on "Assessing Radiation Effects in Polymers" in Vol. I of this series (1).

G. Thermogravimetric Analysis

1. Historical

The development of thermogravimetric analysis (TGA) with a review of its applications to inorganic chemistry has been presented by Duval (116) and Rocchiccioli (117). Apparently, the first application of thermogravimetry to a polymer can be attributed to Duval (118) in 1948 when he

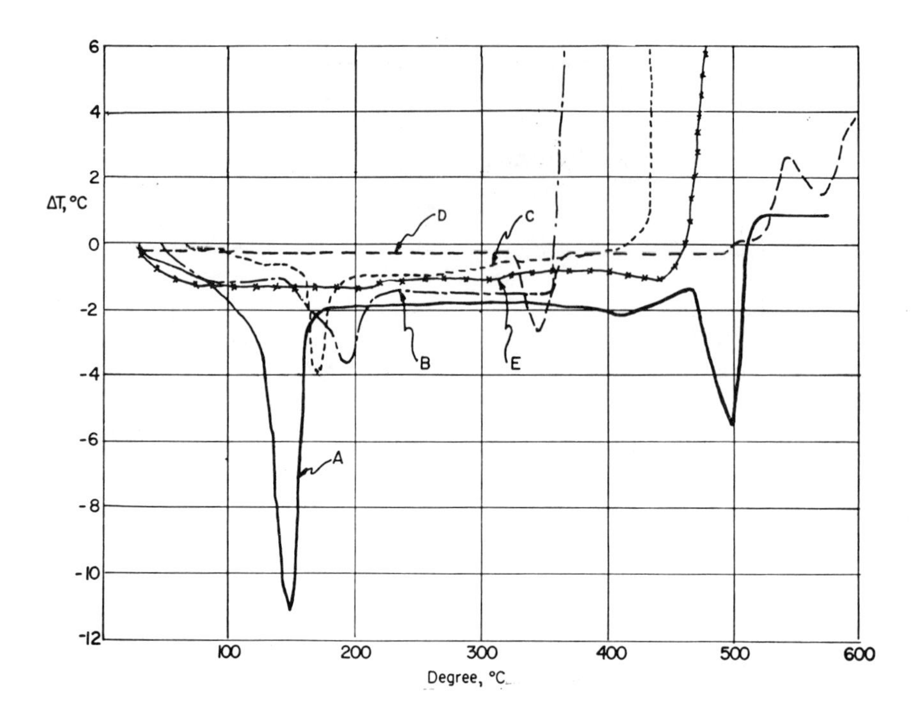

FIG. 24. Thermograms of (A) polyethylene, (B) polyvinyl fluoride), (C) poly(vinylidene fluoride), (D) polytetrafluorethylene, and (E) vinylidene fluoride-perfluoropropene copolymer. [Reprinted from Ref. (112), p. S42, by permission of Wiley.]

pyrolyzed filter paper in a Chevenard thermobalance. In 1949 Jellinek (119) used a quartz spring balance to study the thermal degradation of several synthetic polymers in vacuo. A most frequent application of the technique has been in the study of the thermal stability of polymers. During the past two decades, progress in instrumentation and data treatment has made thermogravimetry, a widely used technique. An excellent review of the application of thermogravimetry in the study of polymer degradation has been given by Reich (120).

2. Instrumentation

The modern thermobalance is designed to produce a continuous record of weight as a function of the temperature. It consists of the following components: (a) recording balance, (b) furnace, (c) temperature programmer,

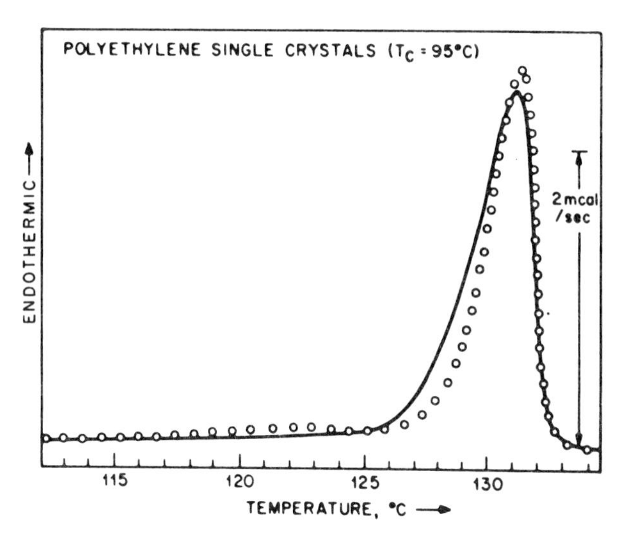

FIG. 25. Comparative melting behavior of unirradiated and irradiated 95°C crystals at 10°C/min (~900μ g).

- Unirradiated crystals
o Irradiated crystals, 25 Mrads

[Reprinted from Ref. (115), p. 36, by permission of Plenum Press.]

and (d) recorder (see Fig. 27). A number of excellent reviews have been written which describe the various commercial and noncommercial recording balances (8, 116, 121, 122). A recent equipment survey (123) gives detailed information on commercial instruments available from 16 manufacturers. Some of the factors to be considered in the construction or purchase of an automatic thermobalance has been given in a review by Lukaszewski and Redfern (124).

The recording balance, being the most important component of the thermobalance, should possess the usual characteristics of a good quality analytical balance. In addition it should have an adjustable range of weight change, rapid response to changes in weight, be relatively unaffected by vibration, and possess a high degree of electronic and mechanical stability (121). The balances may be divided into two general classifications, based on their mode of operation. They are the null and the deflection type; these are schematically illustrated in Figs. 28 and 29. The automatic null-type balance incorporates a sensing device that detects a deviation of the

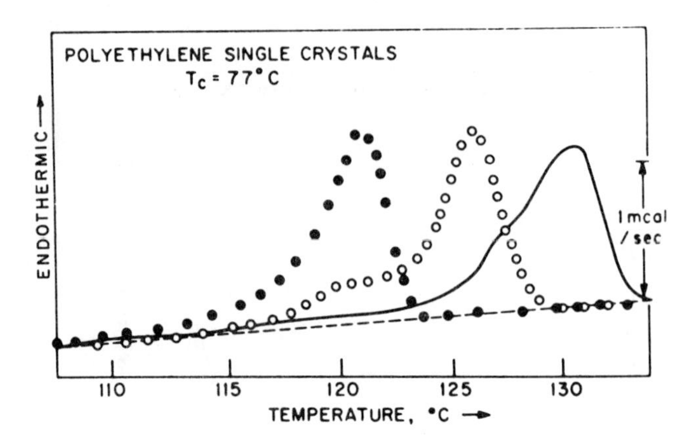

FIG. 26. Comparative melting behavior of unirradiated and irradiated 77°C crystals at 10°C/min (~750μg).

- Unirradiated crystals.
o Irradiated crystals, 25 Mrads
• Irradiated crystals, 100 Mrads.

[Reprinted from Ref. (115), p. 36, by permission of Plenum Press.]

balance beam from its null position. Through appropriate servolinkage, a recorded restoring force is then applied to the beam, returning it to its null position. Deflection balances on the other hand involve the conversion of balance beam deflections into an electrically recorded signal by means of an appropriate displacement measuring transducer.

3. Factors Affecting Readout

The principal problems in thermogravimetry are those associated with heat-transfer effects. These are substantially more serious than those associated with differential thermal analysis because of the need to isolate the sample mechanically from its immediate surroundings. Some sources of errors inherent in thermogravimetry are presently discussed.

a. Buoyancy and convective forces. Common to most thermogravimetric traces is a base-line drift, arising mainly from a density change in the gaseous environment and an increase in the convective force. The drift, which is due to the decrease in buoyancy, may be approximated by the expression (125),

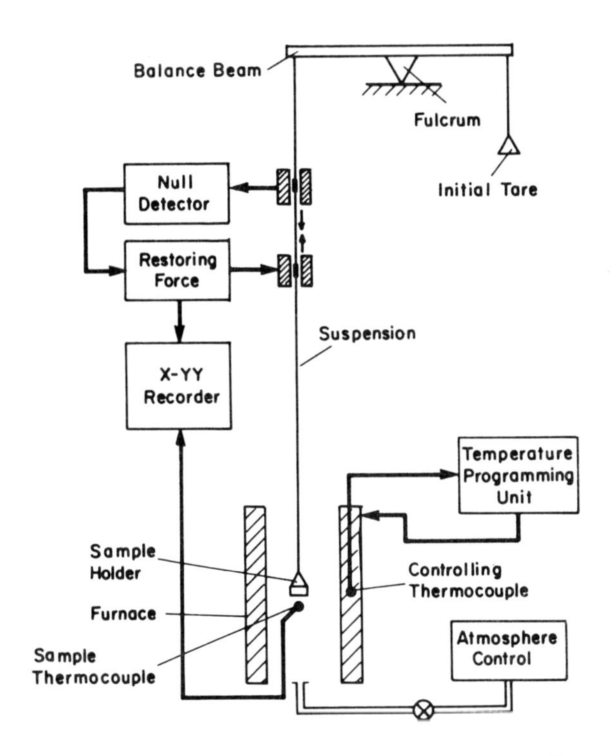

FIG. 27. Schematic representation of an automatic null-type recording thermobalance. For vacuum work, the entire balance is enclosed.

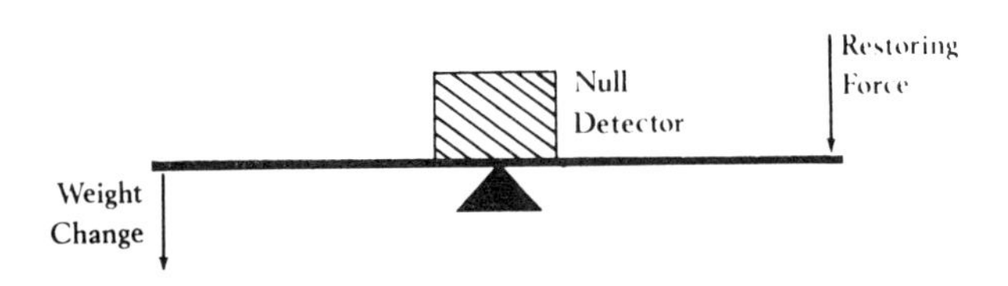

FIG. 28. Null-type balance.

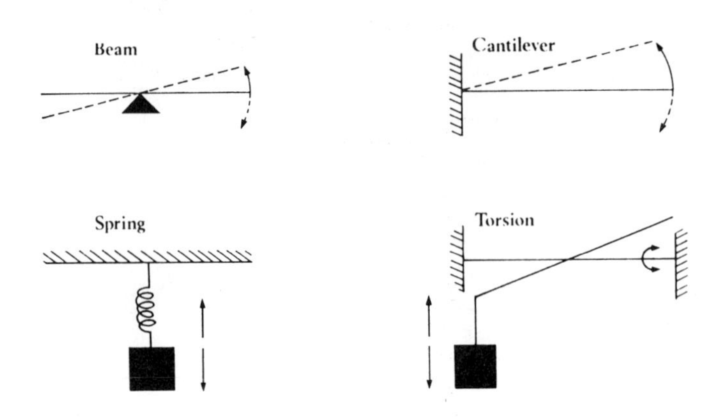

FIG. 29. Deflection-type balance. [Reprinted from Ref. (121), p. 272R, by permission of The American Chemical Society.]

$$\Delta W = Vd\left(1 - \frac{273}{T}\right) \tag{14}$$

where ΔW = apparent weight increase
V = volume of sample, crucible, and holder
d = density of gaseous environment at 273°K
T = Kelvin temperature

The effect of air on the sample container and support mechanism has been the subject of numerous studies with more than one type of thermobalance (116, 126-129). The apparent weight-gain curve for the Chevenard thermobalance, as a function of temperature, is given in Fig. 30. Various methods have been suggested for minimizing drift (128-132). Empirical equations have also been given (127, 133, 134) for compensating for drift. The apparent weight change is caused by the interplay of several factors, such as buoyancy, convection effects within the furnace, crucible geometry, radiation effects, nature of the furnace atmosphere, temperature gradient within the furnace, with most corrections differing with the rate of heating. It appears that the simplest way to achieve a correction to drift is the use of an average curve drawn for several blank runs, conducted under experimental conditions as nearly identical as possible to those of the actual experiment (135).

b. Considerations of aerodynamic forces and Knudsen's effect. Aerodynamic forces as applied to thermogravimetry may be divided into two

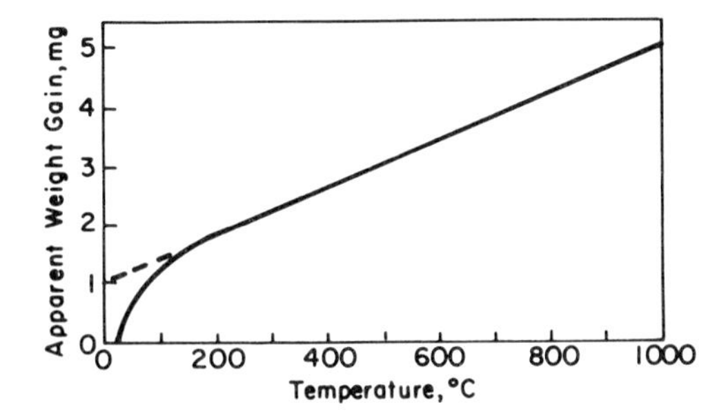

FIG. 30. Apparent weight gain as a function of temperature for the Chevenard thermobalance. [Reprinted from Ref. (127), p. 49, by permission of The American Chemical Society.]

components: a constant offset and a varying component (129). The varying component always looks like the "noise" seen on spectrophotometer traces, and is characterized by the peak-to-peak (pp) variation. These forces are noticeable only with balances of high sensitivity. It has been found that aerodynamic noise is strongly affected by hangdown tube diameter and pressure, is proportional to the horizontal area of the sample container, is only slightly affected by changes in sample temperature, and is independent of sample weight. The effect of air pressure with a 100-mg sample is shown in Fig. 31. Up to 150 torr, the noise is not more than 1 μg (pp) and it is apparent that the utilization of pressures below this value substantially eliminates this type of noise.

In the pressure range from about 10^{-3} to 1 torr there is another phenomenon, thermolecular flow, which also exerts real forces on heated samples and hangdown wires. The mass variations observed in vacuum thermobalance techniques have been ascribed to longitudinal Knudsen forces (136, 137). The apparent increase of weight before an actual weight loss in vacuum work is presumably due to the pressure changes caused by gas molecules being rapidly released by the sample. In order to minimize this effect the system should be kept below 10^{-4} torr during the course of the experiment. This necessitates adjusting the sample size to the pumping capacity of the vacuum system.

c. Temperature measurement and heating rate. Generally the temperature of the sample has been taken as the temperature measured by a thermocouple located either immediately above, below, or adjacent to the sample holder. In definitive work and especially in the determination of kinetic parameters from thermogravimetric data, the temperature so obtained may be grossly inadequate, even for a properly calibrated measuring

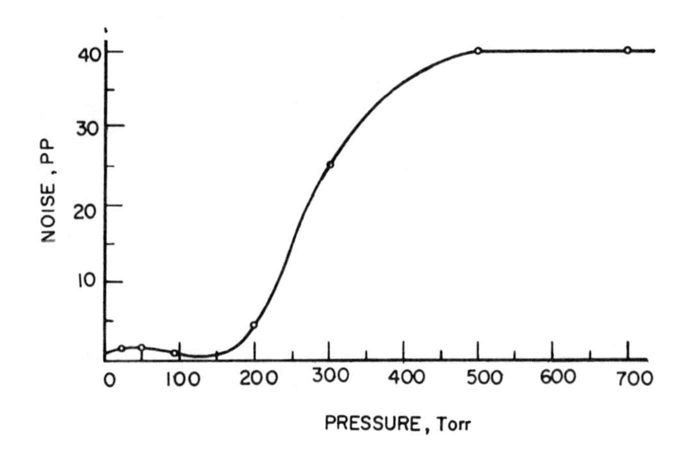

FIG. 31. Effect of air pressure on aerodynamic noise as a function of absolute pressure. [Reprinted from Ref. (129), p. 1730, by permission of The American Chemical Society.]

system, because of temperature gradients within the sample and between the sample and thermocouple. This aspect of thermogravimetry has been the subject of several studies (126, 138, 139). The temperature difference between a control thermocouple in the furnace and a thermocouple in the crucible for various sample sizes is illustrated in Fig. 32 (126). The temperature lag varies most at the temperature of greatest interest, that is, in the reaction region, and the effect is accentuated by the larger sample. The effect of heating rate on crucible temperature was also investigated in the same apparatus and is illustrated in Fig. 33. Here the lag varies from 3°C to 14°C and is roughly proportional to the heating rate. These studies indicate that it is best to use low heating rates and as little sample as possible.

Temperature calibration is usually carried out with a dummy analysis by inserting a thermocouple in the sample holder and measuring this temperature and that of the monitor. A correction curve is constructed and is subsequently applied to the sample. Another procedure for obtaining the sample temperature consists of inserting a temperature probe into the sample without impairing the sensitivity of the weighings. A number of investigators have made instrumental modifications using very fine connecting wires to conduct the signal outside of the suspended portion of the balance (40, 140-146). These modifications are best used in null-type instruments, although it is doubtful that this coupling does not affect the performance of a highly sensitive balance. The ideal solution for accurate determination of sample temperature is to have a fast-response transducer in

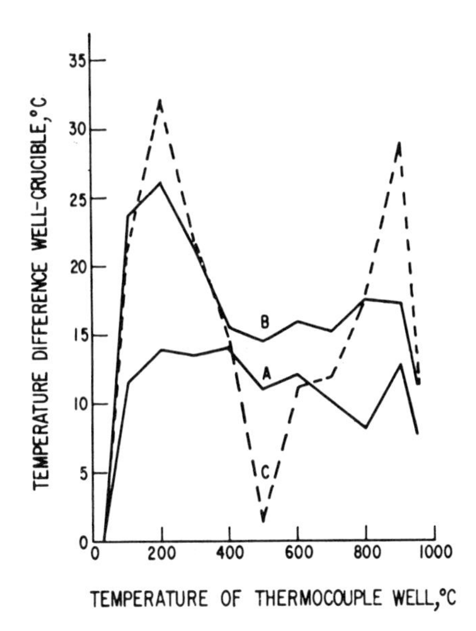

FIG. 32. Temperature differences between sample and thermocouple well in a Chevenard thermobalance for different weights of $CaC_2O_4H_2O$ heated at 10°C/min. (A) Crucible only; (B) crucible + 200 mg sample; (C) crucible + 600 mg of sample. [Reprinted from Ref. (126), p. 1562, by permission of the American Chemical Society.]

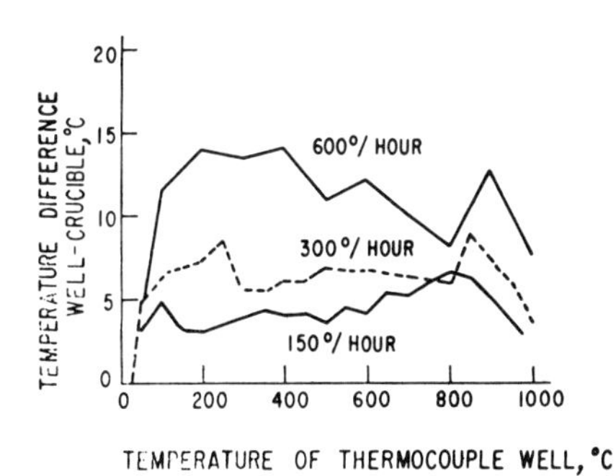

FIG. 33. Effect of heating rate on temperature lag of the specimen holder. [Reprinted from Ref. (126), p. 1559, by permission of The American Chemical Society.]

physical contact with the sample, its information being instantaneously relayed to an indicating device with a nonmechanical coupling. To this end Manche and Carroll (146) have constructed a simple semiconductor thermometer consisting of a unijunction transistor relaxation oscillator with a thermistor as the resistive part. The entire primary circuit including the power supply is made part of the balance suspension and is weighed along with the sample. Since the thermistor only is in contact with the sample, the frequency of oscillation that is a function of sample temperature is relayed to an events-per-unit-time meter, and subsequently to a digital printer, via a mutual inductance between two suspended coils.

More recently two additional publications have appeared describing means to measure the temperature of a sample suspended in a thermobalance without interfering with the balance operation. Schnizlein et al. (147) utilize a magnetic amplifier to sense a current caused by the potential generated by a thermocouple in the sample. A dc potential linear with temperature is recorded. This technique necessitates provision for maintaining a constant balance temperature or compensation for any such variation. Terry (148), on the other hand, describes a system whereby a miniature reflecting galvanometer is suspended from the beam of the thermobalance. The signal from the thermocouple, which resides in the sample, is transmitted optically in conjunction with a conventional lamp and scale. The accuracy of this system also depends on the balance temperature. Chatfield (149) has described a technique that is elegant because of its simplicity. Passive elements are used. Instead of relaying the signal to the recording system via fine wires, split planes and knife edges made of stainless steel provide a conducting path from a noble metal thermocouple that is in contact with the sample.

d. Sample holder. Two serious errors may result from the use of an improper sample holder. The errors may be caused by faulty shape and the susceptibility of the holder to chemical attack. The effect of crucible geometry has been studied (150, 151). An example of the absence of chemical inertness is the case of polytetrafluoroethylene, which should not be analyzed in silicaceous containers since it forms volatile silicon compounds quite rapidly (152, 167).

e. Condensation of volatiles. Special precautions should be taken when volatile decomposition products may condense on the cooler parts of the weighing mechanism (138).

The magnitude of this effect can be ascertained by weighing the sample holder, sample, and the support assembly both before and after each run.

Decrepitation and spattering are common occurrences in polymer pyrolysis. The use of small sample or tall crucibles, perhaps with cover, may be necessary.

f. Effect of sample weight and shape. Sample weight can affect the thermogram in a number of ways. It has been already seen in Fig. 32 that the greater the sample weight, the steeper the thermal gradients within it. This will cause the temperature of the sample to deviate from the desired program. The diffusion of a gaseous product or controlled atmosphere through the sample will be governed in part by the size of the sample (129). Figure 34 shows the effect of sample size on the position and shape of the thermogram for copper sulfate pentahydrate. Not only are the transitions better defined but the values obtained using the smaller sample are generally more accurate. Although small sample sizes invariably introduce fewer experimental artifacts, in some cases a small difference between two samples can be magnified by using a large sample size. For example, Chiu (153) was able to detect an initial weight loss of about 0.4% for a low-pressure polyethylene which was not observed for a high-pressure polyethylene when a sample size of 5 g was used. This difference was not observable when 10-mg samples were used and was attributed to the volatilization of low molecular weight polyethylenes.

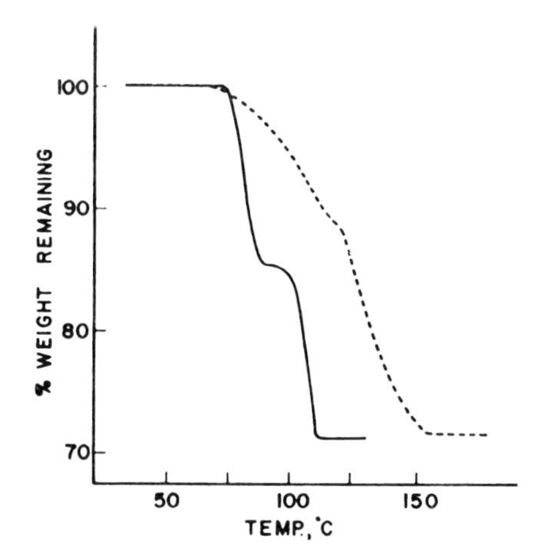

FIG. 34. Thermogravimetric curves for copper sulfate pentahydrate in air at atmospheric pressure with samples of 0.426 mg (solid line) and 18.00 mg (dashed line) under identical conditions. [Reprinted from Ref. (129), p. 1729, by permission of The American Chemical Society.]

The physical state and shape of the sample need careful consideration. For example, Vassallo (154) showed the effect of film thickness on weight loss for polypropylene at 260°C in air as shown in Fig. 35. The thinner samples degraded much more rapidly than the thicker ones. However, in this study it was found that for films of 2 mils and thinner there was little or no effect on weight loss. In a subsequent study on the rate of thermal degradation of thin polymer films, Barlow et al. (155) showed that the rate of degradation was independent of film thickness when the sample was below 250 Å. Thus if the film is taken sufficiently thin, the whole regime can be changed from diffusion control of the rate of weight loss to control by the rate of decomposition.

The degree of subdivision of a powdered sample has been shown to have a profound effect on the rate parameters of solid state reactions (156). It can be generally stated that the smaller the particle size, the lower the apparent decomposition temperature. Grassie and Melville (157) reported that the rate of monomer evolution from polymethymethacrylate at 260°C depended on particle size unless the mesh size was made smaller than 50 mesh.

g. Heating rate. The heating rate is also a major procedural variable. The effect of heating rate on the thermogram is shown in Fig. 36. Newkirk (126) obtained thermograms for the pyrolysis of polystyrene at a number of different heating rates. At any given temperature it is seen that the extent of decomposition is greater at a slow heating rate than for a similar sample heated at a faster rate.

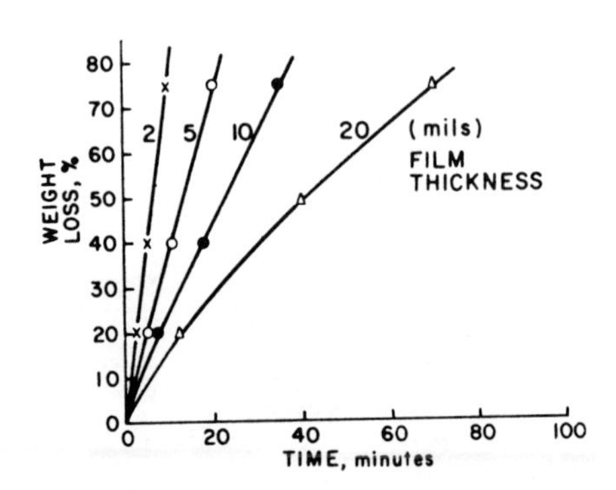

FIG. 35. Effect of film thickness on weight loss for polypropylene heated at 260°C in air. [Reprinted from Ref. (154), p. 1823, by permission of The American Chemical Society.]

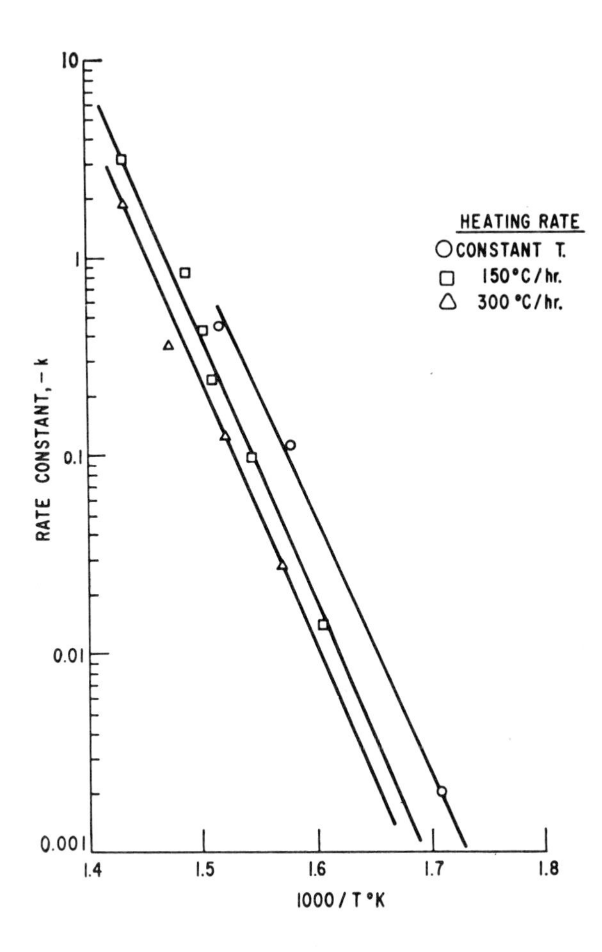

FIG. 36. Thermograms of polystyrene in nitrogen at different heating rates. [Reprinted from Ref. (126), p. 1562, by permission of The American Chemical Society.]

h. Atmosphere control. Another important factor is the type of atmosphere that the sample is subjected to in the course of an analysis. The effect of different atmospheres on the thermogram has been the subject of numerous investigations (135). For the case of a reversible solid-gas reaction, Garn and Kessler (150, 151) demonstrated the pronounced differences in thermograms for samples obtained in the absence and presence of a self-generated atmosphere, the latter achieved by simply using a cover on the sample crucible. The apparent decomposition temperature range can thus be varied by supplying a reversibly reactive gas in the vicinity of the sample. An example that illustrates this effect is shown in Fig. 37. Here the decomposition of linear isotactic polypropylene was studied under varying oxygen partial pressure as well as in vacuum (158). The thermograms reveal that the polymer is stable in vacuum up to about 400°C, where it loses about 8% of its original weight and is completely volatilized at 455°C. Upon the introduction of oxygen, the sample gains about 1% in weight between 150 to 180°C, due to oxygen uptake, followed by rapid decomposition, the rate of which increased with an increase in oxygen pressure.

Although the use of controlled atmospheres has been a well established practice in the thermogravimetric investigation of polymeric materials (159), the general practice has been limited to noncorrosive gases such as N_2, H_2, O_2, CO_2, and the noble gases. The use of "wet" corrosive gas atmospheres, however, has been severely restricted by the danger of damaging the measurement apparatus and, in particular, the weight change measurement mechanism. The problem of condensation on the cooler parts of the thermobalance is real, since some of these gaseous atmospheres are essentially high-boiling-point liquids. A technique has been developed (160) that protects the instrument from corrosive effluent gases and permits the deliberate use of dynamic corrosive gas atmospheres such as the halogen gases, SO_2, NH_3, HCl, etc. Figure 38 shows a schematic diagram of such an instrument which is available commercially (162). The corrosive gas atmosphere may be introduced through a separate gas inlet directly into sample chamber. Back flow to the TGA measurement unit is

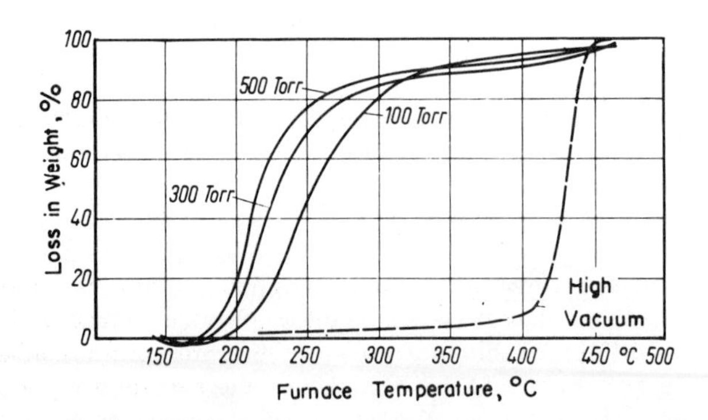

FIG. 37. Thermogravimetric traces of polypropylene in vacuum and in the presence of air at various pressures. [Reprinted from Ref. (158), p. 949, by permission of Chemie-Ingenieur-Technik.]

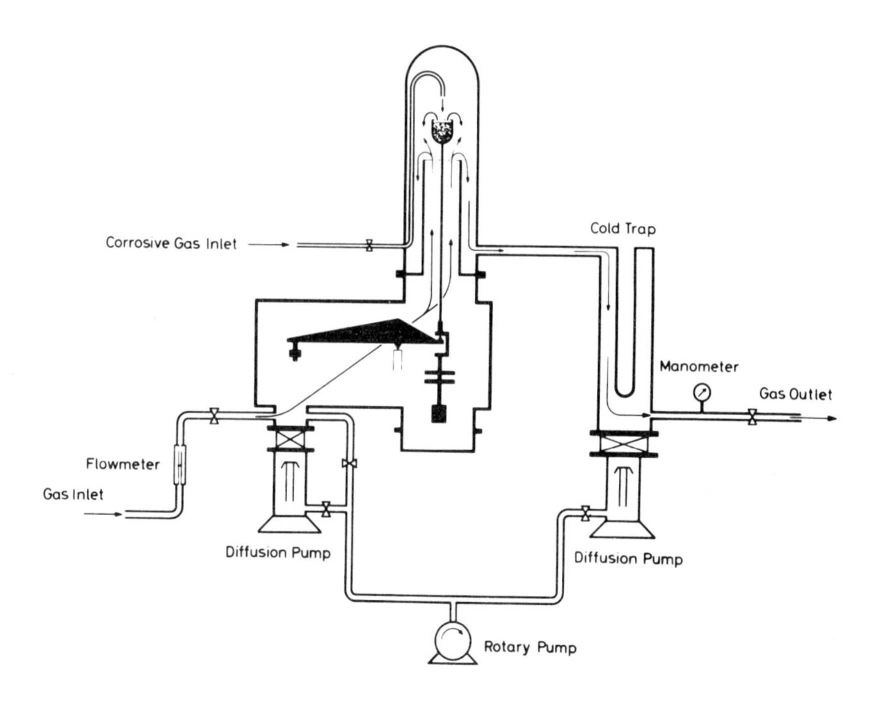

FIG. 38. Schematic diagram of thermobalance capable of operation in a dynamic flow of corrosive atmospheres. Arrows indicate gas flow through the apparatus. [Reprinted from Ref. (161), p. 13, by permission of Plenum Press.]

prevented by an opposing noncorrosive gas flow introduced from the main gas inlet and a baffle system. The instrument is equally effective in high vacuum. An additional feature of interest is its simultaneous capability for DTA in corrosive atmospheric conditions.

i. Note on temperature calibration. A note is necessary for thermobalances in which sample temperature is not measured directly. Although much has been said on the methods to achieve calibration, including the use of thermal standards (163), an alternative solution to the problem of temperature calibration is the use of the ferromagnetic-paramagnetic transition of selected metals and alloys (164). The technique works most conveniently with thermobalances that have small furnace assemblies, because a permanent magnet must be located near the sample holder. When milligram samples of ferromagnetic materials are placed in the sample holder, a downward component of magnetic force, in addition to mass, is recorded by the balance. As the sample temperature is programmed upward, each sample will abruptly lose its ferromagnetism at its Curie point. The calibration is made from the sharp apparent weight-loss traces that correspond to each of the standards used.

4. Application of Thermogravimetry to Macromolecular Systems

The greatest interest in the application of thermogravimetry to the study of polymeric materials has been in the assessment of their relative thermal stabilities and, more recently, the evaluation of kinetic parameters. In addition thermogravimetry has a broader scope of application which has not been fully appreciated. It includes the analysis of additives, characterization of polymer blends and copolymers, and sorption-desorption studies, and the application to biological systems.

a. Thermal stability. Arbitrary methods for the assessment of the thermal stability of polymers from thermogravimetric traces have been suggested and briefly reviewed by Reich and Levi (159). These methods can be broadly classified as qualitative and semiquantitative. The former compares stabilities by mere visual observation. The latter may be evaluated by the method of Doyle (165). He proposed indices of thermal stability by using a defined procedural decomposition temperature. One of these, the "integral procedural decomposition temperature" (ipdt), was devised as a way of summing up the whole shape of the "normalized" curve which is readily available and is highly reproducible. To place all materials on an equal procedural footing, the curve is integrated on the basis of the total experimentally accessible range from 25°C to 900°C. The integral procedural decomposition temperature, designated at T*, is obtained from the expression,

$$T^* = 875A^* + 25$$

where A* is the total area under the thermal curve normalized with respect to both residual weight and temperature. This is obtained by dividing the area of all the cross-hatched region by the total rectangular area bounded by 25°C and 900°C and 0 and 1.0 (see Fig. 39). The integral procedural decomposition temperatures for a few familiar polymers, analyzed in an inert atmosphere, are listed in Table 3.

An example where the relative thermal stability of various polymers has been ascertained by the mere visual observation of its thermal curve is shown in Fig. 40. Here it may be seen that the order of thermal stability is as follows: polytetrafluoroethylene > polyethylene > polypropylene > polyhexafluoroethylene. A further observation may be made. Fully fluorinated structures are generally found to be the more stable. However, some polymers containing branched perfluorinated structures may be less stable than their hydrocarbon analogs (166).

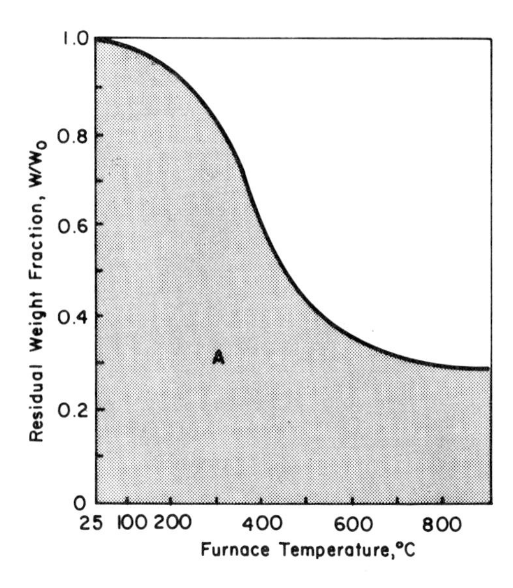

FIG. 39. Thermogram areas used in determining integral procedural decomposition temperatures. A* equals shaded area under curve divided by area of rectangle. [Adopted from Ref. (165), p. 78 by permission of The American Chemical Society.]

TABLE 3

Integral Procedural Decomposition Temperature for Some Polymers (165)

Polymer	ipdt, °C
Polystyrene	395
Maleic-hardened epoxy	405
Plexiglass	345
Nylon 66	419
Teflon	555
Kel-F	410
Viton A	460
Silicone resin	505

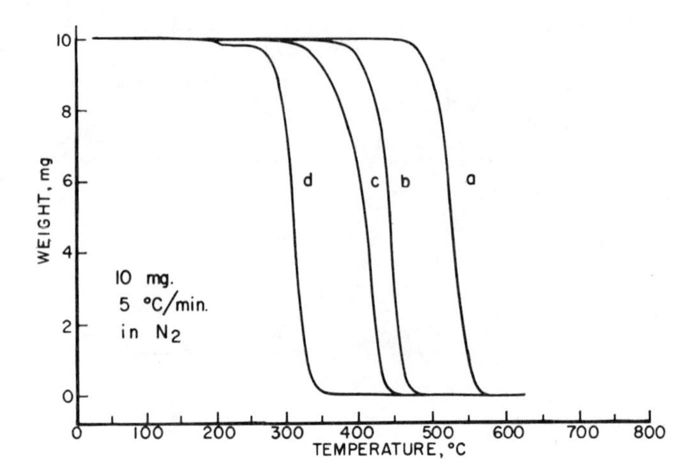

FIG. 40. Relative thermal stability of polymers by dynamic thermogravimetry. (a) Polytetrafluoroethylene; (b) polyethylene; (c) polypropylene; (d) polyhexafluoroethylene. [Reprinted from Ref. (166), p. 30, by permission of Wiley.]

b. Characterization of polymer blends and copolymers. In general the thermal stability of a copolymer falls between those of the two homopolymers and changes in a regular fashion with the copolymer composition (167). This is substantiated by a number of thermogravimetric studies on copolymer systems. An example is the study of Baer (168) on block copolymers, random copolymers, and homopolymers of styrene and α-methylstyrene and is illustrated in Fig. 41.

c. Analysis of additives. Thermogravimetry can be more than a convenient analytical tool for the analysis of polymer additives, either organic or inorganic. For example, Light et al. (152) used this technique for the analysis of silica-filled polytetrafluoroethylene. Figures 42 and 43 represent thermograms obtained on the pure polymer and on three mixtures with colloidal silica in the amount of 10%, 25%, and 50%. One set of experiments was carried in air and the other in helium. When heated in air, mixtures containing up to 23% silica are totally volatilized at 600°C. When heated in inert atmosphere, only the polymer is volatilized at 600°C, leaving the theoretical value of added silica as a residue. In order to account for this interaction between the polymer and silica when analyzed in air, the following stoichiometric equation was proposed,

$$(-C_2F_4-)_n + SiO_2 + 2O_2 \rightarrow 2COF_2 + SiF_4 + 2CO_2 + (n-2)C_2F_4 + C_2F_4 \text{ (recombination products)} \quad (15)$$

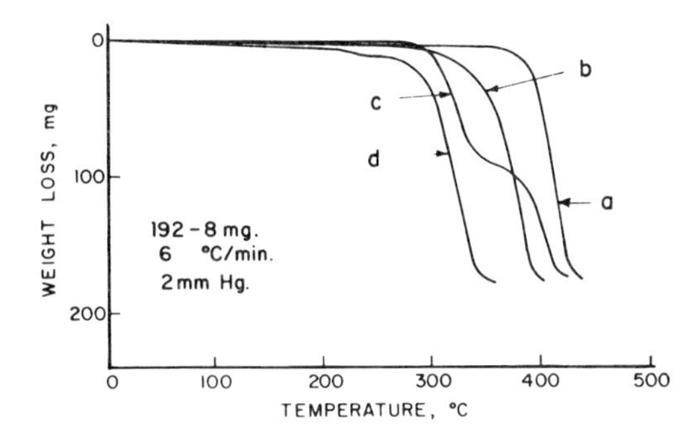

FIG. 41. Thermogravimetric study of styrene-α-methylstyrene copolymers. (a) Polystyrene; (b) styrene-α-methylstyrene (random); (c) styrene-α-methylstyrene (block); (d) polystyrene-α-methylstyrene. [Reprinted from Ref. (168), p. 417, by permission of Wiley.]

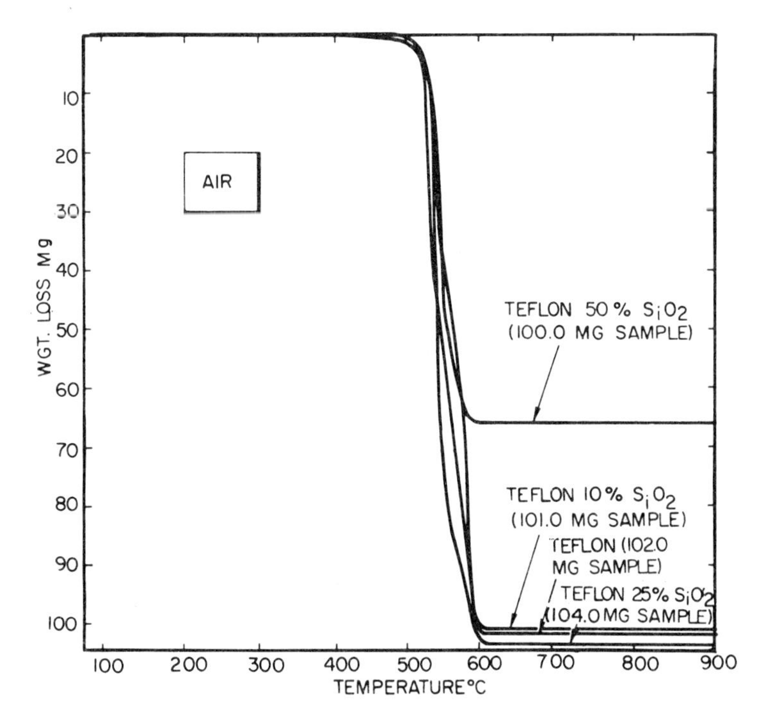

FIG. 42. Thermogravimetric analysis (TGA) curves of polytetrafluoroethylene (PTFE, Teflon) and polytetrafluoroethylene-silica mixtures in air. [Reprinted from Ref. (152), p. 80, by permission of The American Chemical Society.]

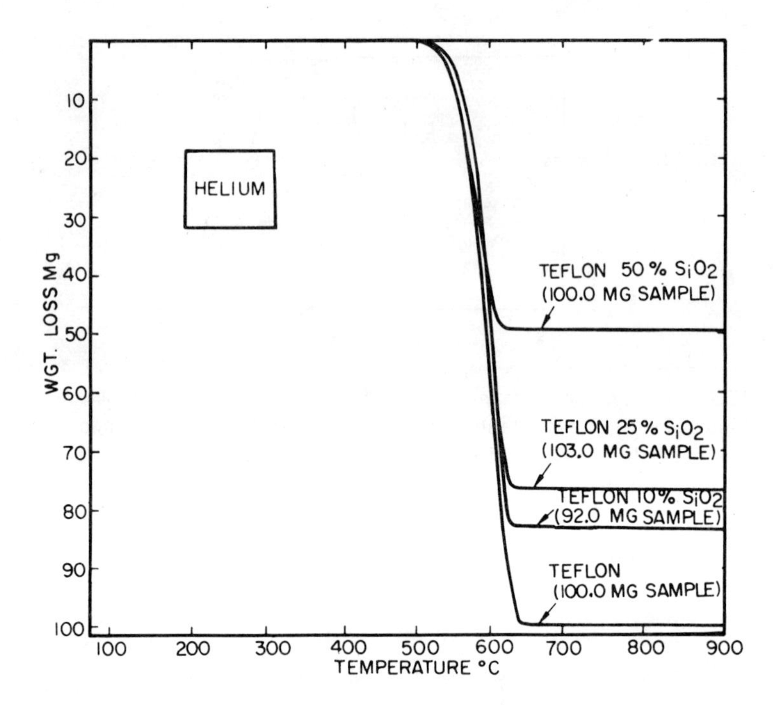

FIG. 43. Thermogravimetric analysis curves of polytetrafluoroethylene and polytetrafluoroethylene-silica mixtures in helium. [Reprinted from Ref. (152), p. 80, by permission of The American Chemical Society.]

The decomposition mechanism, as supported by the stoichiometry of the reaction, the need for oxygen to be present, and the identified decomposition products, suggests several steps whose summation leads to Eq. (15). The first step is a simple decomposition (unzipping) of the polymer which occurs in either inert or oxygen atmosphere.

$$(-C_2F_4-)_n(s) \rightarrow C_2F_4(g) + C_3F_6(g) + \text{cyclo } C_4F_8(g) \tag{16}$$

In an inert atmosphere no further reaction is expected. In air, however, the following two steps were postulated,

$$C_2F_4 + O_2 \rightarrow 2COF_2 \tag{17}$$

and

$$2COF_2 + SiO_2 \rightarrow SiF_4 + 2CO_2$$

d. Sorption-desorption studies. Sorption is the process in which vapor molecules are attached to sites in solids. This process has been used in the study of natural and synthetic polymers. An excellent review of the subject, including methods of data gathering, has been given by McLaren and Rowen (169), who stated that the sorption-solution theories had far outstripped acquisition of data. Thermogravimetric apparatus for the automatic measurement of sorption and desorption isotherms has been described by Hofer and Mohler (170). They studied the sorption kinetics of starch by means of a fully automatic Cahn-Electrobalance.

Theoretical equations have been developed for analyzing the thermodesorption data (171). The cases considered are those for first- and second-order rate processes when desorption occurs in vacuo for a linear temperature program.

H. Simultaneous Use of Thermal Techniques

The addition of differential thermal analysis to a thermogravimetric procedure so as to include information for those enthalpic changes that are unaccompanied by a weight change greatly extends the usefulness of thermal analysis. Acquisition of additional information such as changes in the electrical property, rate of expansion, or the detection and identification of the composition of the gases evolved by the sample greatly increases the utility and application of thermoanalytical techniques to the solution of complex chemical problems, and frequently facilitates the elucidation of their reaction mechanism. The most important instrumental consideration is the design of apparatus for obtaining complementary information simultaneously on the same sample under identical experimental conditions.

1. Simultaneous Thermogravimetric Analysis and Differential Thermal Analysis

By far the most common scanning thermal analytical combination has been simultaneous TGA-DTA, where the thermograms for the same sample are usually displayed on the same recorder chart. A variety of such instruments have been described by Powell (140), Blažec (172), Papailhau (141), Paulik et al. (173), Reisman (40), Krüger and Bryden (145), McAdie (143), and Wiedemann (174). These instruments are modifications of the conventional thermogravimetric apparatus. They feature an additional holder for the reference material which is located near the sample holder. The modification must include provision for the simultaneous measurement of differential temperatures and sample weight. The most important consideration in the design of these instruments is the avoidance of interference on the balance performance by the electric connections necessary to measure the differential temperatures.

2. Thermal Analysis—Effluent Gas Analysis

A polymer may be pyrolyzed on a heated wire or in an oven-type unit, and the pyrolyzate subsequently analyzed with a gas chromatograph, a mass spectrometer, or an infrared spectrophotometer. This technique has been well described, with widely different experimental arrangements, by Madorsky (175), and others (176-179).

For maximum information the effluent gases should be identified continuously with the programmed temperature of the sample. Various arrangements have been described whereby established scanning thermal analytical techniques, such as DTA, DSC, and TGA, have been coupled to simple continuous gas detectors or gas chromatographs, mass spectrometers, and infrared spectrophotometers so that one or more of the processes such as desorption, vaporization, or degradation can be followed. Continuous measurement of evolved unidentified gases, although a common practice, will not be treated here.

a. Mass spectrometry. Langer and Gohlke (180) first used a mass spectrometric thermal analyzer (MTA), which was subsequently modified for simultaneously recording of DTA data (MDTA) (181). A modification of Gohlke's procedure was made by Shulman (182), who apparently was the first to report the use of MTA as a tool in the study of the thermal degradation of polymers.

The coupling of thermogravimetry and mass spectrometry has been described recently by Zitomer (182a), whereby the weight changes of a sample are continuously monitored as the volatiles are collected at various stages in the course of the experiment and intermittently analyzed with a mass spectrometer having rapid-scan capabilities. Figure 44 is a schematic representation of a TGA unit coupled to a time-of-flight mass spectrometer. The purge gas serves the dual function of controlling the atmosphere in the furnace cell and sweeping the volatiles into the mass spectrometer. Noble gases and most fixed gases may be used. The design provides for negligible dead time by using short tubing between the effluent gas port and the metering valve. Temperature is regulated on the entire interfacing unit with variable control heating lines. Figure 45 shows the thermogram of a sample of polymethylene sulfide that was contaminated with residual monomer. Mass spectra for this experiment were scanned magnetically at the numbered points. In Fig. 46 are the mass spectra, in bar form, which were recorded at points 0, 1, 7, 11 of Fig. 45. The mass spectrum at point 0 shows only low levels of air and water which are the results of instrument background and residual air in the furnace area. At point 1, the molecular ions of the first two volatiles evolved, trithiane and o-dichlorobenzene, can be seen at m/e 138 and 146, the former due to

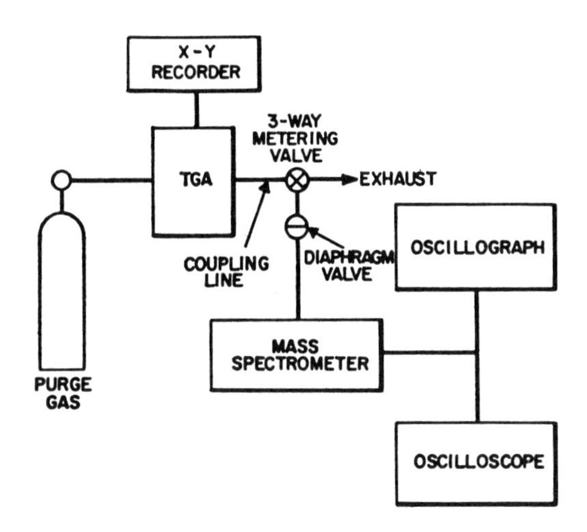

FIG. 44. Block diagram of TGA-MS instrumentation. [Reprinted from Ref. (182a), p. 1092, by permission of The American Chemical Society.]

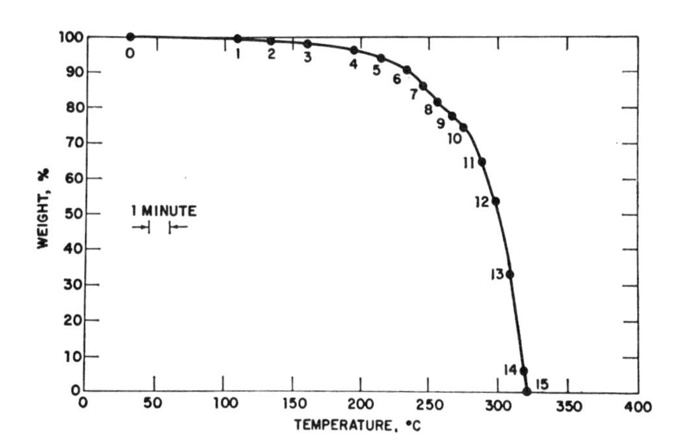

FIG. 45. Thermogram of polymethylene sulfide containing solvent impurities and residual monomer. [Reprinted from Ref. (182a), p. 1092, by permission of The American Chemical Society.]

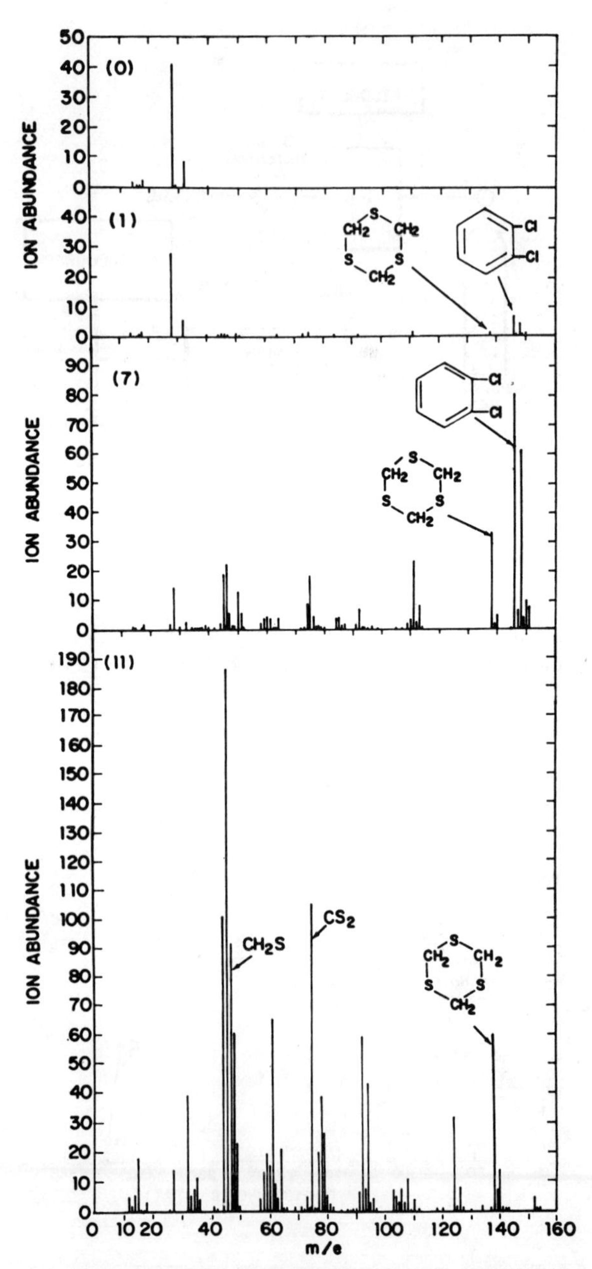

FIG. 46. Mass spectra of volatiles generated from corresponding points shown in the thermogram of Fig. 45. [Reprinted from Ref. (182a), p. 1094, by permission of The American Chemical Society.]

unpolymerized starting material and the latter due to the solvent used. At point 7, just before degradation, the mass spectra of both compounds are strongly evident. At point 11, after the start of polymer degradation, it is seen that the two major degradation products are thioformaldehyde and carbon disulfide. In general TGA-Mass spectrometry should be particularly useful in the investigation of trace solvent or impurities in addition to kinetic and mechanism studies (182b).

b. Gas chromatography. One of the earliest attempts to combine thermogravimetry and gas-effluent analysis was reported by Blažek (172), who also included DTA in the simultaneous analysis. The possibility of coupling TGA with gas chromatography was reported by Cano (183).

A detailed description of a TGA-Gas chromatographic apparatus was recently reported by Chiu (184). The usefulness of the technique is illustrated in the study of polymer blends. For example, the results for a blend of polystyrene and poly-(α-methylstyrene) is shown in Fig. 47. The upper diagram shows the thermogram for each of the polymers and for a one-to-one physical mixture. The lower diagram represents pyrograms, in the form of bar graphs, for the solvent and model substances. The last three rows correspond to pyrograms for the blend at points 1, 2, and 3 in the thermogram. It can be seen that a stepwise analysis may provide both qualitative and quantitative information of each component in the sample in a manner somewhat similar to that for mass spectrometric techniques. However, the latter technique has the advantage of the relatively short dead time, that is, the time between product evolution and the readout. Gas chromatography is basically a batch method. In the case in which a reaction is associated with evolution of more than one gas, gas chromatography may greatly simplify identification, especially if the eluted gases are further analyzed by mass or infrared spectroscopy.

c. Infrared absorption. A number of rapid-scan spectrometers have been described in the literature (185-188), some with scan speeds up to 20,000 scans/sec. Clearly instruments of this type can be adapted to gas-effluent analysis. The various infrared techniques adaptable to this analysis have been reviewed by Low (189).

d. Simultaneous thermogravimetric, derivative thermogravimetric, differential thermal and electrothermal analysis. A system that simultaneously records changes in weight (TGA), rate of change in weight (DTG), enthalpy change (DTA), and electric conductivity (ETA) on the same sample has been described by Chiu (10). The system has been designed around a commercial thermal analyzer. This technique has been shown to be valuable in studying complex systems involving consecutive or overlapping thermal events. The features of the technique have been illustrated on studies including polyethylene oxide and polyacrylonitrile.

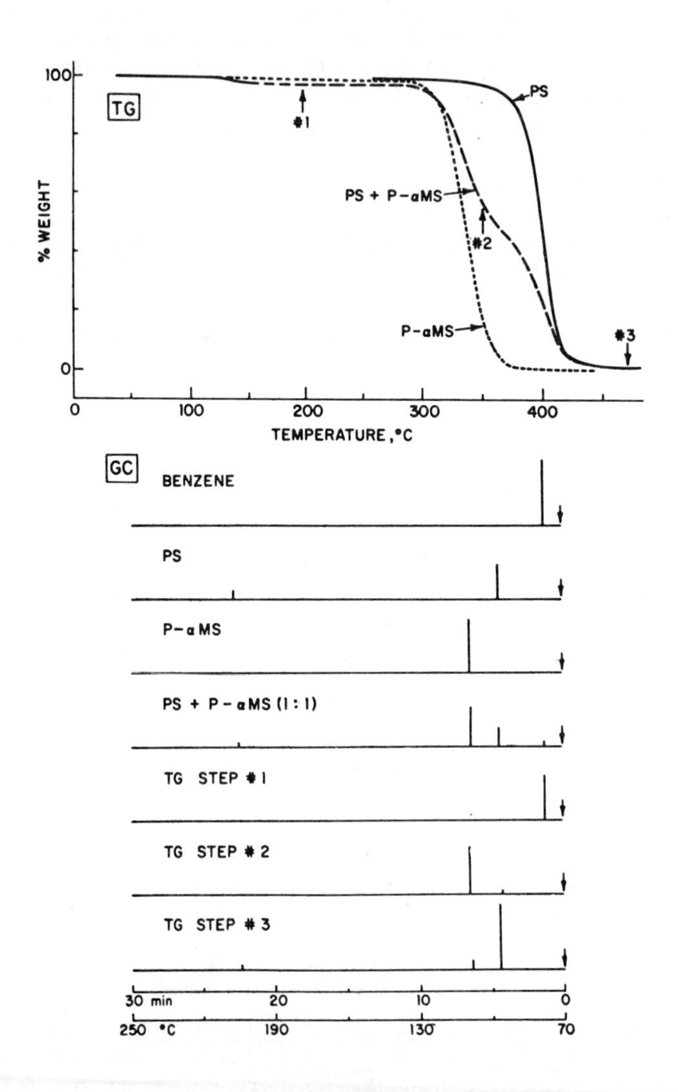

FIG. 47. Thermogravimetric analysis-gas chromatography of polystyrene (PS), poly (α-methylstyrene) (P-αMS), and their blend; TGA conditions: sample size, 2 mg; helium flow, 160 ml/min; heating rate, 5°C/min. [Reprinted from Ref. (184), p. 1518, by permission of The American Chemical Society.]

I. Miscellaneous Thermal Techniques

1. Thermodilatometric Analysis

Thermodilatometric analysis (TDA) is the technique that involves the continuous measurement of the sample length as it is heated or cooled at a constant rate. The information derived from it is especially useful in the determination of glass transition temperatures. A number of commercial instruments are available. However, a simple modification of DTA apparatus and the addition of an electromechanical amplifier for detection of sample expansion was described by Barrall et al. (190). The apparatus has a maximum sensitivity of 2.5×10^{-5} in. of sample expansion per inch of recorder deflection and is capable of operation down to liquid nitrogen temperatures. Figure 48 shows the expansion record for a sample of polypropylene, in addition to quartz and copper used to check the sensitivity. Through the use of expansion coefficients above and below T_g, measurement of the crystallinity of the polymer is also possible using the correlation of Beck et al. (191).

2. Thermal Volatilization Analysis

A new method for the study of polymer degradation has been recently described by McNeill (192-194). Thermal volatilization analysis (TVA) depends on the fact that when a polymer degrades to volatile products in a

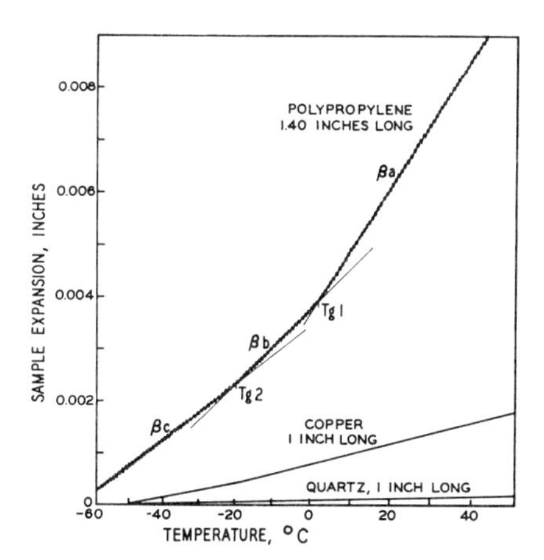

FIG. 48. Thermal expansion records for polypropylene, quartz, and copper. [Reprinted from Ref. (190), p. 2317, by permission of The American Chemical Society.]

continuously evacuated system, a small pressure will exist between the hot sample and a cold trap placed some distance away due to the time taken by the products to distil from the hot to the cold surface. The magnitude of the pressure developed varies with the rate of volatilization of the sample and sample size, which normally falls in the range 10^{-5} to 10^{-1} torr. The sample is temperature programmed at a linear rate, while the pressure is continuously recorded with a Pirani gauge. Figure 49 is a schematic of the apparatus required for this technique. A flat-bottomed glass tube containing the polymer sample powder or film is inserted in a temperature programmed oven. The Pirani gauge, E, which is between the tube and the trap senses the changes in pressure. The gauge control unit provides an output for a potentiometric recorder so that a continuous trace of pressure versus time (temperature) may be obtained.

The apparatus has several advantages in that it is simple and sturdy, having no moving parts. Perhaps of greater importance is that, unlike TGA, the rates of volatilization are measured, thus giving what would be essentially the first derivative of a TGA trace.

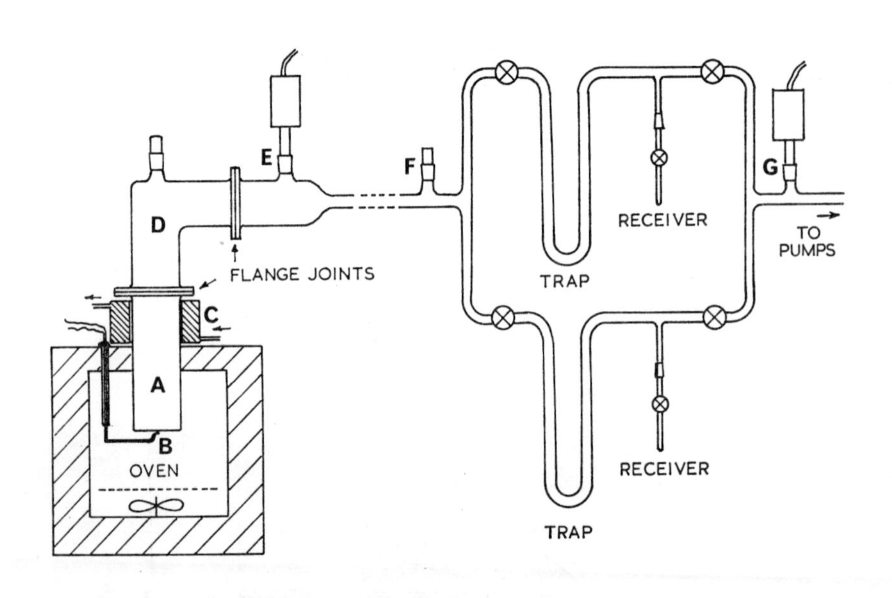

FIG. 49. Layout of TVA apparatus. (A) Pyrex glass degradation tube; (B) chromel-alumel thermocouple; (C) cold water jacket; (D) removable lid; (E), (F), and (G), attachment points for Pirani gauge heads. [Reprinted from Ref. (193), p. 411, by permission of Pergamon Press.]

Figure 50 shows the TVA curves obtained for three samples of poly(methyl methacrylate) of different initial molecular weights. The thermograms show that the rate of degradation at a given temperature between 220°C to 300°C is lower for the samples of higher molecular weight (first large peak), but higher at a given temperature between 320°C and 350°C. These observations are consistent with the accepted view that the first stage in the degradation of poly(methyl methacrylate) is end initiated and the second is initiated by random chain scission.

3. Thermoparticulate Analysis

A recent technique, thermoparticulate analysis (TPA), for the thermal analysis of polymeric materials (195) is based on the detection of condensation nuclei evolved from a polymer sample that is temperature programmed. Condensation nuclei are particles that have radii of 10^{-7} to 10^{-5} cm. When humidified to approximately 100% RH, and allowed to expand adiabatically, moisture condenses on the nuclei, effecting a particle growth to

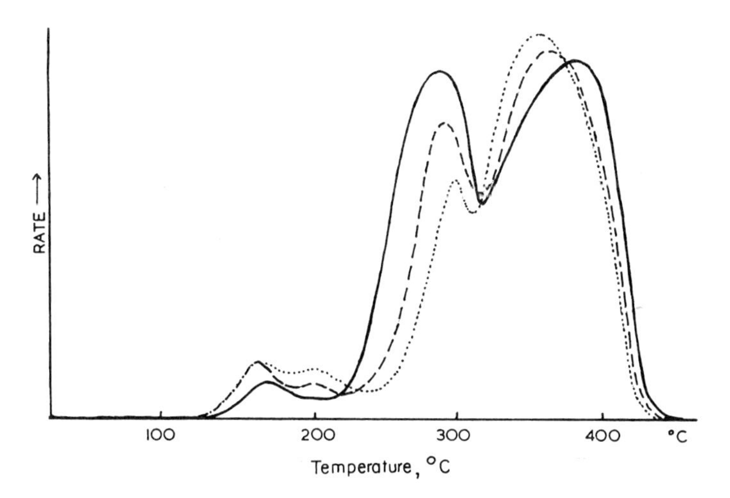

FIG. 50. The effect of initial molecular weight on the thermal stability of poly(methyl methacrylate) samples prepared by the free radical mechanism at 60°C.

Dotted line, $\overline{M}_n = 480,000$

Dashed line, $\overline{M}_n = 100,000$

Full line, $\overline{M}_n = 20,000$

[Reprinted from Ref. (194), p. 24, by permission of Pergamon Press.]

micron size in a few milliseconds. The particulate matter so developed can scatter light in a dark field optical system. A photodetector registers the intensity that is directly proportional to the number of condensation nuclei originally present in the gas phase. The essential equipment for TPA is the condensation-nuclei counter (196). In addition to the direct application of the technique, indirect applications involving the selective conversion of gaseous decomposition products to condensation nuclei is also possible. An excellent review of this technique with applications to the polymer field has been given by Murphy (197).

4. Thermoluminescence

Thermoluminescence analysis is the technique that involves the measurement of the emitted light energy of a sample as a function of temperature. The sample is normally heated at a linearly programmed rate to a temperature well below that of incandescence. A few polymers have been studied by this method, these being polyethylene (198-200) as well as fluorinated polymers (200, 201).

In this technique one measures three quantities, namely, total light emission as a function of time, temperature as a function of time and spectral composition of the glow. Figure 51 shows the schematic arrangement of an apparatus that consists of a sample holder and heater, a thermocouple, a microphotometer to measure total intensity, and a monochromator to analyze the emitted radiation.

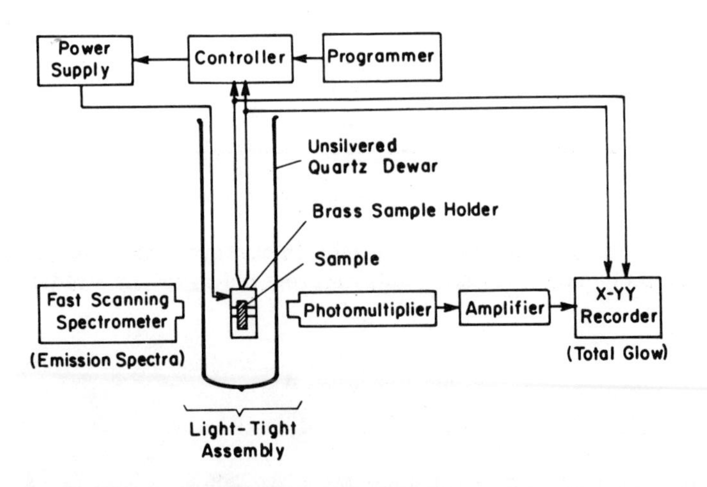

FIG. 51. Schematic arrangement of thermoluminescence apparatus. [Adapted from Ref. (202), p. 460, by permission of Pergamon Press.]

In the experiments on γ-ray irradiated polyethylenes at liquid nitrogen temperature and in the absence of oxygen, three distinct peaks were revealed (α, β, γ), which have different relative heights. A high degree of crystallinity favored the α peak at the expense of the β and γ peaks. However, the peaks were quite unaffected by the various impurities present in commercial polyethylenes (199). This offers the possibility of comparing samples of polyethylenes that are chemically identical but of different degrees of crystallinity. Additional findings report the coincidence of glow peaks with the appearance of molecular motions. These results have been interpreted on the basis of electron-trap theory.

5. Oxyluminescence

When some polymers are heated in the presence of oxygen to temperatures in the neighborhood of 200°C, luminescence results. This phenomenon was investigated by Ashby (203) who ascribed the name "oxyluminescence." The presence of oxygen was found to be necessary, the intensity of emission being proportional to the percentage of oxygen in contact with the polymer surface as illustrated in Fig. 52. Experimentally the same apparatus used for thermoluminescence, shown in Fig. 51, may be used in oxyluminescence to measure the low light levels of 10^{-10} to 10^{-8} lumens emitted. This technique appears to be useful in the study of antioxidant action of compounds.

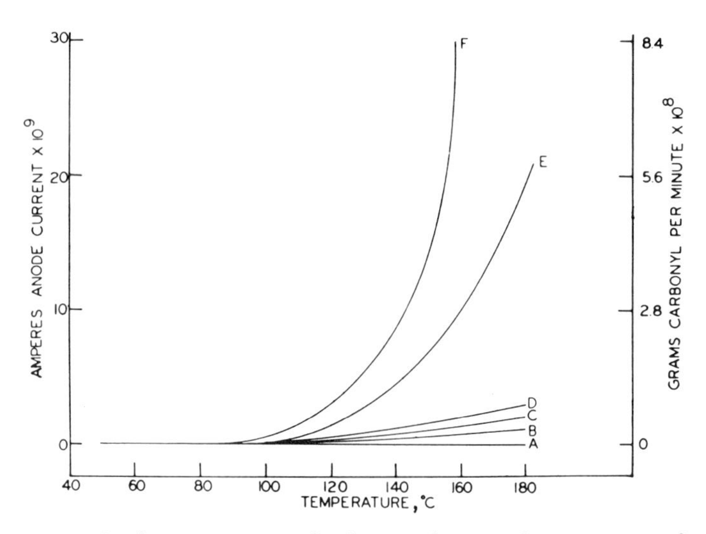

FIG. 52. Oxyluminescence of polypropylene in the presence of varying amounts of oxygen. (A) Argon; (B) 2.6% oxygen in nitrogen; (C) 6.4% oxygen in nitrogen; (D) 11.0% oxygen in nitrogen; (E) air; and (F) oxygen. [Reprinted from Ref. (203), p. 101, by permission of J. Wiley.]

III. KINETIC ANALYSIS

A. Differential Thermal Analysis

Considerable attention has been directed toward the kinetic analysis of thermal traces (204). The methods have been numerous, each method offering some particular advantage in obtaining numerical values for the specific rate constant k, the activation energy E, the order of reaction n, and the pre-exponential factor Z, as defined by the Arrhenius equation,

$$k = Z \exp(-E/RT) \tag{19}$$

Although kinetic analysis is relatively simple for isothermal procedures, there are advantages in using programmed heating. One nonisothermal experiment may take the place of many isothermal ones. The nonisothermal procedure may be used as a convenient scanning method. Most important, the continuous temperature changes may reveal the existence of a temperature range of special interest that could be overlooked if the conventional isothermal procedure were used. The differential thermal trace of a system undergoing two or more simultaneous reactions, each having its individual set of kinetic parameters, may exhibit resolved peaks as the entire temperature spectrum is scanned. Subsequent use of the usual and more precise isothermal procedures may then be undertaken at temperatures that are most meaningful.

Most polymer reactions examined by thermal techniques have involved thermal degradation (111). These reactions are considered essentially heterogeneous. Factors such as particle size and shape as well as the mode of sample packing may be found to be significant. The latter factor may not only affect the temperature gradient but also the retention of a reaction products. Other reactions that have been investigated to a minor extent have been polymerization reactions (205a) and the denaturation of some bipolymers (205b). A wide variety of kinetic analysis methods are available for each thermal technique. Our discussion will be restricted to DTA, DSC, and TGA. With minor modifications, the kinetic analysis for these methods may usually be adapted to other thermal methods.

We may write that the time rate of conversion $d\alpha/dt$ is given as

$$d\alpha/dt = kf(\alpha) \tag{20}$$

where α is the degree of completion of the reaction with respect to the initial sample. It should be noted that the specific rate constant k as defined in Eq. (20) has the physical dimensions of a first-order rate constant, that is, reciprocal time, regardless of the order or type of reaction.

We shall consider DTA traces first. As has been made clear in the earlier section of this chapter, the DTA cell may appear as a simple device but in fact is a complicated reactor. Though no analytical model has been devised to take all phenomena into account, three basic models which overlap to some degree have been used by most workers and will be discussed here.

1. Model Based on Similarities of Differential Thermal Analysis and Differential Thermogravimetric Analysis

In this case one assumes that the rate of heat absorption or liberation is proportional to the rate of reaction so that the peak differential temperature T_m occurs at the maximum reaction rate. Although this assumption has been questioned by Reed and co-workers (206, 207) more recently; Akita and Kase (208) have indicated that any discrepancy between peak temperature of the DTA trace and the maximum reaction rate will probably be negligible under the usual experimental conditions.

It appears that the differential equation for many reactions at a given temperature may be written as

$$d\alpha/dt = k(1-\alpha)^n \tag{21}$$

The substitution of Eq. (19) in Eq. (21) when the heating rate is $dT/dt = \Phi$ yields

$$d\alpha/dT = (Z/\Phi)\exp(-E/RT)(1-\alpha)^n \tag{22}$$

Setting $d^2\alpha/dT^2 = 0$ for the maximum reaction rate and marking the corresponding temperature as T_m yields

$$E/RT_m^2 = n(Z/\Phi)(1-\alpha_m)^{n-1}\exp(-E/RT_m) \tag{23}$$

Here α_m refers to α at T_m. For the frequently occurring case where $n = 1$, Eq. (23) yields

$$E/RT_m^2 = (Z/\Phi)\exp(-E/RT_m) \tag{24}$$

The above equation is inconvenient to use. However, Eq. (24) may be rewritten and differentiated so that

$$\frac{d[\ln(\Phi/T_m^2)]}{d(1/T_m)} = -\frac{E}{R} \tag{25}$$

Assuming then that $n = 1$, a series of traces at various linear heating rates, each trace yielding a particular value for T_m provides the information for the extraction of the activation energy.

In general for $n = 1$ or 0 it can be shown (67) that

$$n(1 - \alpha_m)^{n-1} \simeq 1 + (n - 1)(2RT_m/E) \tag{26}$$

In view of the fact that n is a small number so that

$$(n - 1)(2RT_m/E) \ll 1 \tag{27}$$

substitution of $n(1 - \alpha_m)^{n-1} \simeq 1$ in Eq. (23) leads to Eq. (25) regardless of the order of the reaction. Equation (25) appears to yield reliable values for the activation energy (67). More recently in the study of the kinetics of a glass transition McMillan (209) obtained a value of $E = 42.4 \pm 1.2$ kcal/mole using Eq. (25) as compared to the literature value of 40.0 kcal/mole obtained by a dielectric relaxation method.

Concerning the order of a reaction, Kissinger observed that the value of $(1 - \alpha_m)$, that is, fraction of undecomposed reactant at T_m, increased with increasing n, as may be inferred from Fig. 53. A shape index, S, was defined as the ratio of the slopes of the tangents to the curve at the inflection points as indicated in Fig. 54. The shape index was found related to n by

$$n \simeq 1.26S^{\frac{1}{2}} \tag{28}$$

The values obtainable for n are not very precise. The Kissinger method has been applied to vinyl polymers (210).

2. Model for Stirred Homogeneous Solutions

a. Method of Borchardt and Daniels. A number of DTA methods are based on the original treatment of Borchardt and Daniels (211, 212). In this treatment the enthalpy of the reaction ΔH may be written as

$$\Delta H = KA \tag{29}$$

where A is the total area under the DTA trace and K is the heat transfer coefficient. The latter quantity depends on cell geometry, thermal conductivity, and heat capacity of the sample. The heat evolved or absorbed should be proportional to the number of moles reacting, that is,

$$dH = -(KA/N_0)\, dN \tag{30}$$

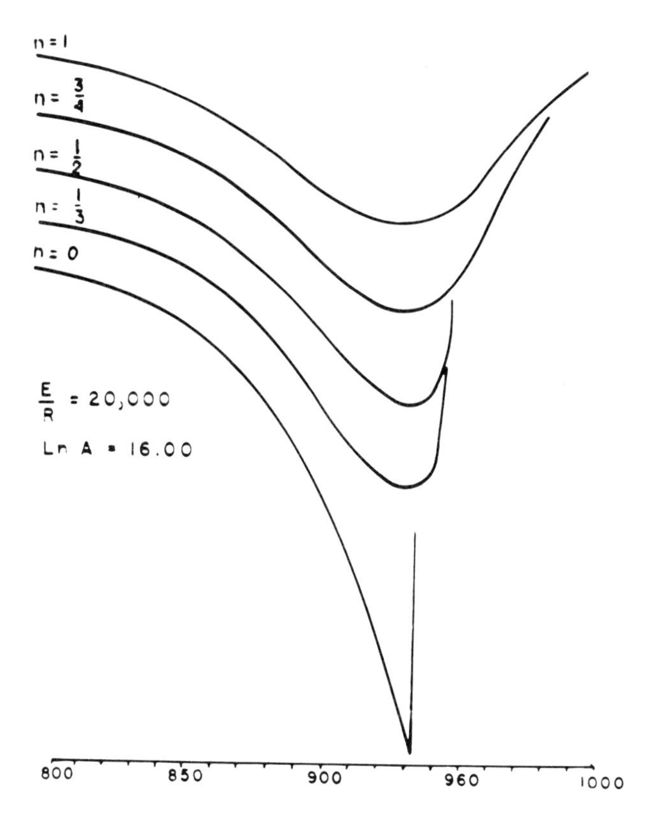

FIG. 53. Effect of order of reaction on plots of reaction-rate versus temperature for constant heating rate, frequency factor, and activation energy. [Reprinted from Ref. (67), p. 1703, by permission of the American Chemical Society.]

Here N_o is the initial number of moles of reactant and $-dN$ is the loss of moles of reactant corresponding to the enthalpy change dH.

Equation (30) may be written in terms of weights so that

$$dH = (KA/w_o)\ dw \tag{31}$$

From the heat balance equation,

$$dH = C_p\ d(\Delta T) + K\ \Delta T\ dt \tag{32}$$

where C_p is the total heat capacity of the sample in the cell (assumed to be

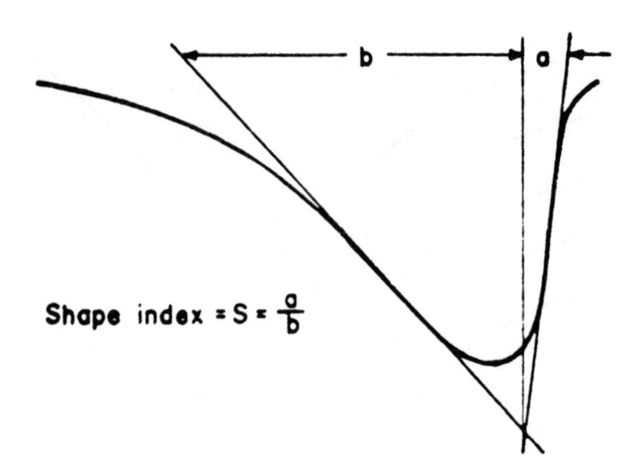

FIG. 54. Method for measuring amount of asymmetry in an endothermic differential thermal analysis peak. [Reprinted from Ref. (67), p. 1703, by permission of the American Chemical Society.]

equal to the heat capacity of the cell containing the inert substance only) and ΔT is the temperature difference between the two cells. Combining Eqs. (31) and (32) we obtain the relation,

$$-dw/dt = (w_o/KA) \quad [C_p d(\Delta T)/dt + K\,\Delta T] \tag{33}$$

Using the Borchardt assumption

$$K\,\Delta T \gg C_p\, d(\Delta T)/dt \tag{34}$$

Equation (33) becomes

$$-(1/w_o)\, dw/dt = d\alpha/dt \simeq \Delta T/A \tag{35}$$

The variable t may be replaced by temperature T, so that

$$-(1/w_o)\, dw/dT \simeq \Delta T/A \tag{36}$$

provided temperature is plotted as the abscissa of the DTA curve. It should be clear that both w and w_o refer to the active sample. These values are obtained by subtracting the inactive residue (at the end of the heating process) from the actual weights of the system. Integrating Eq. (36), that is,

$$-\int_{w_o}^{w} \frac{dw}{w_o} \simeq \int_{T_o}^{T} \frac{\Delta T}{A}\, dT$$

$$w/w_o \approx \tilde{a}/A \tag{37}$$

where

$$\tilde{a} = A - \int_{T_0}^{T} \Delta T \, dT \tag{38}$$

or

$$\tilde{a} = A - a \tag{39}$$

as illustrated in Fig. 55. Both Eqs. (36) and (37) make possible a variety of kinetic methods similar to those for TGA. Obviously these equations contain the assumption that DTA and DTG curves are similar.

b. Method of Freeman and Carroll (214). Consider a simple reaction of the form

$$d\alpha/dt = k(1 - \alpha)^n$$

so that

$$d\alpha/dt = (Z/\Phi) \exp(-E/RT) (1 - \alpha)^n \tag{23}$$

Taking the logarithm of both sides of Eq. (23), differentiating, and then integrating, we obtain for a particular fixed linear heating rate

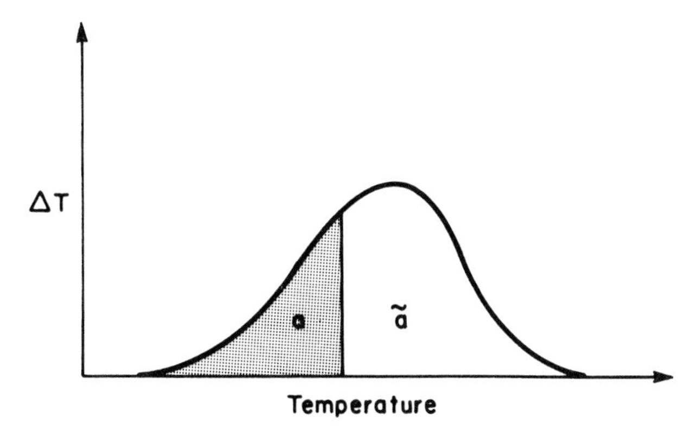

FIG. 55. Simple exothermic DTA curve schematically showing a and $\tilde{a}$ as used in Eq. (39).

$$\Delta \ln(d\alpha/dT) = n\Delta \ln(1-\alpha) - (E/R)\,\Delta\,(1/T) \tag{40}$$

providing we assume the frequency factor Z is temperature independent. In terms of the DTA trace, Eqs. (35), (38), and (40) may be combined as

$$\Delta \ln(\Delta T) = n\Delta \ln\tilde{a} - (E/R)\Delta\ (1/T) \tag{41}$$

At this point there is a choice of procedures in order to obtain both E and n from a single DTA trace. We may plot

$$\Delta \ln(\Delta T)/\Delta \ln \tilde{a} \text{ vs } \Delta(1/T)\Delta \ln \tilde{a} \tag{42}$$

the slope leading to a value of E/R and the intercept yielding a value for the order of the reaction n. An alternative procedure (215, 216) is to divide the thermal trace into segments so that interval $\Delta\,(1/T)$ may be held constant. Under these conditions a plot of

$$\Delta \ln\,(\Delta T) \text{ vs } \Delta \ln \tilde{a}$$

should yield a linear relationship whose slope is n and intercept is (E/R). Knowing E and n, it is a simple matter to use the original differential rate equation and the Arrhenius equation to obtain the specific rate constant k and the frequency factor Z.

All of the above kinetic equations have been given on a weight basis instead of the usual concentration basis. This has been done since it is questionable in the case of reactions in the solid state, in which the reactions are heterogeneous as a rule, whether concentration is meaningful. Furthermore, seldom, if ever, are volumes experimentally determined during a kinetic run. However, in the case of homogeneous reactions such as polymerizations in solution, concentration terms should be used. Setting up the differential rate equation in terms of moles of reactant per unit volume will lead to the Borchardt-Daniels expressions. Starting with

$$-(1/V)\, dN/dt = k(N/V)^n \tag{43}$$

where V is the volume, Eq. (33) may be written as

$$k = \frac{(KAV/N_o)^{n-1}[C_p d(\Delta T)/dt + K\,\Delta T]}{[K(A - a) - C_p\,\Delta T]^n} \tag{44}$$

Using the assumption contained in Eq. (34) leads to

$$k = (AV/N_o)^{n-1}\,\Delta T/(A - a)^n \tag{45}$$

where a is the area under the DTA curve from time 0 to t. Equations (44) and (45) may be treated in the previous manner, yielding values for E and n (214) provided V is constant.

c. Method of Piloyan et al. (218) The general expression for a reaction rate may be written as

$$d\alpha/dT = (Z/\Phi)f(\alpha)\exp(-E/RT) \tag{46}$$

where $f(\alpha)$ is a function of extent of the reaction. Substituting Eq. (35) in Eq. (46) and taking logarithms, the following equation is obtained

$$\ln \Delta T = c + \ln f(\alpha) - E/RT \tag{47}$$

where

$$c = \ln A + \ln (Z/\Phi)$$

On the assumption that under usual laboratory conditions, particularly at the early stage of reaction, ΔT changes more rapidly than α, we may write

$$\ln \Delta T = c = E/RT \tag{48}$$

A plot of $\ln \Delta T$ versus $1/T$ from approximately the point of maximum curvature to about the peak of the trace should provide the information required for the determination of activation energy. Comparison of the author's data for several inorganic compounds with data from other source indicates that this simple procedure of data handling yields good results. In this method the order of reaction is lost. However, Reich has suggested a novel procedure to obtain the order for the case of a simple reaction.

d. Method of Reich (217). Rewriting Eq. (47) so that

$$f(\alpha) = (1 - \alpha)^n$$

we have

$$\ln(\Delta T) = c + n \ln(1 - \alpha) - E/RT \tag{49}$$

For any two temperatures prior to or after the DTA peak, which possess the same value for ΔT, we have according to Eq. (41)

$$n = -(E/R)\Delta(1/T)/\Delta \ln \tilde{a} \tag{50}$$

Thus n may be obtained in a simple manner and subsequently used in Eq. (49). The procedure may be summarized in the following equations,

$$\ln(\Delta T) = \ln c - V(E/R) \tag{51}$$

where

$$V = (1/T) - [\Delta(1/T)/\Delta \ln \tilde{a}] \ln \tilde{a} \tag{52}$$

This equation differs slightly from that given by L. Reich. Since in Eq. (52) both $\Delta(1/T)$ and $\Delta \ln \tilde{a}$ refer to some sequence of temperature, that is,

$$\Delta(1/T) = (1/T_2) - (1/T_1) \text{ and } \Delta \ln \tilde{a} = \ln \tilde{a}_{T_2} - \ln \tilde{a}_{T_1}$$

the Reich equation is equivalent to the one given in Eq. (52).

It will be seen that when n = 0, Eq. (49) becomes identical to Eq. (48). It is interesting to note that Reich tested both methods using the decomposition of benzene-diazonium chloride in aqueous solution. A value of E of 30.5 kcal/mole was obtained for Eq. (51). This is about 2 kcal higher than the value given by Borchardt and Daniels. A plot of the data according to Eq. (48) yielded a value of 22 kcal/mole. However, a value of 26 kcal/mole was obtained when the initial experimental points were considered. The most reliable value obtained by the conventional isothermal method appears to be 27.025 kcal as obtained by Moelwyn-Hughes and Johnson (219). The value for n which is unobtainable by the Piloyan method was found by Reich to be 1.08. The accepted value for the order of benzene-diazonium chloride decomposition is unity.

Reich has suggested several alternative methods for determining kinetic parameters from thermal methods (111, 220). An interesting method is based on the use of two DTA traces obtained at different heating rates. Here it is not necessary to evaluate the area under the DTA traces in order to obtain E and n. Using an approximate procedure whereby DTA traces are "linearized" values of E and n have been estimated for polyethylene, polypropylene, and polystyrene in fair agreement with other methods.

B. Differential Scanning Calorimetry

Whereas in DTA a temperature change is measured between base line and trace as the sample is being heated or cooled, in DSC the power required to prevent such temperature change is measured. Thus dH/dt, the time rate of change of the enthalpy of the reaction, is a direct observable. The simplicity and exactness of this recently developed instrument should make it widely applicable for kinetic analysis.

The area under the thermal trace is a measure of the enthalpy of the reaction, and

$$dH/dt = (\Delta H/N_o)\ (dN/dt) \tag{53}$$

Also

$$N/N_o = (A - a)/A \tag{54}$$

where N_o is the initial number of moles of reactant and N is the number of moles of reactant remaining at time t. We consider a simple reaction again. Since DSC apparatus is frequently used in solution kinetics, we write

$$dc/dt = kc^n \tag{55}$$

where c represents the molar concentration of the reactant. Equation (55) may be written as

$$dN/dt = KN^n/V^{n-1} \tag{56}$$

Substitution of Eq. (53) in Eq. (56) yields

$$dH/dt = k(N^n/V^{n-1})(\Delta H/N_o) \tag{57}$$

or

$$k = (V^{n-1}/N^n)(N_o/A)(dH/dt) \tag{58}$$

upon setting $\Delta H = A$. Equation (58) may be simplified by making use of Eq. (54). The relationship needed is

$$N^n/N_o{}^n = (A - a)^n/A^n \tag{59}$$

Substitution of Eq. (59) in Eq. (58) leads to

$$k = (AV/N_o)^{n-1}(dH/dt)/(A - a)^n \tag{60}$$

Equation (60) may be used isothermally to obtain the specific rate constants at various temperatures. A subsequent plot of the rate constants versus (1/T) yields the activation energy.

An alternative procedure is to use a linear heating rate along with the Freeman and Carroll method. The method may be applied to Eq. (60) as follows

The Arrhenius equation is substituted in Eq. (60) so that

$$(Z/\Phi)\ \exp\ (-E/RT) = (AV/N_0)^{n-1}(A - a)^n(dH/dt) \tag{61}$$

where Φ is the linear heating rate and the abscissa of the trace is on a temperature basis. Taking logarithms of both sides of Eq. (61), differentiating, and then integrating, we obtain

$$\Delta \ln\ (dH/dT) = -\ (E/R)\,\Delta\,(1/T)\ + n\Delta\ \ \ln\ (A - a) \tag{62}$$

provided that it is assumed that the volume does not change. Thus, a plot of Eq. (62), such as

$$\Delta \ln\ (dH/dt)/\Delta\ \ \ln\ (A - a)\ \text{vs}\ \Delta\,(1/T)/\Delta\ \ \ln\ (A - a) \tag{63}$$

will lead to an intercept for n; the slope will yield a value of (E/R). The procedure of dividing up the trace into constant values for $\Delta\,(1/T)$ as was suggested earlier in the case of DTA may be found convenient here also (222). Because the ordinate displacement between the base line and thermal curve is equal to dH/dT, no mathematical approximations are necessary to yield kinetic parameters. In regard to the volume being constant, this assumption does not appear to be relevant to solid state reactions since the appropriate rate equation is of the form given in Eq. (21). The above procedure was recently suggested and used by Ellerstein on a number of polymers (222). Other procedures have been suggested for DSC (223). Also the treatment of base-line deviations, particularly for the case of fast scanning rates, has been considered in detail by Heuvel and Lind (224).

C. Thermogravimetric Analysis

The application of TGA in determining the kinetics of polymer systems has been reviewed critically by Flynn and Wall (225). The subject has also been reviewed by Reich and Levi (159) with emphasis on specific polymers. A wide variety of methods have been developed, assessed, and reassessed. Clearly the balance is capable of enormous accuracy and precision and in the case of the thermal decomposition of polymers the use of TGA has been extremely widespread. Most methods may be classified as derivative (182b) or integral.

1. Derivative Methods

a. Method of Freeman and Carroll (214). Freeman-Carroll method has been widely used in spite of its limited precision which some have reported (226). Defining the decomposition rate in terms of weight lost per unit time we write

$$-dw/dt = kf(w) \tag{64}$$

where w is the weight of the reactive portion of the sample, that is, the actual weight at time t minus the final weight of the char. If we assume $f(w) = w^n$, then

$$-dw/dt = Z \exp(-E/RT)w^n \tag{65}$$

provided that f(w) is a constant at constant w and is independent of the temperature. Equation (65) may be written as

$$-dw/dT = (Z/\Phi) \exp(-E/RT)w^n \tag{66}$$

where Φ is the linear heating rate, so that

$$\Delta \ln(-dw/dt) - (E/R)\,\Delta(1/T) + n\Delta \ln w \tag{67}$$

The above equation is usually written as

$$\Delta \log(-dw/dT)/\Delta \log w = [(-E/R)\,\Delta(1/T)/2.303\,\Delta \log w] + n \tag{68}$$

and leads to values for E and n by plotting $\Delta \log(-dw/dT)/\Delta \log/\Delta \log w$ vs $\Delta(1/T)/\Delta \log w$.

It has been found convenient to have the logarithm of the experimental rates plotted initially against 1/T so that the smoothed-out curve may be used to obtain a set of values for constant increments in 1/T. Keeping $\Delta(1/T)$ constant a subsequent plot of $\Delta \log(-dw/dT)/\Delta \log w$ versus $1/\Delta \log w$ leads to a value of n as the intercept. The activation energy is obtained from the slope. The reliability of this procedure is illustrated for the case of the vaporization of octamethylcyclotetrasiloxane. Here it is known that the vaporization is a zero-order rate process and that the activation energy should be equal to the thermodynamic latent heat of vaporization, this being 10.9 kcal $mole^{-1}$. Identical values are obtained from the analysis of the thermal trace (216).

b. Method of multiple heating rates (159, 182b). In the Freeman-Carroll method an analytical form for f(w) or $f(\alpha)$ has to be assumed.

However, it is possible to obtain the activation energy without this assumption if the thermal traces are obtained for several different constant linear heating rates. In this case we write Eq. (66) as

$$\ln [\Phi(-dw/dT)] = \ln[Zf(w)] - E/RT \tag{69}$$

Since f(w) is assumed to be constant at constant w, a plot of $\ln[\Phi(-dw/dT)]$ versus (1/T) for a given value of w at various values of Φ will lead to a value of E. This method may be used to check changes in activation energy as a function of the progress of the reaction. The value for n may be obtained by setting the intercepts of the above plots, that is, $\ln[Zf(w)]$ equal to $\ln Z + n \ln w$. A plot of the intercepts against ln w will lead to values for n and Z.

c. Method of maximal rates. The method of maximal rates is applicable to reactions showing one or more successive thermogravimetric inflection points. The location of the inflection point and the slope of the curve at that point must be determined. The order is unavailable; the value of (E/n) instead may be calculated. This will be seen from the following: At the point of inflection $(d^2w/dT^2)_i = 0$; also $(d \ln (dw/dT)/dT)_i = 0$, from Eq. (66) it follows that $\ln (dw/dT) = -\ln (Z/\Phi) - E/RT + n \ln w$; taking the derivative of the above equation with respect to T at the inflection point yields $0 = (E/R)(1/T_i^2) + n(d \ln w/dT)$ or

$$E/n = (RT_i^2/w_i)\ (dw/dT)_i \tag{70}$$

Frequently the order is obtained in some other way so that the maximal rate may be used to obtain the activation energy. However, the precision of this method which is based on locating a single point will be seen to be rather limited.

d. Method of Achar, Brindley, and Sharp. A derivative method has recently been suggested (226). If the correct a priori analytical form $f(\alpha)$ is assumed, then a linear plot may be obtained. This may be seen on rearranging the terms in Eq. (69) and replacing w with α

$$\ln[1/f(\alpha)\ (d\alpha/dT)] = \ln (Z/\Phi) - (E/RT) \tag{71}$$

Thus a plot of the left-hand side of Eq. (71) versus 1/T should result in a straight line and yield E and Z from values of the slope and intercept.

Chemical intuition guides the selection of the form for $f(\alpha)$ with modifications of the order n so that a linear plot may be achieved.

2. Integral Methods

A variety of integral methods are available. These methods depend in part on whether a single or several thermogravimetric traces are available for kinetic analysis. The use of a single trace is based on a trial and error procedure. A general analytical form for the reaction is assumed. Successive values for the order are tried; the equation usually selected is $f(\alpha) = (1 - \alpha)^n$. The criterion for the proper kinetic parameters are those that lead to a calculated thermal trace that duplicates the experimental trace.

In 1961 Doyle (227) presented a formal method for the precise curve fitting of a single thermogram. He also presented an approximation procedure to reduce the tediousness of this method. Several approximation methods have appeared since then, each method claiming special advantages. In this category appear the names of Flynn and Wall (225), Zsakó (228), Ozawa (229), Cameron and Fortune (229a), MacCallum and Tanner (230), and Coates and Redfern (231).

The variables for the general expression [Eq. (46)] for a reaction rate may be separated,

$$d\alpha/f(\alpha) = (Z/\Phi) \exp(-E/RT)\, dT \tag{72}$$

We define

$$F(\alpha) = \int_0^{\alpha} d\alpha/f(\alpha) \tag{73}$$

so that

$$F(\alpha) = (Z/\Phi) \int_{T_0}^{T} \exp(-E/RT)\, dT \tag{74}$$

where T_0 is the value of T at the beginning of the reaction. Ordinarily the reaction rate is negligible at low temperatures so that

$$\int_{T_0}^{T} \exp(-E/RT) = \int_{0}^{T} \exp(-E/RT)$$

Integration of Eq. (74) yields

$$F(\alpha) = (ZE/\Phi R)[(-e^x/x) + \int_{-\infty}^{x} (e^x/x)\,dx] = (AE/\Phi R)p(x) \tag{75}$$

where $x \equiv -E/RT$.

If $f(\alpha)$ is of the form $(1 - \alpha)^n$ as given in Eq. (21), then

$$[(1 - \alpha)^{1-n} - 1]/(1 - n) = (ZE/\Phi R)p(x) \tag{76}$$

for $n \neq 1$, and

$$\ln(1 - \alpha) = (ZE/\Phi R)p(x) \tag{77}$$

for n = 1. Two series expansions that have been used in many of the kinetic methods are asymptotic expansion,

$$p(x) = (e^x/x^2)\,[1 + (2!/x) + (3!/x^2) + \ldots] \tag{78}$$

and the Schlömilch expansion,

$$p(y) = [e^{-y}/y(y + 1)]\;[1 - 1/(y + 2) + 2/(y + 2)(y + 3) - \ldots] \qquad y = -x \tag{79}$$

The values for p(x) have also been tabulated for limited ranges (225, 227, 228). The difficulty in applying Eq. (75) is that p(x) depends on both temperature and activation energy.

a. Methods of Doyle and Zsakó. Doyle evolved an approximation for p(x) when E/RT exceeds 20, a most frequent case for the thermal degradation of polymers, viz.,

$$\log p(x) = -2.315 - 0.4567E/RT \tag{80}$$

This approximation is based on the first two terms of the asymptotic expansion. Equation (80) along with an assumed analytical form for $f(\alpha)$ lead to a trial-and-error curve-fitting method. The approximation affords a good first estimate of E. Progressively better estimates of E can be made by modifying the presumed value of E so that agreement between the calculated and experimental thermogravimetric trace is achieved.

Zsakó tried to simplify Doyle's trial-and-error method in the following way. Equation (75) may be written as

$$\log (ZE/R\Phi) = \log F(\alpha) - \log p(x) = B \tag{81}$$

where B for a given reaction will depend only on the heating rate and not on the temperature. The value of $F(\alpha)$ for a given temperature can be calculated from the experimental data if an expression for $f(\alpha)$ is assumed. Similarly, p(x) for the same temperature can be found if the value for E is known. The constancy of B permits a quantitative method for testing the validity of different kinetic equations and of determining the apparent activation energy consistent with a given analytical form for $f(\alpha)$. The Zsakó paper contains a comparison of some data for the modified Doyle method and the Freeman-Carroll method. These values are shown in Table 4.

b. Method of MacCallum and Tanner (230). The MacCallum-Tanner method avoids the need for successive approximations. The authors have shown that for a given value of E, -log p(x) is linear with 1/T so that

$$-\log p(x) = X + a/T$$

They also found that a plot of E versus a was linear, that is, $a = Y + bE$, so that

$$-\log p(x) = X + (Y + bE)/T \tag{82}$$

Though X is not linear with respect to E, log X is linear,

$$\log X = \log A + c \log E$$

or

$$X = AE^{c}$$

TABLE 4 (228)

Method	Complex I[a]		Complex II[b]	
	n	E (kcal/mole)	n	E (kcal/mole)
Doyle-Zsakó	1	28.9	1	28.3
Freeman-Carroll	1.16	29.2	0.82	28.7

[a]Complex I [Co(dimethylglyoxime)$_2$ (p-ethylamine)$_2$] NCS

[b]Complex II [Co(dimethylglyoxime)$_2$ (γ-picoline)$_2$] NCS

The value for n is assumed to be unity for the Doyle-Zsakó method.

Equation (82) may then be written as

$$-\log p(x) = AE^{c} + (Y + bE)/T \tag{83}$$

Substituting the evaluated constants from the table of values for log p(x) we have

$$-\log p(x) = 0.4828E^{0.4351} + (0.449 + 0.217E)/T \cdot 10^{-3} \tag{84}$$

MacCallum and Tanner found that the values of -log p(x) so deduced agreed with Zsakó's tables to about 1% or better. Thus the complete integrated rate equation may be expressed as

$$\log F(\alpha) = \log (ZE/\Phi R) - 0.48E^{0.44} - (0.45 + 0.22E)/T \cdot 10^{-3} \tag{85}$$

The correct form for $F(\alpha)$ and the experimental values for α (as a function of the temperature) permit the numerical evaluation of $F(\alpha)$ and its log. A plot of log $F(\alpha)$ against 1/T should result in a linear plot and lead to values for E and Z.

c. Method of Coats and Redfern (231). If the first three terms of the asymptotic approximation are used and if $f(\alpha) = (1 - \alpha)^n$, then it can be shown that

$$[1 - (1 - \alpha)^{1-n}]/(1 - n) = (ZRT^2/\Phi E)[1 - (2RT/E)] \exp(-E/RT) \tag{86}$$

or

$$\log \{[1 - (1 - \alpha)^{1-n}]/T^2(1 - n)\} = \log \{(ZR/\Phi E)[1 - (2RT/E)]\} - E/2.3RT \tag{87}$$

for all values of n except n = 1. In the latter case

$$\log \{-\log [(1 - \alpha)/T^2]\} = \log (ZR/\Phi E) - E/2.3RT \tag{88}$$

Thus a plot of either

$$\log \{[1 - (1 - \alpha)^{1-n}]/T^2(1 - n)\} \text{ vs } 1/T$$

or when n = 1

$$\log \{-\log [(1 - \alpha)/T^2]\} \text{ vs } 1/T$$

should result in a straight line of slope $-E/2.3R$ for the correct value of n. This is due to the almost constant value for $\log \{(ZR/\phi E)[1 - (2RT/E)]\}$ for most values of E and T over which reactions generally occur.

In both the Doyle and the Coats and Redfern procedure the problem of determining order may be circumvented by going to low values for α. Assuming all reactions behave as zero order as $\alpha \rightarrow 0$ Eq. (87) becomes

$$\log (c/T^2) = \log \{(ZR/\phi E)[1 - (2RT/E)]\} - E/2.3RT \tag{89}$$

and a plot for $\log (c/T^2)$ versus $1/T$ should be a straight line yielding values for E and Z.

d. Method of varied heating rates. The above methods almost always involve the assumption that the reaction follows a simple order. The question arises as to whether the activation energy can be obtained without this assumption. The use of thermal traces at different heating rates makes this possible. A procedure of this type has been described in the previous section on derivative methods. In the present method the magnification of the experimental scatter due to slope measurements may be avoided. Variations of this basic method have been described (225, 229, 232, 233).

Substituting the Doyle approximation [Eq. (80)] in Eq. (75) we may write

$$\log F(\alpha) \simeq \log (ZE/R) - \log \phi - 2.315 - (0.457\ E/RT)$$

so that at constant values of α for several heating rates a plot of $\log \phi$ versus $1/T$ will have a slope of

$$d(\log \phi)/d(1/T) \simeq -0.457\ E/R \tag{90}$$

or

$$E = -4.35\ d(\log \phi)/d(1/T); \quad (1 - \alpha) = \text{constant} \tag{91}$$

By determining E at various constant conversion values, the possible variation of the activation energy during the course of the reaction may be tested.

Once the value for E has been obtained via Eq. (91), the constant -4.35 in the equation can be readily recalculated for the appropriate value of E/RT listed in the tables for p(x). These new values may be utilized to obtain more accurate activation energies. Convenient tables for this purpose have been published by Flynn and Wall (232).

IV. CONCLUSIONS

The question has been raised whether kinetic parameters obtained via nonthermal or dynamic methods are the same as those obtained by the conventional isothermal methods (234). At least in some cases MacCallum and Tanner claim to have found very poor agreement for the numerical values of the kinetic parameters for the thermal decomposition of polymers and have suggested that the basic equations in the dynamic methods may be inaccurate.

If we re-examine Eq. (20) using the Arrhenius equation,

$$d\alpha/dt = Z[\exp(-E/RT)]f(\alpha)$$

where α is the degree of conversion at time t, it will be seen that among the many assumptions, it is inferred that the function $f(\alpha)$ is independent of the temperature so that at a given degree of conversion the function $f(\alpha)$ is uniquely defined. It has been suggested that it would be more accurate to write

$$d\alpha/dt = (\partial\alpha/\partial t)_T + (\partial\alpha/\partial T)_t\Phi \tag{92}$$

so that only when the heating rate Φ is zero will the dynamic and isothermal rates be identical. For those cases in which both methods yield the same results it would appear that the last term in Eq. (92) is negligible. Unfortunately, MacCallum and Tanner at the time of writing have not published numerical values. However, Audebert and Aubineau (235) have published values on the activation energies for the thermal decomposition of some polymers comparing isothermal and dynamic values. These are shown in Table 5. It may be seen that, within the precision of the measurements, the values for both methods appear to be in fair agreement. In general it has been the experience of the authors, particularly in the thermal decomposition of inorganic solids (236), that the dynamic and isothermal methods yield the same kinetic parameters within the experimental error.

The correlation of kinetic parameters and mechanism of chemical changes has not attracted much effort. Probably with the availability of instrumentation and valid procedures for kinetic analysis, this situation may be altered.

TABLE 5

Comparison of Activation Energies Obtained by Isothermal and Dynamic Thermogravimetry on the Pyrolysis of Three Different Polymers (235)

Polymer	Isothermal E (kcal/mole)	Dynamic E (kcal/mole)
Polytetrafluroethylene	81 ± 9	74 ± 5
Poly(dichloropiperidyltriazine-bisphenol - A)	39 ± 12	35 ± 2.5
Poly(ethylene oxide)	45 ± 5	48.5 ± 3

APPENDIX I

Recommendations for Reporting Thermal Analysis Data*

In 1965 the First International Conference on Thermal Analysis (ICTA) established a Committee on Standardization charged with the task of studying how and where standardization might further the value of these methods. One area of concern was with the uniform reporting of data, in view of the profound lack of essential experimental information occurring in much of the thermal analysis literature. The following recommendations have been put forward by the Committee on Standardization.

To accompany each DTA or TGA record, the following information should be reported:

1. Identification of all substances (sample, reference, diluent) by a definitive name, an empirical formula, or equivalent compositional data.

2. A statement of the source of all substances, details of their histories, pretreatments, and chemical purities, so far as these are known.

*Reprinted in part from Analytical Chemistry, Vol. 39, No. 4, April 1967 with permission of Analytical Chemistry and the copyright holder, the American Chemical Society.

3. Measurement of the average rate of linear temperature change over the temperature range involving the phenomena of interest.

4. Identification of the sample atmosphere by pressure, composition, and purity; whether the atmosphere is static, self-generated, or dynamic through or over the sample. Where applicable, the ambient atmospheric pressure and humidity should be specified. If the pressure is other than atmospheric, full details of the method of control should be given.

5. A statement of the dimensions, geometry, and materials of the sample holder; the method of loading the sample where applicable.

6. Identification of the abscissa scale in terms of time or of temperature at a specified location. Time or temperature should be plotted to increase from left to right.

7. A statement of the methods used to identify intermediates or final products.

8. Faithful reproduction of all original records.

9. Wherever possible, each thermal effect should be identified and supplementary supporting evidence stated.

In the reporting of TGA data, the following additional details are also necessary:

10. Identification of the thermobalance, including the location of the temperature-measuring thermocouple.

11. A statement of the sample weight and weight scale for the ordinate. Weight loss should be plotted as a downward trend and deviations from this practice should be clearly marked. Additional scales (for example, fractional decomposition, molecular composition) may be used for the ordinate where desired.

12. If derivative thermogravimetry is employed, the method of obtaining the derivative should be indicated and the units of the ordinate specified.

When reporting DTA traces, these specific details should also be presented:

13. Sample weight and dilution of the sample.

14. Identification of the apparatus, including the geometry and materials of the thermocouples and the locations of the differential and temperature-measuring thermocouples.

15. The ordinate scale should indicate deflection per degree Centigrade at a specified temperature. Preferred plotting will indicate upward deflection as a positive temperature differential, and downward deflection as a negative temperature differential, with respect to the reference. Deviations from this practice should be clearly marked.

APPENDIX II

Compounds Useful for Temperature or Heat of Reaction Calibration for DTA and DSC [Barshad (237, 238) and Others]

Substance	Point[a]	Temperature, °C	Heat of Fusion, cal/g
NH_4NO_3	I	32	19.1
	I	85	12.3
	I	125	
	F	170	
Stearic Acid	F	69.4	
m-Dinitrobenzene	F	90	24.7
o-Dinitrobenzene	F	117	32.3
$BaCl_2 \cdot 2H_2O$	D	120	119
Benzoic Acid	F	122.4	33.9
KNO_3	I	128	
$CaSO_4 \cdot 2H_2O$	D	140	160
In	F	156.6	6.8
$AgNO_3$	I	160	
	F	212	16.7
	Dec.	400	210
AgI	I	147	
	F	552	

Substance	Point[a]	Temperature, °C	Heat of Fusion, cal/g
Sn	F	231.9	14.5
AgC1	F	307	
Zn	F	419.5	
$NaNO_3$	F	314	45.3
	Dec.	667	776
Pb	F	327.4	
Ag_2SO_4	I	412	
	F	652	
Quartz (SiO_2)	I	573	
K_2SO_4	I	583	
$NaMo_2O_4$	I	642	
	F	687	
$CaCO_3$	Dec.	787	468
NaC1	F	800.4	
$SrCO_3$	F	925	

[a]I = inversion; F = fusion; D = dehydration; Dec = decomposition.

Precisely defined temperature standards for a number of organic compounds covering the range of 50°C to 285°C are available commercially[a]. The triple point of each lot for each standard is specified on the label and certified accurate to within ± 0.05°C.

Standard	Typical Triple Point, Centigrade
p-Nitrotoluene	51.50°
Naphthalene	80.25°
Benzoic Acid	122.35°
Adipic Acid	151.40°
Anisic Acid	182.50°
2-Chloroanthroquinone	209.00°
Carbazole	245.00°
Anthraquinone	284.50°

[a]Fisher TherMetric Standards, Fisher Scientific Company, Pittsburgh, Pennsylvania.

REFERENCES

(1) P. Y. Feng and E. S. Freeman, in Physical Methods in Macromolecular Chemistry, (B. Carroll, ed.), Vol. 1, Dekker, New York, 1969, Chap. 4.

(2) H. Le Chatelier, Bull. Soc. Fr. Mineral., 10, 204 (1887).

(3) W. C. Roberts-Austen, Proc. Inst. Mech. Engrs. (London), 1, 35 (1899); and Metallographist, 2, 189 (1899).

(4) P. Brasseur and G. Champetier, Bull Soc. Chim. Fr. 1947, p. 117.

(5) R. C. MacKenzie and F. R. Farquharson, Congr. geol. intern., Compt. rend. 19th Sess., Algiers, 1952, No. 18, p. 183-200 (1953).

(6) H. G. McAdie, Anal. Chem., 39, 543 (1967).

(7) P. D. Garn, Thermoanalytical Methods of Investigation, Academic, New York, 1965.

(8) W. W. Wendlandt, Thermal Methods of Analysis, Wiley-Interscience, New York, 1964.

(9) J. D. Walton, J. Am. Ceram. Soc., 38, 438 (1955).

(10) J. Chiu, Anal. Chem., 39, 861 (1967).

(11) E. M. Bollin, J. A. Dunne, and P. F. Kerr, Science, 131, 661 (1960).

(12) E. M. Barrall, II, R. S. Porter, and J. F. Johnson, J. Appl. Polymer Sci., 9, 3061 (1965).

(13) B. Wunderlich and D. M. Bodily, J. Polymer Sci., C6, 137 (1964).

(14) M. Inoue, J. Polymer Sci., A1, 2697 (1963).

(15) D. J. David, Anal. Chem., 36, 2162 (1964).

(15a) S. J. Strella, J. Appl. Polymer Sci., 7, 569 (1963).

(16) C. B. Murphy, J. A. Palm, C. D. Doyle, and E. M. Curtiss, J. Polymer Sci., 28, 447 (1958).

(17) H. J. Donald, E. S. Humes, and L. W. White, J. Polymer Sci., C6, 93 (1964).

(18) E. M. Barrall, II, and L. B. Rogers, Anal. Chem., 34, 1101 (1962).

(19) D. A. Vassallo and J. C. Harden, Anal. Chem., 34, 132 (1962).

(20) P. G. Harold and T. J. Planje, J. Am. Ceram. Soc., 31, 20 (1948).

(21) C. Mazières, Anal. Chem., 36, 602 (1964).

(22) H. Rase and R. L. Stone, Anal. Chem., 29, 1273 (1955).

(23) R. L. Stone, Anal. Chem., 32, 1582 (1960).

(24) Robert L. Stone Co., Austin Texas, Bulletin, DTA-656.

(25) D. B. Gasson, J. Sci. Instrum., 39, 78 (1962).

(26) P. D. Garn and J. E. Kessler, Abstract Papers, 140th Meeting, American Chemical Society, Illinois, Sept. 1961, p. 2B.

(27) A. D. Russell, J. Sci. Instrum., 44, 399 (1967).

(28) R. L. Bohon, Anal. Chem., 33, 1451 (1961).

(29) J. L. Kulp and P. F. Kerr, Science, 105, 413 (1947).

(30) D. B. Cox and J. F. McGlynn, Anal. Chem., 29, 960 (1957).

(31) The Deltatherm, Technical Equipment Corp., Denver, Colorado.

(32) W. H. King, Jr., C. T. Camilli, and A. F. Findeis, Anal. Chem. 40, 1330 (1968).

(33) G. N. Rupert, Rev. Sci. Instrum., 34, 1183 (1963).

(34) G. N. Rupert, Rev. Sci. Instrum., 36, 1629 (1965).

(35) E. Rudy, Aerometrics, Aeroject-General Corp. brochure "Differential Thermal Analysis."

(36) J. A. Hill and C. B. Murphy, Anal. Chem., 31, 1443 (1959).

(37) L. Brewer and P. Zavitsanos, J. Phys. Chem. Solids, 2, 284 (1957).

(38) C. Campbell and G. Weingarten, Trans. Faraday Soc., 55, 222 (1959).

(39) C. Geacintov, R. S. Schotland, and R. B. Miles, J. Polymer Sci., C6, 197 (1964).

(40) A. Reisman, Anal. Chem., 32, 1566 (1960).

(41) R. D. Garwood, J. R. Moon, and B. F. Thrusth, Rev. Sci. Instrum., 37, 108 (1960).

(42) J. Chiu, Anal. Chem., 34, 1841 (1962).

(43) P. D. Garn, Paper 19, Pittsburgh Conf. Analytical Chemistry Appl. Spectroscopy, Pittsburgh, Pa., Feb. 21-25, 1966.

(44) J. Simura and H. Noda, Kogyo Kagaku Zasshi, 69, 1650 (1966).

(45) E. Sturm, J. Phys. Chem., 65, 1935 (1961).

(46) Engelhard Industries, Inc., East Newark, New Jersey, USA.

(47) W. Lodding and E. Sturm, Am. Mineralogist, 42, 78 (1957).

(48) M. J. Joncich and D. R. Bailey, Anal. Chem., 32, 1578 (1960).

(49) J. M. Pakulak and G. W. Leonard, Anal. Chem., 31, 1037 (1959).

(50) E. E. Weaver and W. Keim, Proc. Indian Acad. Sci., 70, 123 (1960).

(51) N. A. Nedumov, Russ. J. Phys. Chem. (English Transl.), 34, 84 (1960).

(52) E. S. Watson, M. J. O'Neill, J. Justin, and N. Brenner, Anal. Chem., 36, 1233 (1964).

(53) R. F. Schwenker, Jr., and R. K. Zuccarello, J. Polymer Sci., C6, 1 (1964).

(54) J. Chiu, J. Polymer Sci., C8, 27 (1965).

(54a) P. D. Garn, Anal. Chem., 37, 77 (1965).

(55) W. Lodding, and L. Hammel, Anal. Chem., 32, 657 (1960).

(56) D. J. David, Anal. Chem., 37, 82 (1965).

(57) R. M. Perkins, G. L. Drake, Jr., and W. A. Reeves, J. Appl. Polymer Sci., 10, 1041 (1966).

(58) F. Danusso and G. Polizzotti, Chim. Ind. (Milan), 44, 241 (1962).

(59) R. H. Still and C. J. Keattch, J. Appl. Polymer Sci., 10, 193 (1966).

(60) F. P. Bundy, W. R. Hibbard, Jr., and H. M. Strong, Progress in Very High Pressure Research, Wiley, New York, 1961.

(61) S. L. S. Thomas, H. S. Turner, and W. F. Wall, High Pressure Engineering Conference, Institute of Mechanical Engineers, London, 1967.

(62) L. H. Cohen, W. Klement, Jr., and G. C. Kennedy, J. Phys. Chem. Solids, 27, 179 (1966).

(63) N. G. Savill and W. F. Wall, J. Sci. Instrum., 44, 839 (1967).

(64) M. Tamayama, N. Andersen, and H. Eyring, Proc. Nat. Acad. Sci., 57, 554 (1967).

(65) T. Davidson and B. Wunderlich, J. Polymer Sci., A2, 7, 377 (1969).

(66) H. E. Kissinger, J. Res. Nat. Bur. Stand., 57, 217 (1956).

(67) H. E. Kissinger, Anal. Chem., 29, 1702 (1957).

(68) S. Speil, L. H. Berkelhamer, J. A. Pask, and B. Davies, U.S. Bur. Mines, Tech. Paper 664, 81 (1945).

(69) H. T. Smyth, J. Am. Ceram. Soc., 34, 221 (1951).

(70) S. Strella, J. Appl. Polymer Sci., 7, 569 (1963).

(71) P. L. Arens, A Study of the Differential Thermal Analysis of Clays and Clay Minerals, Excelsior Foto-Offsets, Gravenhage, Wageningen, Netherlands, 1951.

(72) M. J. Vold, Anal. Chem., 21, 683 (1949).

(73) J. L. Soulé, J. Phys. Radium, 13, 516 (1952).

(74) B. Ke, J. Polymer Sci., A1, 1453 (1963).

(75) R. C. Mackenzie and V. C. Farmer, Clay Minerals Bull., 1, 262 (1952).

(76) E. M. Barrall, II, and L. B. Rogers, Anal. Chem., 34, 1106 (1962).

(77) H. Morita and H. M. Rice, Anal. Chem., 27, 336 (1955).

(78) See Ref. 53.

(79) K. G. Kumanin, Zh. Prikl. Khim., 20, 1242 (1947).

(80) C. Eyraud, Compt. rend., 238, 1511 (1954).

(81) C. Eyraud, Compt. rend., 240, 862 (1955).

(82) D. M. Speros and R. L. Woodhouse, J. Phys. Chem., 67, 2164 (1963).

(83) Perkin-Elmer Corporation, Norwalk, Conn., USA.

(84) R. A. W. Hill, and R. P. Slessor, Trans. Faraday Soc., 65, 340 (1969).

(85) A. P. Gray, in Analytical Calorimetry, (R. S. Porter and J. F. Johnson, eds.), Plenum Press, New York, 1968.

(86) M. L. Keith and O. F. Tuttle, Am. J. Sci., 1, 203 (1952).

(87) G. L. Driscoll, I. N. Duling, and F. Magnotta, see Ref. 85.

(88) W. Kauzman, Chem. Rev., 43, 219 (1948).

(89) J. D. Ferry, Viscoelastic Properties of Polymers, Wiley, New York, 1961.

(90) F. J. Hybart and J. D. Platt, J. Appl. Polymer Sci., 11, 1449 (1967).

(91) J. J. Keavney and E. C. Eberlin, J. Appl. Polymer Sci., 3, 47 (1960).

(92) S. S. Rogers and L. Mandelkern, J. Phys. Chem., 61, 985 (1957).

(93) N. Hirai and H. Erying, J. Appl. Phys., 29, 810 (1958).

(94) N. Hirai and H. Eyring, J. Polymer Sci., 37, 51 (1959).

(95) B. Wunderlich, J. Phys. Chem., 64, 1052 (1960).

(96) B. Ke, J. Polymer Sci., 42, 15 (1960).

(97) F. P. Price (ed.), The Meaning of Crystallinity in Polymers, (J. Polymer Sci., Symposium No. 18) Wiley-Interscience, New York, 1967.

(98) G. Gee, Proc. Chem. Soc., 1957, 111.

(99) M. Dole, J. Polymer Sci., 18C, 57 (1967).

(100) M. J. O'Neill, Anal. Chem., 38, 1331 (1966).

(101) J. R. Knox, see Ref. 85.

(102) B. Ke, J. Appl. Polymer Sci., 6, 624 (1962).

(103) R. W. Ford, J. D. Ilavsky, and R. A. Scott, see Ref. 85, P. 41.

(104) B. Ke, J. Polymer Sci., 61, 47 (1962).

(105) P. J. Flory, Principles of Polymer Chemistry, Cornell Univ. Press, Ithaca, New York, 1953.

(106) B. Ke, J. Polymer Sci., 50, 79, (1961).

(107) J. R. Knox, see Ref. 85.

(108) B. Ke, Polymer Letters, 1, 167 (1963).

(109) J. H. Griffith and B. G. Rånby, J. Polymer Sci., 44, 369 (1960).

(110) S. Igarashi and H. Kambe, Bull. Chem. Soc. Jap., 37, 176 (1964).

(111) L. Reich, in Macromolecular Reviews (A. Peterlin, M. Goodman, S. Okamura, B. H. Zimm, and H. F. Mark, eds.), Vol. 3, Wiley-Interscience, New York, 1968, p. 49.

(112) K. L. Paciorek, W. G. Lajiness, R. G. Spain, and C. T. Lenk, J. Polymer Sci., 61, S41 (1962).

(113) C. B. Murphy and J. A. Hill, Nucleonics, 18, 78 (1960).

(114) J. Fock, Polym. Letters, 5, 635 (1967).

(115) H. E. Bair, T. W. Huseby, and R. Salovey, see Ref. 85, P. 31

(116) C. Duval, Inorganic Thermogravimetric Analysis, 2nd ed., Elsevier, New York, 1963.

(117) C. Rocchiccioli, Mikrochim. Acta, 6, 1017 (1962).

(118) C. Duval, Anal. Chim. Acta, 2, 92 (1948).

(119) J. H. G. Jellinek, J. Polymer Sci., 4, 13 (1949).

(120) L. Reich, see Vol. 1 of Ref. 111.

(121) S. Gordon and C. Campbell, Anal. Chem., 32, 271R (1960).

(122) L. Jacqué, G. Guiochon, and P. Gendrel, Bull. Soc. Chim. Fr., 1961, 1061.

(123) Survey of thermogravimetry instruments in Scientific Research, 4, 31 (1969).

(124) G. M. Lukaszewski and J. P. Redfern, Lab. Practice, 10, 469 (1961).

(125) F. Claisse, F. East, and F. Abesque, Use of the Thermobalance in Analytical Chemistry, Dept. of Mines, Province of Quebec, P. R. 305, 1954.

(126) A. E. Newkirk, Anal. Chem., 32, 1558 (1960).

(127) E. L. Simons, A. E. Newkirk, and I. Aliferis, Anal. Chem., 29, 48 (1957).

(128) G. M. Lukaszewski, Nature, 194, 959 (1962).

(129) L. Cahn and H. Schultz, Anal. Chem., 35, 1729 (1963).

(130) L. Cahn and N. C. Peterson, Anal. Chem., 39, 403 (1967).

(131) C. Duval, Mikrochim. Acta, 1958, 705.

(132) P. D. Garn, C. R. Geith, and S. DeBala, Rev. Sci. Instrum., 33, 293 (1962).

(133) H. Peters and H. G. Wiedemann, Z. Anorg. Chem., 300, 142 (1959).

(134) H. G. Wiedemann, Z. Anorg. Chem., 306, 84 (1960).

(135) A. W. Coats and J. P. Redfern, Analyst, 88, 906 (1963).

(136) J. A. Poulis and J. M. Thomas, J. Sci. Instrum., 40, 95 (1963).

(137) J. A. Poulis, B. Pelupessy, C. H. Massen, and J. M. Thomas, J. Sci. Instrum., 41, 295 (1964).

(138) J. R. Soulen and I. Mockrin, Anal. Chem., 33, 1909 (1961).

(139) G. M. Lukaszewski, Lab. Practice, 13, 32 (1964).

(140) D. A. Powell, J. Sci. Instrum., 34, 225 (1957).

(141) J. Papailhau, Bull. Soc. Fr. Mineral. Cristallogr., 82, 367 (1959).

(142) F. Paulik, S. Gal, and L. Erdey, Anal Chim. Acta, 29, 381 (1963).

(143) H. G. McAdie, Anal. Chem., 35, 1840 (1963).

(144) F. W. Wilburn and J. R. Hesford, J. Sci. Instrum., 40, 91 (1963).

(145) J. E. Krüger and J. G. Bryden, J. Sci. Instrum., 40, 178 (1963).

(146) E. P. Manche and B. Carroll, Rev. Sci. Instrum., 35, 1486 (1964).

(147) J. G. Schnizlein, J. Brewer, and D. F. Fischer, Rev. Sci. Instrum., 36, 591 (1965).

(148) D. R. Terry, J. Sci. Instrum., 42, 507 (1965).

(149) E. J. Chatfield, J. Sci. Instrum., 44, 649 (1967).

(150) P. D. Garn and J. E. Kessler, Anal. Chem., 32, 1563 (1960).

(151) P. D. Garn and J. E. Kessler, Anal. Chem., 32, 1900 (1960).

(152) T. S. Light, L. F. Fitzpatrick, and J. P. Phaneuf, Anal. Chem., 37, 79 (1965).

(153) J. Chiu, Applied Polymer Symposia No. 2, 25 (1966).

(154) D. A. Vassallo, Anal. Chem., 33, 1823 (1961).

(155) A. Barlow, R. S. Lehrle, and J. C. Robb, Makromol. Chem., 54, 230 (1962).

(156) M. D. Karkhanavala and S. G. Rege, J. Indian Chem. Soc., 40, 459 (1963).

(157) N. Grassie and H. W. Melville, Proc. Roy. Soc., Ser. A, 199, 1 (1949).

(158) B. Kaesche-Krischer, Chem.-Ing.-Tech., 37, 944 (1965).

(159) L. Reich and W. Levi, in Macromolecular Reviews (A. Peterlin, M. Goodman, S. Okamura, B. H. Zimm, and H. F. Mark, eds.), Vol. 1, Wiley-Interscience, New York, 1968, p. 173.

(160) H. Jucker, H. G. Wiedemann, and H. P. Vaughan, paper presented at the Pittsburgh Conference on Analytical Chemistry and Applied Spectroscopy, March 2, 1965.

(161) H. P. Vaughan and H. G. Wiedemann, in Vacuum Microbalance Techniques, (P. M. Waters/ed.), Vol. 4, Plenum Press, 1965, p. 1.

(162) Mettler Instrument Corporation, 20 Nassau St., Princeton, N.J., U.S.A.

(163) C. J. Keattch, Talanta, 14, 77 (1967).

(164) A. P. Gray, paper presented at the Third Toronto Symposium on Thermal Analysis, Toronto, Canada, February 25-26, 1969.

(165) C. D. Doyle, Anal. Chem., 33, 77 (1961).

(166) J. Chiu, Applied Polymer Symposia No. 2, 25 (1966).

(167) H. H. G. Jellinek and H. Kachi, Polym. Eng. Sci., 5, 1 (1965).

(168) M. Baer, J. Polymer Sci., A2, 417 (1964).

(169) A. D. McLaren and J. W. Rowen, J. Polymer Sci., 7, 289 (1951).

(170) A. A. Hofer and H. Mohler, Helv. Chim. Acta, 45, 1415 (1962).

(171) A. W. Czanderna, in Vacuum Microbalance Techniques, (A. W. Czanderna, ed.), Vol. 6, Plenum Press, New York, 1967, p. 129.

(172) A. Blažek, Hutnicke listy, 12, 1096 (1957).

(173) F. Paulik, J. Paulik, and L. Erdey, Z. Anal. Chem., 160, 241 (1958).

(174) H. G. Wiedemann, Chem.-Ing.-Tech., 36, 1105 (1964).

(175) S. L. Madorsky, Thermal Degradation of Organic Polymers, Wiley-Interscience, New York, 1964.

(176) A. S. Kenyon, in Techniques and Methods of Polymer Evaluation, (P. E. Slade and L. I. Jenkins, eds.), Vol. 1, Dekker, New York, 1966, p. 217.

(177) B. Groten, in Gas Effluent Analysis (W. Lodding, ed.), Dekker, New York, 1967, Chap. 4.

(178) R. L. Levy, J. Gas Chromatog., 5, 107 (1967).

(179) G. Guiochon, (ed.), Pyrolysis and Reaction Gas Chromatography, Preston Abstracts, Evanston, Illinois, 1968.

(180) H. G. Langer and R. S. Gohlke, Anal. Chem., 35, 1301 (1963).

(181) H. G. Langer, R. S. Gohlke, and D. H. Smith, Anal. Chem., 37, 433 (1965).

(182) G. P. Shulman, Polymer Letters, 3, 911 (1965).

(182a) F. Zitomer, Anal. Chem., 40, 1091 (1968).

(182b) H. L. Friedman, J. Macromal. Sci. (Chem.), A1, 57 (1967).

(183) G. Cano, Bull. Soc. Chim. Fr., 1963, 2540.

(184) J. Chiu, Anal. Chem., 40, 1516 (1968).

(185) W. T. Smith, Modern Optical Engineering, McGraw-Hill, New York, 1966, pp. 385 and 393.

(186) S. A. Dolin, H. A. Kruegle, and G. J. Penzias, Appl. Opt., 6, 267 (1967).

(187) R. M. Hexter and C. W. Hand, Appl. Opt. 7, 2161 (1968).

(188) P. Buchhave and C. H. Church, Appl. Opt., 7, 2200 (1968).

(189) M. J. D. Low, in Gas Effluent Analysis, (W. Lodding, ed.), Dekker, New York, 1967, Chap. 5.

(190) E. M. Barrall, II, R. S. Portor, and J. F. Johnson, Anal. Chem., 36, 2316 (1964).

(191) D. L. Beck, A. A. Hiltz, and J. R. Knox, SPE Trans., 3, 279 (1963).

(192) I. C. McNeill, J. Polymer Sci., Part A-1, 4, 2479 (1966).

(193) I. C. McNeill, Eur. Polym. J., 3, 409 (1967).

(194) I. C. McNeill, Eur. Polym. J., 4, 21 (1968).

(195) C. D. Doyle and C. B. Murphy, in Thermal Analysis 1965, (J. P. Redfern, ed.), Proceedings of First International Congress, Aberdeen, Macmillan, London, 1965, p. 49.

(196) F. W. VanLuik, Jr., and R. E. Rippere, Anal. Chem., 34, 1617 (1962).

(197) C. B. Murphy, in Gas Effluent Analysis, (W. Lodding, ed.), Dekker, New York, 1967, Chap. 7.

(198) A. Charlesby and R. H. Partridge, Proc. Roy. Soc., Series A, 271, 170, 188 (1963).

(199) R. H. Partridge and A. Charlesby, J. Polymer Sci., 1, 439 (1963).

(200) A. Charlesby and R. H. Partridge, Proc. Roy. Soc., Series A, 283, 312 (1965).

(201) A. Mele, A. Delle Site, C. Bettinali, and A. DiDomenico J. Chem. Phys., 49, 3297 (1968).

(202) I. Boustead and A. Charlesby, Eur. Polym. J., 3, 459 (1967).

(203) G. E. Ashby, J. Polymer Sci., 50, 99 (1961).

(204) H. L. Friedman, "Thermal Aging and Oxidation" in Treatise on Analytical Chemistry, (I. M. Kolthoff, P. J. Elving, and F. H. Stross, eds.), Wiley-Interscience New York, Part III (in press).

(205a) J. W. Liskowitz and B. Carroll, J. Macromol. Sci. (Chem.), A2, 1139 (1968).

(205b) H. W. Hoyer, J. Am. Chem. Soc. 90, 2480 (1968).

(206) R. L. Reed, L. Weber, and B. S. Gottfried, Ind. Eng. Chem. Fundamentals, 4, 49 (1968).

(207) R. L. Reed, L. Weber, and B. S. Gottfried, Ind. Eng. Chem. Fundamentals, 5, 287 (1968).

(208) K. Akita and M. Kase, J. Phys. Chem., 72, 906 (1968).

(209) J. A. McMillan, J. Chem. Phys, 42, 3497 (1965).

(210) R. N. Young, "DTA of Some Polymer Decompositions," in Thermal Analysis 1965, (J. P. Redfern, ed.), MacMillan, London, 1965, p. 70.

(211) H. J. Borchardt and F. Daniels, J. Am. Chem. Soc., 79, 41 (1957).

(212) H. J. Borchardt, J. Inorg. Nucl. Chem. Soc., 79, 41 (1960).

(213) H. L. Friedman, Polymer Letters, 7, 41 (1969).

(214) E. S. Freeman and B. Carroll, J. Phys. Chem., 62, 394 (1958).

(215) D. A. Anderson and E. S. Freeman, J. Polymer Sci., 54, 253 (1961).

(216) B. Carroll and E. P. Manche, J. Appl Polymer Sci., 9, 1895 (1965).

(217) L. Reich, Die Macromol. Chemie, 123, 42 (1969).

(218) G. O. Piloyan, I. D. Ryabchikov, and O. S. Novikova, Nature, 212, 1229 (1966).

(219) E. A. Moelwyn-Hughes and P. Johnson, Trans. Faraday Soc., 36, 948 (1940).

(220) L. Reich, J. Appl. Polymer Sci., 10, 1033 (1966).

(221) K. E. J. Barret, J. Appl. Polym. Sci., 11, 1617 (1967).

(222) S. M. Ellerstein, see Ref. 85, p. 279-287.

(223) A. A. Duswalt, see Ref. 85, p. 313-317.

(224) H. M. Huevel and K. C. J. B. Lind, Anal. Chem., 42, 1044 (1970).

(225) J. F. Flynn and L. A. Wall, J. Res. Natl. Bur. Std., 70, A487 (1966).

(226) J. H. Sharp and S. A. Wentworth, Anal. Chem., 41, 2060 (1969).

(227) C. D. Doyle, J. Appl. Polymer Sci., 15, 285 (1961).

(228) J. Zsakó, J. Phys. Chem., 72, 2406 (1968).

(229) T. Ozawa, Bull. Chem. Soc. Jap., 38, 1881 (1965).

(229a) G. G. Cameron and J. D. Fortune, Eur. Polym. J., 4, 333 (1968).

(230) J. R. MacCallum and J. Tanner, Eur. Polym. J., 6, 1033 (1970).

(231) A. W. Coats and J. P. Redfern, Nature, 201, 68 (1964).

(232) J. F. Flynn and L. A. Wall, J. Polymer Sci., B4, 323 (1966).

(233) L. Reich, Polymer Letters, 2, 621 (1964).

(234) J. R. MacCallum and J. Tanner, Nature, 225, 1127 (1970).

(235) R. Audebert and C. Aubineau, J. Chim. Phys., 67, 617 (1970).

(236) W. Fisco, Doctoral Dissertation, Rutgers, the State Univ., New Jersey, (1969).

(237) I. Barshad, Am. Minerologist, 37, 667 (1952).

(238) Selected Values of Chemical Thermodynamic Properties, Natl. Bur. Standards Circ. 500, Washington, D.C. (1952) and Supplements.

AUTHOR INDEX

Numbers in parentheses are reference numbers and indicate that an author's work is referred to although his name is not cited in the text. Underlined numbers show the page on which the complete reference is listed.

G

S

T

U

V

W

Y

Z

SUBJECT INDEX